STEPHEN BLAIR, PHILLIP DAVIES, TERRY GRIMWOOD, ANDY JEFFERY

PLUMBING STUDIES

PEARSON

Published by Pearson Education Limited, Edinburgh Gate, Harlow, Essex, CM20 2JE.

www.pearsonschoolsandfecolleges.co.uk

Text © Pearson Education Ltd; JTL for content within ISBN 978 0 435 03131 2
Edited by Maria Anson
Typeset by Tek-Art, Crawley Down, West Sussex
Original illustration © Pearson Education 2013
Illustrated by Tek-Art, Crawley Down, West Sussex
Cover images: *Front:* **Getty Images:** Vetta

The rights of Stephen Blair, Phillip Davies, JTL, Terry Grimwood and Andy Jeffery to be identified as authors of this work have been asserted by them in accordance with the Copyright, Designs and Patents Act 1988.

First published 2013
16 15 14 13
10 9 8 7 6 5 4 3 2 1

British Library Cataloguing in Publication Data
A catalogue record for this book is available from the British Library.

ISBN 978 1 447 94034 0

Printed in the UK by Butler, Tanner and Dennis

Acknowledgements
The author and publisher would like to thank the following individuals and organisations for permission to reproduce photographs.

(Key: b-bottom; c-centre; l-left; r-right; t-top)

Alamy Images: Adrian Sherratt 4, Arcaid Images 284c, Catchlight Visual Service 401, DW Images 379, fotofacade 91, Ingram Publishing 378, Jim West 39, Judith Collins 124 (Blowtorch), Justin Kase 189, Marshall Ikonography 285, Niels Poulsen std 263, Razorpix 307 (Mirror), Rosemary Owen 284r, Zoonar GmbH 92; **Comstock Images:** 360b; **Construction Photography:** Jean-Francois Cordella 372; **Corbis:** 35, Fancy / Veer 50; **Digital Vision:** 43; **Fotolia.com:** fasphotographic 326bl, fefufoto 322, Maksym Dykha 371, scaliger 324l, Silvano Rebai 71 (Timer); **Getty Images:** 389; **Hepworth Building Products:** 166; **Imagemore Co., Ltd:** 29; **Imagestate Media:** John Foxx Collection 373, 380; **PhotoDisc:** 49, 70b, 98, 139 (Screw), 360c, 377; **Photolibrary.com:** Moodboard 375; **Photos.com:** Bob Dorn 1, Jupiterimages 47; **Science Photo Library Ltd:** Pekka Parviainen 85; **Shutterstock.com:** Dmitry Bruskov 206, Dmitry Naumov 281, Zhukov Oleg 121; **SuperStock:** F1 Online 337; **Surestop Ltd:** 202b; **Veer/Corbis:** Barney Boogles 326br, Blue Wren 38, Ivonne 181, Keerati Thanitthitianant 357 (Hose), Keith Wilson 126 (Mini tube cutter), LCS 65t, leungchopan 93, naumoid 309, Paul Littler 145 (Pencils), rodho 290, Tom Gowanlock 289; **Wavin Limited:** Hepworth 141 (Plastic saddle clip), 142 (Squareline clip), 343cl, 343bl, 343br; **www.imagesource.com:** 2

All other images © Pearson Education Ltd: Clark Wiseman, Studio 8; Gareth Boden; Jules Selmes; Naki Photography; Stuart Cox and Trevor Clifford

In some instances we have been unable to trace the owners of copyright material, and we would appreciate any information that would enable us to do so.

Websites
Pearson Education Limited is not responsible for the content of any external internet sites. It is essential for tutors to preview each website before using it in class so as to ensure that the URL is still accurate, relevant and appropriate. We suggest that tutors bookmark useful websites and consider enabling students to access them through the school/college intranet.

Contents

Introduction

This book supports the Level 2 Diploma in Plumbing Studies (6035) currently offered by City & Guilds although, at the time of writing, other awarding organisations, including EAL, offer similar qualifications. The diploma has been prepared by working with the awarding organisations to provide a qualification for those seeking a career in the plumbing mechanical services industry.

The diploma has been approved on the Qualifications and Credit Framework (QCF), the government framework that regulates all vocational qualifications. The QCF ensures that qualifications are structured and titled consistently and that they are quality assured.

Who the qualification is aimed at

The standard industry plumbing courses, the Level 2 and 3 NVQ Diplomas, are funded only for those who are working as apprentices in the plumbing and mechanical services industries. The Diploma in Plumbing is suitable for people who are:

- not currently working in the sector but looking for a qualification to support their efforts to find employment as plumbing improvers or trainees
- working in the industry but who do not have the academic qualifications, or breadth of experience, required to undertake the Level 2 or 3 NVQ Diplomas.

In essence:

- the Level 2 qualification is designed for people who are new to the industry. It provides the basic skills and knowledge they will need to go on to Level 3
- the Level 3 qualification is aimed at those who have already completed the Level 2 qualification or have some relevant experience and knowledge of the industry.

This qualification tests both practical and knowledge-based skills, but it will not qualify you as a plumber. In order to become fully qualified, you will need to meet certain performance criteria which can be found in the National Occupational Standards which were created and structured by Summit Skills. Successful completion of the Level 2 NVQ Diploma will meet the requirements necessary for entry into the profession. It is possible to map the qualifications gained in this Level 2 Diploma across to the Level 2 NVQ Diploma, so that you only have to complete the outstanding performance units.

Your tutor or assessor will be able to explain how you can progress onto the Level 2 and 3 NVQ Diplomas. However, you should be aware that the relevant performance units will need to be carried out in industry.

About this book

This book supports the Level 2 Diploma in Plumbing Studies. The chapters match the qualification units as follows: the unit numbering refers to the City & Guilds theory and practical units. The content is also appropriate for the Level 2 EAL qualification in Plumbing and Heating 600/6725/1.

Chapter 1	Unit 201/501 Health and safety in building services engineering
Chapter 2	Unit 202/502 Electrical principles and processes for building services engineering
Chapter 3	Unit 203 Scientific principles for domestic, industrial and commercial plumbing
Chapter 4	Unit 204/504 Common plumbing processes
Chapter 5	Unit 205/505 Cold water systems
Chapter 6	Unit 206/506 Domestic hot water systems
Chapter 7	Unit 207 Sanitation
Chapter 8	Unit 208/508 Central heating systems
Chapter 9	Unit 209/509 Drainage systems
Chapter 10	Unit 210 Understand how to communicate with others within building services engineering

Each unit consists of a set of outcomes. As far as possible, these outcomes have been presented in order within each chapter. There are occasions when the order has been changed to create a logical path through the chapter, but the entire content of each unit is explored within the book.

There are progress checks and knowledge checks throughout the book, which will enable you to assess your own level of knowledge and understanding at various stages of the course.

The book has been written by vocational lecturers with many years of experience in the plumbing and heating trade as well as in further education, where they currently teach plumbing and heating qualifications to wide and diverse groups of students.

Using this book

This book is to be used as part of your training and, beyond that, as a reference work to support your career in the plumbing and heating industry. It is not intended as a handbook to an actual installation. Nor is it a code of practice or guidance note. British Standards, manufacturer's data, and HSE documents are among the materials to be used for actual installation work.

Features of this book

This book has been fully illustrated with artworks and photographs. They will help to give you more information about certain concepts and procedures, as well as helping you to follow step-by-step processes or identify particular tools or materials.

This book also contains a number of different features to help your learning and development.

Key term

These are new or difficult words. They are picked out in **bold** in the text and defined in the margin.

Remember

This highlights key facts or concepts, sometimes from earlier in the text, to remind you of important things you will need to think about

Did you know?

This feature gives you interesting facts about the plumbing trade.

 Safe working

Red safety tips remind you of things you **should not** do.

 Safe working

Blue safety tips remind you of things you **should** do.

 Safe working

Yellow safety tips indicate warnings of **hazards** or danger.

Case study

These features highlight real life events or situations which are relevant to the plumbing and heating industry.

Working practice

These features give you a chance to read about and debate real-life work scenarios or problems. Why has the situation occurred? What would you do?

Activity

These features suggest short activities and research opportunities, designed to help you gain further information about, or increase your understanding of, a topic area

Progress check

These features contain a series of short questions and usually appear at the end of each learning outcome. They give you the opportunity to check and revise your knowledge.

Knowledge check

This is a series of multiple choice questions at the end of each unit. Answers to the questions are supplied on the training resource disk.

ACKNOWLEDGEMENTS

Pearson would like to thank all those who have contributed to the development of this book making sure that standards and quality remained high through to the final product.

Phillip Davies would like to thank all his teaching and work colleagues. Stephen Blair would like to thank his wife Teresa for her encouragement and support throughout the process of researching and writing his contributions to this book.

Thanks to Richard Dixey and Ross Barson at Leicester College; Glen Lambert, Thomas Hendriks, Lee Burd and Carl Weymouth at Oaklands College; and Clark Wiseman at Studio 8, Kidlington for their patience, assistance, advice and support during the photo shoot.

Thanks also to the National Association of Plumbing Teachers for additional help with photos and technical skills.

Health and safety in building services engineering

Chapter 1

This chapter will cover the following learning outcomes:

- Know health and safety legislation
- Know how to handle hazardous situations
- Know electrical safety requirements when working in the building services industry
- Know the safety requirements for working with gases and heat producing equipment
- Know the safety requirements for using access equipment in the building services industry
- Know the safety requirements for working safely in excavations and confined spaces in the building services industry
- Be able to apply safe working practice

Introduction

Whether it is an office or a construction site, the workplace is a dangerous environment. Equipment of all kinds is in constant use; there are electrical supplies, cables and other tripping hazards, dangerous tools, chemicals, many of which are toxic or flammable, and people – all busy in whatever jobs and roles they are employed to do.

Accidents happen and can result in injury, disability and even death. While this is, of course, terrible for the victims, it is also traumatic for their families and loved ones. Accidents at work can lead to loss of earnings, personal problems and drastic changes to your life.

For the company, there may be a heavy financial cost resulting from lost hours, lost business, possible legal action and fines to pay. Added to this is the loss of credibility if your company is seen to be one where health and safety is not the main priority.

Because of this, successive governments are constantly introducing and updating regulations and processes to make sure that, as far as possible, working life, and the workplace itself, are safe and healthy.

The building services engineer is expected to have a general knowledge of health and safety laws and practices. Ignorance is no defence, so it will be pointless to tell a judge 'I didn't know I wasn't supposed to work in that way.' This chapter is intended to help you gain the health and safety awareness you will need in the workplace.

KNOW HEALTH AND SAFETY LEGISLATION

Legislation is a set of rules put in place by the government. It is debated and refined until it can be voted in and become an act of parliament. This legislation is intended to make sure that the workplace is safe and healthy, not only for the workers themselves, but for anyone in the vicinity or connected with the work. This includes everyone from site visitors to the general public.

Health and safety legislation is enforced by inspection, and prosecution if any serious breaches of the law are discovered. Serious accidents may be followed by an inquiry, both to investigate what went wrong and to establish how to prevent such accidents from happening again.

Figure 1.01: The Houses of Parliament

The responsibilities of individuals under health and safety legislation

Everyone who works on, visits or passes near to a work area is responsible for their own health and safety and for that of others.

Employers

Employers must make sure that the workplace and equipment they own (or are responsible for) are **fit for purpose**, safe and healthy. Employers are also responsible for providing safety equipment such as hard hats, ear defenders and safety features on tools and machinery. They must appoint people who are responsible for introducing and maintaining health and safety standards and processes, from managers who are responsible for the health and safety of the whole organisation, to first aiders who are there to help individual staff members if they are injured.

Employees

It's no use saying 'Well, it isn't my job to worry about health and safety, I only work here'. The law doesn't see it that way and has placed the following responsibilities on the employee.

- Take care of yourself and of any other person who may be affected by your actions.
- Report any equipment or other issues that pose risks to health and safety.
- Cooperate with your employer in relation to health and safety issues.
- Use any safety equipment provided – for example, you MUST wear a hard hat on a construction site, even if you don't think it suits you!
- Do not interfere with or misuse anything provided in the interests of health and safety.

Contractors

A contractor is someone who has been taken on by an employer to do a specific job, but who is not an employee of that company. It is vital that:

- the right contractor is used for the job
- there is a clear understanding of what the contractor is expected to do and what is involved in the work, including any risks
- the contractor receives training before starting work
- the contractor provides training to everyone who will use any specialist installation or equipment that they set up.

Visitors to site

A visitor to a workplace can be an inspector, the customer, a repair engineer, a delivery driver or anyone else who needs to enter the area on official business. Once on site they are exposed to many of the same hazards and risks as employees and contractors.

- Training – make sure the visitors are aware of how to conduct themselves safely in your workplace. This may involve a short training talk, for example.
- Visitors must be provided with any PPE needed in that area.
- Visitors are responsible for wearing the PPE and using any other safety equipment or procedures that are in place.

> **Key term**
>
> *Fit for purpose* – using the right tool for the right job and the right environment for the work. For example, try assembling delicate electronic components in a dirty, tumbledown barn. The environment is simply not right for that job. It is not 'fit for purpose' and will lead to a faulty product.

Figure 1.02: A work area protected by barriers and warning signs

General public

Although the general public is banned from most construction sites, there are times when building services engineering work takes place in a public area, for example a shop, school, office or in the street. In this case, the work area must be protected by barriers and warning signs making it clear that:

- the work area is hazardous
- only the engineer and other authorised people are allowed to cross the barriers.

Barriers and signs must force people to avoid the area, for example by diverting people a safe distance around the base of a scaffold tower so that they will not be injured by falling objects or bump into the tower and cause the engineer to fall off.

Members of the public have their own responsibilities to heed barriers and warning signs and not enter a work area.

Statutory health and safety materials

Workplace health and safety law is explained in a series of documents issued by the government. These are **statutory** documents, which means that you must adhere to them. If you don't, you are breaking the law and can be prosecuted.

There are two main types of regulations:

- general – these apply to everyone in the workplace, whether it is a construction site or an office
- specific – these deal with specific types of work, for example working at height or in a confined space.

This section will look at the general statutory documents used in the building services engineering workplace.

Progress check 1.01

1 What is legislation?
2 What are an employee's main health and safety responsibilities?
3 A customer wants to visit your site. How will you keep the customer safe?

Key term

Statutory – statutory documents are those that have been debated and issued by the government. The word 'statutory' comes from the fact that they are on the Statute Book, which means that they are part of the law of the land.

Health and Safety at Work etc. Act 1974

The Health and Safety at Work etc. Act 1974 is the central document on which all other statutory health and safety regulations are based. Its basic message is that:

- employers have a duty to keep their employees, contractors and visitors safe in the workplace
- employees must take care of their own health and safety and cooperate with their employers with regards to health and safety.

The Health and Safety Executive (HSE) has powers to enforce health and safety law on a company and can prosecute them if they do not cooperate.

Reporting of Incidents, Diseases and Dangerous Occurrences Regulations (RIDDOR) 1995

RIDDOR is a regulation that makes employers, or anyone else in charge of a workplace, responsible for reporting:

- serious workplace accidents
- occupational diseases
- **dangerous occurrences**.

Accidents and other incidents that cause people harm in the workplace need to be reported as soon as possible. This is because an investigation may be required. It will also provide a record to help with treatment if there are any long-term effects from the incident. There is an online RIDDOR form on the HSE website.

Types of reportable injury include:

- fracture (except to fingers, thumbs and toes)
- loss of a limb
- dislocated joint
- loss of sight (temporary or permanent)
- burns
- electric shock serious enough to cause unconsciousness and require resuscitation or hospital admittance.

Electricity at Work Regulations 1989 (revised in 2002)

Although the title of these regulations contains the word 'electricity', the Electricity at Work Regulations do not just apply to people working on electrical systems. The intention of the Electricity at Work Regulations is to make sure that workplace electrical installations and equipment are safe for all the people working on the premises. Some of the main points are that:

- electrical systems must not cause any risk to anyone working near to them or working with them
- electrical systems must be regularly maintained
- electrical equipment will be strong and capable enough of doing its job without endangering anyone
- there must be a means of cutting off the supply, both automatically if there is a fault and manually if the system or equipment has to be worked on.

> **Key term**
>
> *Dangerous occurrence* – a near miss. This is when an accident occurs that could have caused serious injury but didn't, for example, if a scaffold collapses after working hours when the entire workforce is off site.

Personal Protective Equipment at Work Regulations 1992

Personal protective equipment, or PPE as it is usually called, is the safety gear used to protect your body while you are at work (see Figure 1.03). PPE includes:

- hard hat
- protective boots
- high-visibility (hi-vis) jacket or waistcoat
- ear defenders
- gloves
- masks and respirators.

The employer should provide necessary PPE to their employees free of charge. The employees must wear, and take good care of, the PPE provided. On most construction sites it is mandatory to wear a hard hat, high-visibility clothing and protective footwear. The rule is: No PPE = No Entry.

If any item of PPE is damaged or faulty, you should not wear it. Report the fault to your employer and dispose of the item.

Goggles protect the eyes from flying objects when using power tools such as drills, saws and cutters.

Hi-vis waistcoat makes sure you can be seen at all times. Compulsory on most construction sites.

Protective footwear protects feet from injury from heavy objects. Compulsory on most construction sites.

Figure 1.03: An employer must issue all employees with PPE

Control of Substances Hazardous to Health (COSHH) 2002

It would be easy to think that 'substances hazardous to health' just means poisons or dangerous liquids such as acid, but these regulations apply to any substance used in the workplace, for example:

- paint
- varnish
- cleaning fluids
- adhesives
- solvents (used to weld certain plastics).

Even the most everyday products contain ingredients that can be harmful, whether by contact with the skin or by inhaling fumes. Prolonged exposure to many substances, such as the glue used by an electrician to fit together parts of a plastic conduit system, can cause illness and injury.

COSHH is really a huge set of regulations covering a vast range of chemicals and substances, from dust to the components of nanotechnology. There are data sheets and guides for specific chemical types. These guides contain instructions and recommendations for the safe handling and use of each chemical. The main requirements that apply to most substances are as follows.

- Use the substance according to the manufacturer's instructions.
- Be aware of the specific hazards before using a substance – this information should be provided by the manufacturer (on the tin or instruction leaflet, for example).
- Prepare a risk assessment (see pages 13–14) before using any type of chemical substance.
- Provide good ventilation in the work area.
- Provide (and wear) suitable PPE such as respirators, gloves and eye protection.

Figure 1.04: A COSHH notice

Provision and Use of Work Equipment Regulations (PUWER) 1998

Nearly every job, whether it is a practical trade or office-based, involves the use of tools or work equipment. This can be anything from hammers and screwdrivers to computers. Tools get a lot of hard use in the workplace and once damaged, worn out or faulty, they can become dangerous. For example, a loose hammer head might fly off while in use and electrical equipment can heat up or cause an electric shock.

The Provision and Use of Work Equipment Regulations provide guidelines on the use and provision of tools and other work equipment. The main message of PUWER is that work equipment should be:

- the right equipment for the right job (do not use a pair of pliers to hammer in a nail, or a hammer to drive in a screw!)
- only used if it is safe
- maintained in a safe condition and, in some cases (e.g. electrical, compressed air and hydraulic equipment), inspected and tested
- accompanied by information, instructions and training
- fitted with safety devices, warning labels and markings.

Electrical equipment, which covers everything from a power tool to a television in a hotel room, must be tested regularly. This type of testing is called Portable Appliance Testing (PAT). You can read more about PAT on page 136.

Control of Asbestos at Work Regulations 2012

Asbestos is a mineral found naturally in certain rocks. When separated from rock, it becomes a fluffy, fibrous material. In the past, it was used for many purposes in the construction industry, including in cement products, roofing, plastics, insulation, floor and ceiling tiling, and fire-resistant board for doors and partitions.

There are three main types of asbestos:

- chrysotile – white (accounts for about 90 per cent of asbestos found)
- amosite – brown or grey
- crocidolite – blue.

Asbestos is a very dangerous substance. Although it comes under the COSHH regulations, it is so hazardous that it also has its own set of regulations. Asbestos must never be drilled, cut or removed by anyone who is:

- not trained to work with asbestos
- not wearing the appropriate PPE – which includes full body protection and respirator equipment.

Asbestos fibres cause a disease of the lungs called asbestosis. It is not only the worker who is at risk from asbestos dust and fibres, but also anyone else in the vicinity. If asbestos has to be removed from a building or area, that area has to be shut down, sealed off and the asbestos itself cleared by fully trained, licensed and equipped professionals.

If you inadvertently drill or cut asbestos, then it must be reported as a dangerous occurrence, covered by the Reporting of Incidents, Diseases and Dangerous Occurrences Regulations (RIDDOR) (see page 5).

The Control of Asbestos at Work Regulations also state:

- If existing asbestos is in good condition and not likely to be damaged, it can be left in place. However, its condition must be monitored and managed to make sure it is not disturbed.
- The presence of asbestos has to be made known to anyone who is going to carry out work in the area.
- Training is mandatory for anyone likely to be exposed to asbestos fibres while they are at work.

Health and Safety (First Aid) Regulations 1981

First aid is the initial treatment and help given to someone who has suffered an accident or been taken ill. In the workplace there should be trained staff who can carry out first aid immediately. The names of trained first aiders and their contact details must be clearly displayed so they can be called as soon as an accident occurs. Any accident must be recorded in an accident book as soon as possible after the incident. It doesn't matter whether the injury or illness is work-related or not.

An employer must provide first aid equipment and facilities. The regulations require:

- a first aid box
- an appointed person who is responsible for all first aid arrangements and facilities
- information for employees about first aid arrangements.

Figure 1.05: A first aid box is essential in all work areas

Workplace (Health, Safety and Welfare) Regulations 1992

Not only must people in the workplace be protected against accidents, it is also important to consider their welfare. This means their health and well-being. For example, proper hygienic toilet and washing facilities must be provided. There must be adequate lighting and working space. People in the workplace should, as far as possible, be comfortable enough to carry out their work without risk to their general health.

Working at Height Regulations 2005

Please refer to pages 33–38 for more information.

Manual Handling Operations Regulations 1992

Please refer to pages 44–45 for more information.

Confined Spaces Regulations 1997

Please refer to pages 41–43 for more information.

Please refer to pages 33–38 for more information.

Please refer to pages 44–45 for more information.

Please refer to pages 41–43 for more information.

Progress check 1.02
1 What is RIDDOR?
2 List five items of PPE.
3 Who is authorised to work with asbestos?

Activity 1.1

Research four more health and safety regulations relevant to the building engineering services sector. Produce a table giving the name of each regulation, its purpose and main points.

Non-statutory health and safety materials

Non-statutory documents are not law but they are based on statutory regulations. They provide guidelines on how to work according to 'best practice'. Even though they are not in themselves mandatory, working according to non-statutory codes of practice and guidelines means that you will be applying safe and effective methods.

Most health and safety at work regulations are available from the Health and Safety Executive (HSE). The HSE also provides sets of non-statutory, easy-to-understand guidelines explaining how the regulations can be applied in everyday work situations.

The different roles enforcing health and safety legislation

In order to manage and maintain health and safety both nationally and within a company, various organisations have been set up and roles created. Some of the main ones are outlined below.

Health and Safety Executive (HSE)

The HSE is a government agency that has the authority to enforce health and safety law. Its inspectors can carry out checks in the workplace and impose improvement notices on any business that does not comply with health and safety law. The HSE can also prosecute a business if it does not comply.

Intimidating as all this sounds, the HSE is also there to advise and help businesses make their workplaces safe for employees, visitors, contractors and the general public.

Safety manager

A safety manager is a member of the management team. He or she has responsibility for making sure that health and safety laws and practices are put in place throughout the company. The safety manager may be the head of the company or someone at senior level. They will not necessarily carry out the work themselves, but assign it to safety officers and representatives. They do, however, bear responsibility for health and safety compliance.

Safety officer

This is a more hands-on role, and a safety officer is the person you are most likely to see regarding keeping your workplace safe. The officer will carry out inspections and has the authority to order improvements. The safety officer may be full time in that role, as is the case in many large companies, or they may have other responsibilities as well as health and safety.

Safety representative

Safety representatives are usually members of the workforce who are responsible for making sure that their particular part of the company or workplace is safe, and also for reporting health and safety matters to the management. For example, if the electricians tell their safety representative that the ladders provided by the company are not safe, the representative will go to the management, usually via the safety officer, bring this to their attention and argue the case for the ladders to be replaced with safer ones.

First aider

If someone has an accident in the workplace, the first person on the scene is a first aider. These are employees who have been trained to deal with the immediate consequences of an accident. For example:

- to control heavy bleeding in the case of a cut
- to provide support for an injured limb, for example a temporary sling or splint
- to attend to an unconscious patient by trying to revive them or making sure they are safe and in the correct position.

The names and contact details of all first aiders must be clearly displayed so that they can be called on quickly if there is an accident (see Figure 1.06).

Many companies give all their staff some basic first aid training on a regular basis, so that they can, at least, keep the victim of an accident safe and calm until a trained first aider can get to the scene.

Fire warden

Fire is an ever present danger in all workplaces. Construction sites contain many fire hazards such as gas bottles, chemicals and naked welding and brazing flames. There may also be paper and plastic packaging and wood and wood shavings everywhere.

The fire warden is a designated employee who has been trained to get you to the correct assembly point if the fire alarm is sounded. Once at the assembly point, they will take a register to make sure everyone is present. If someone is missing, the fire warden must not go back into the building to try to find them. Instead, they will report the missing person to the fire service and to the relevant safety officer.

Figure 1.06: A first aider contact notice

The fire warden also has the authority to inspect the work area and order the removal of any fire hazards.

Progress check 1.03

1 What is the HSE?
2 Who represents the workforce in matters of health and safety?
3 Who is the first person to call when someone is injured in the workplace?

KNOW HOW TO HANDLE HAZARDOUS SITUATIONS

The workplace contains many **hazards**. Some hazards are more immediately apparent and dangerous than others: a construction site, for example, is filled with hazards, whereas a shop or office is relatively safe. Inevitably, accidents will happen and hazardous situations will arise. You need to know how to deal with them. This section will tell you both how to avoid and how to handle hazardous situations.

Key term

Hazard – a situation that poses a threat. For example, a drill with a damaged power lead is a hazard because if you attempt to use it, the drill could give you an electric shock.

Hazardous situations on site

Table 1.01 lists some of the typical hazards found in the kind of workplace building services engineers such as electricians and plumbers will encounter. This is by no means a complete list, but shows the main hazards. Some of these will be covered in more detail later in the chapter.

Hazard	Possible results	Causes

Tripping hazards / trailing leads (see Figure 1.07 on page 12)	• Falls resulting in:
	  o sprains and broken bones
	  o head injury
		• Trailing leads and hoses
		• Untidy workplace

Slippery floors and surfaces	• Falls resulting in:
	  o sprains and broken bones
	  o head injury
		• Wet floors
		• Newly cleaned and polished floors

Electric shock (see page 26)	• Heart failure • Burns • Unconsciousness	• Faulty power tools and equipment • Damaged cables • Incorrect wiring and connections • Exposed cable joints and connections • Water around electrical fittings and supplies • No earthing arrangements
Fire (see pages 30–32)	• Burns • Explosion • Smoke inhalation	• Incorrect handling of gas bottles and other flammable substances • Smoking in a no smoking area • Presence of flammable materials such as paper and wood shavings • Carrying out welding or gas cutting operations near to flammable materials or in a flammable atmosphere
Dust and fumes	• Respiratory problems • Eye and skin problems	• Cutting and drilling masonry • Sweeping up • Use of chemicals – paints, cleaning fluids, etc. • Fibreglass insulation

Table 1.01: Common hazards on a construction site

Continued ▼

Hazard	Possible results	Causes
Contaminants and irritants	• Respiratory problems • Eye and skin problems	• Use of chemicals – paints, cleaning fluids, etc.
Asbestos (see page 8)	• Major respiratory problems	• Cladding found in older buildings

Working at height and in confined spaces (see pages 33–43)	• Falls resulting in:
	  o sprains and broken bones
	  o head injury
	• Dropped materials and equipment causing head injuries to people below
	• Breathing problems and even asphyxiation
		• Ladders
		• Towers
		• Scaffolding
		• Working above trenches and other excavations
		• Working on a roof
		• Working in a confined space such as a duct or cramped plant room with insufficient ventilation

Gas bottles (see pages 28–30)	• Explosions • Fire • Asphyxiation • Low temperature burns	• Welding and cutting • Brazing • Laying bitumen on a flat roof • Paint stripping

Malfunctioning and damaged tools and equipment	• **Hazardous malfunction** resulting in:
	  o eye injury
	  o cuts, bruises and general injuries
	  o electric shock
	  o burns
		• Damaged electric equipment and leads
		• Poorly maintained equipment
		• Using damaged or worn out tools
		• Using the wrong tool for the job

Manual handling and lifting (see pages 44–45)	• Back injuries • Sprains and strain injuries • Cuts to hand • Foot injury	• Incorrect lifting technique • Attempting to manually lift something that is too heavy or awkward • Not using lifting equipment • Failing to wear PPE such as gloves and protective footwear

Table 1.01: Common hazards on a construction site (continued)

Key term

Hazardous malfunction – this occurs when a tool or item of equipment goes wrong and does not injure anyone, but could have caused injury to the person using it or to people in the vicinity.

Figure 1.07: Tripping hazards

Progress check 1.04

1 What is a hazard?

2 What are the possible results of an electric shock?

3 What can cause manual handling and lifting injuries?

Safe systems of work

Risks can be reduced by using safe systems of work. These are procedures that allow you to think about and prepare for the possible hazards presented by a particular job.

Risk assessment

Most risk assessments are an automatic process. You look at a job activity, work out what tools you will need and how you will protect yourself from the hazards involved in the work. We do it all the time – before crossing the road, for example. The first thing we do is check that there is no traffic and that it is safe to cross. We are, in fact, assessing the risks involved in stepping out onto the road and trying to reduce them by crossing when there are no cars coming our way.

For larger, more complex work, a written risk assessment has to be produced. This will be a form with a set of headings. Typical headings are:

1 Tasks – Break down the project into a series of tasks.

2 Hazards – List the hazards associated with each task.

3 Control – How will the risks be alleviated? For example, what PPE is needed?

4 Who is affected – Who will be affected if an accident actually happens?

5 Seriousness – This will be coded using letters or numbers. For example, using the scale in Table 1.02, if the hazard could end up in the loss of a limb or eye, then it will be graded as a 4. Something that results in a bruise but is not serious or life-threatening will be graded as a 1.

6 Likelihood – How likely is it that the hazard will actually cause an accident? Again, this is graded using a number or letter system (see Table 1.03).

Injury or loss	Scale value
No injury or loss	0
Treated by first aid	1
Up to 3 days off work	2
Over 3 days off work	3
Specified major injury	4
Fatality	5

Table 1.02: Maximum consequences (seriousness) of an accident

Likelihood	Scale value
No likelihood	0
Very unlikely	1
Unlikely	2
Likely	3
Very likely	4
Certainty	5

Table 1.03: Likelihood of an accident occurring

TASK	
Manual handling of loads	
Specialist tools	
APPLICATION OF EQUIPMENT	**APPLICATION OF SUBSTANCE**
Pipe-bending machines, stilsons, ropes, lead dressers, bending springs, block and tackle, spanners, etc.	N/A

ASSOCIATED HAZARDS
Risk of muscle strains
Risk of sprains
Risk of musculo-skeletal injury

LIKELIHOOD	CONSEQUENCE	RISK FACTOR
3	3	9

RISK EXPOSURE	SAFEGUARDS HARDWARE
Employees	Nil

CONTROL MEASURES
1 Specific training and instruction to employees – kinetic lifting
2 Individual assessment to be performed for all tasks
3 Workplace inspections conducted at 3-month intervals
4 Random safety inspections
5 Suitable and sufficient personal protective equipment
6 Medical screening for staff at risk

Figure 1.08: Risk assessment form

Figure 1.08 on page 13 shows an example of a generic risk assessment form for working with specialist tools. Most of the form is self-explanatory but you should focus on the following items:

- likelihood of accident occurring (see Table 1.03)
- maximum consequences (severity) of an accident (see Table 1.02)
- risk factor (likelihood × consequence = risk).

The risk factor calculation will give a result between 1 and 25. These are categorised as follows:

- minor risk (risk factor between 1 and 7): can be disregarded but closely monitored
- significant risk (risk factor between 8 and 15): requires immediate control measures
- critical risk (risk factor between 16 and 25): activity must cease until risk is reduced.

Other sections on the form are:

- risk exposure: the individuals or groups of people who may be affected by the work activity or process. Control measures must take account of these people
- safeguards hardware: the in-built safety features of work equipment
- control measures: additional safeguards that underpin your arrangements. Where these are identified, they must be followed through and a record of any outcomes must be kept.

Method statements

Method statements describe, in detail, the way a work task is to be carried out. The method statement should outline the hazards involved and include a step-by-step guide on how to do the job safely.

Method statements often form part of a **tendering** process and can give a potential customer an insight into your company and how it operates. Table 1.04 outlines the information needed for a method statement.

> **Key term**
>
> *Tendering* – the tendering process is the stage of a project when companies are invited to submit a price and a proposal to the customer. The customer will choose the most suitable offer and appoint that company to carry out the work.

Section	What it contains
Section 1	• A title • A brief description of the work to be carried out • Your company details • Start date and completion date for the work • Site address and contact details, including emergency numbers, etc.
Section 2	• Summary of the main hazards that are present and the measures put in place to control them • Personal protective equipment • Environmental or quality procedures
Section 3	• Staff and the training they will need • Permits to work (see page 15) • Machinery shutdown and lock off procedure • Site access • Welfare and first aid
Section 4	• Step-by-step guide to how the work is to be carried out

Table 1.04: What is contained in each section of a method statement

Permit to work

If a particularly hazardous job is to be carried out, then a permit to work should be issued. Jobs and situations that require a permit to work include:

- working on, or near, live electrical equipment or systems
- working in a confined space
- working in deep trenches
- working with or near corrosive or toxic substances
- working near machinery in production areas such as assembly lines.

The permit to work must be authorised by a **competent person**. This is someone who fully understands the dangers associated with the job. In addition:

- the work itself must be carried out by a trained and competent person
- the work itself must be detailed on the permit
- the permit must have a start and finish time.

The finish time is important because if the engineer has not returned by the time stated on the permit, it acts as a warning that there may be something wrong.

A permit to work covers one specific job and cannot be used again.

The layout of a typical permit to work will be as follows:

- Description and scope of work to be carried out
- The location of the work
- Who is responsible
- Who will actually do the work
- Precautions needed
- Hazards identified
- PPE required
- Authorisation to start work
- Necessary checks to be carried out
- Emergency arrangements
- What to do in unusual circumstances
- The time limit of the permit
- Completion signatures
- Cancellation signatures.

Safety signs and notices

Signs are an important part of workplace safety. They are not there simply as decoration; you must heed them. If a sign instructs you to wear ear defenders, then you must wear them. A standard colour scheme has been established for warning signs. Table 1.05 on page 16 lists these standard colours and gives example of signs.

> **Key term**
>
> **Competent person** – BS7671: 2008 and the associated guidance notes and codes of practice define a competent person as someone with the technical knowledge or experience to carry out electrical work without risk of injury to themselves or others.

	Prohibition signs	Mandatory signs	Warning signs	Information or safe condition signs
Shape	Circular	Circular	Triangular	Square or rectangular
Colour	Red border and cross bar Black symbol on white background	White symbol on blue background	Yellow background with black border and symbol	White symbol on green background
Meaning	Shows what you must not do	Shows what you must do	Warns of hazard or danger	Gives information about safety provision
Example				

Table 1.05: Safety signs

Hazardous substance signs and labels

Symbols are also used to denote harmful substances (see Table 1.06). One place you will see these is on the back of a tanker lorry. They give immediate information on whether the contents of the tanker are corrosive, flammable, or dangerous to health.

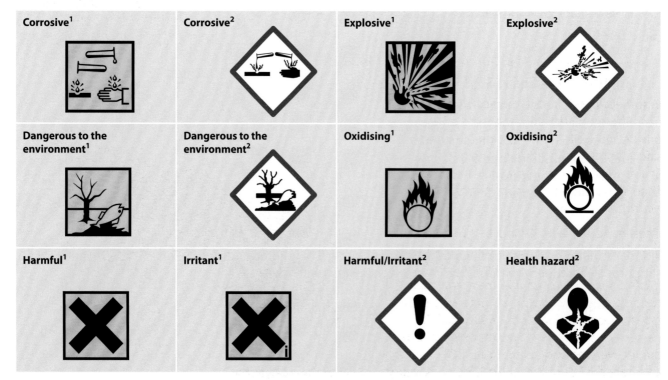

Table 1.06: Hazardous substance signs (1 European label; 2 International Hazardous Substance label)

Toxic[1]	Toxic[2]	Very toxic[1]	Gas under pressure[2]

Highly flammable[1]	Highly flammable[2]	Extremely flammable[1]

Table 1.06: Hazardous substance signs (1 European label; 2 International Hazardous Substance label) (continued)

Actions when an accident or emergency is discovered

Imagine you are working with another plumber who has just cut her hand. There is a lot of blood and your workmate is in distress. Your *first* actions should be to:

- seek or administer immediate first aid
- get help if necessary (e.g. phone for an ambulance)
- report the accident to the site supervisor.

The order in which you carry out these actions will depend on several things:

- Are you a qualified first aider?
- How severe is the injury?
- Is your supervisor or a qualified first aider immediately available?

Later, you will need to:

- write down the details in an accident report book
- complete a company accident report form.

In some situations – for example, if a colleague has suffered an electric shock – your first priority will be to make sure that the area is safe. In this case, you need to isolate the supply first and call for an ambulance if necessary. Then someone can attend to the casualty and inform the supervisor.

If someone has fallen from a height, do not move them. If they have a serious back injury, your actions could make things worse. If the injured person is conscious, make them as comfortable as possible until help arrives. This may include treating minor wounds. If the person is unconscious, they may be put into the recovery position (see Figure 1.09), but only if you are confident that there is no back injury.

> **Remember**
>
> People with back injuries should only be moved by trained personnel.

Figure 1.09: The recovery position

If someone is injured in an accident, the details must be entered in the accident book. If the accident involved a piece of faulty equipment, do not tamper with it as it may be subject to an investigation by the Health and Safety Executive.

Accidents that result in three or more days' absence from work must be reported to the Health and Safety Executive. An example of an accident report form is shown in Figure 1.10.

Other emergencies

Emergencies other than a fire or an accident could include a bomb threat or a chemical spillage. These events are rare, but you still need to be aware of the correct procedures for dealing with them.

- Find a telephone in a safe environment well away from the emergency. Make sure you are not going to become trapped while making the call.
- Dial the emergency services number – usually 999. If you are on an internal exchange, you may need to call a different number, so make sure you know of any special procedures.

> **Did you know?**
>
> Any act of aggression by a colleague at college or work that causes a situation of conflict is considered a dangerous occurrence.

Full name of injured person:	
Home address:	**Sex:** Male / Female
	Age:
Status: Employee Contractor Visitor	
Date of accident:	
Time of accident:	
Precise location:	
What was the accident and its cause? (You may have to give a detailed written description.)	
Name and address of witness, if any:	
Details of apparent injuries:	
Summary:	

Figure 1.10: A typical accident report form

- Keep calm and listen to what the emergency services operator has to say.
- When asked, give your name and the name of the service you require.
- Once you have been connected to the relevant emergency service, stay calm, listen to what the operator has to say and answer their questions as fully as possible. You will need to explain the nature of the emergency and where it occurred. Try to give the operator the exact location of the incident, for example the name and address of the company.
- When you have completed the call, arrange for someone to meet the emergency services and show them where the incident has occurred.

Evacuation procedures

If you are working in an environment where there is an increased risk of an emergency situation occurring, your employer should provide an evacuation procedure that gives details of where to leave the building and which assembly point to go to once you are outside. Here, a specified warden will check that everyone who should be at the meeting point is there. If someone is missing, report it to the emergency services but do not attempt to go back into the building to find them.

Dealing with injuries

Treatment of cuts

Assess the injury. If the cut is minor:

- clean your own hands and put on disposable gloves
- gently clean the cut in running water if it is dirty
- pat the area dry using a sterile dressing or clean cloth
- apply a plaster or sterile dressing.

Severe bleeding

For more severe bleeding, follow the steps below:

- Assess the injury. Is loss of blood the main issue or could there be other medical problems?
- Control the bleeding. This should be your first priority.
- Apply pressure to the injury, using a clean cloth or sterile dressing. If the wound is large, you may have to bring the edges of the wound together.
- Keep applying pressure to the wound and raise the injured area. This reduces blood flow to the wound.
- Ask the injured person to lie down. This brings down the heart rate, reduces blood flow and reduces the risk of shock.
- Cover the wound with a dressing (preferably sterile) and bandage this firmly in place. Make sure the bandage is not too tight – you do not want to cut off circulation to fingers or toes.
- After bandaging, monitor the dressing to check that blood is not seeping through. If it is, apply another dressing on top. Do not remove the first dressing, as this can disturb any blood clot that is forming.
- Treat for shock.
- Dial 999 for an ambulance or seek other urgent medical attention.

Treatment of eye injuries

All eye injuries are potentially serious. The casualty will be experiencing intense pain in the affected eye, with spasms of the eyelids.

Before attempting to treat the injury, wash your hands. If there is something in the eye, irrigate with cool, clean water or sterile fluid from a sealed container to remove loose material. Do not attempt to remove anything that is embedded. If chemicals are involved, flush the open eye with water or sterile fluid for at least 10–15 minutes. Apply an eye pad and send the injured person to hospital.

Exposure to fumes

If you find someone suffering from the effects of fume inhalation or asphyxiation, get them outside into the fresh air as soon as possible, provided it is safe to do so. Loosen any clothes around their neck or chest that may impair their breathing.

Call the emergency services. Remember that some fumes, such as welding fumes, contain carcinogens (cancer-producing substances) so there is a need for urgent first aid and medical management following accidental exposure.

Chapter

1

Progress check 1.05

1   State three headings found on a risk assessment form.

2   What is the purpose of a method statement?

3   Why is the finish time on a permit to work so important?

Treatment of broken bones

If a broken bone is suspected, get expert help. Do not move the casualty, unless they are in a position that exposes them to immediate danger.

Unconsciousness

Any unconscious person should be placed in the recovery position (see Figure 1.09 on page 17), as long as there is no obvious back injury. This position prevents the tongue from blocking the throat and allows liquids to drain from the mouth, reducing the risk of the casualty choking on their own vomit. The head, neck and back are kept in a straight line, while the bent limbs keep the body propped in a secure and comfortable position.

KNOW ELECTRICAL SAFETY REQUIREMENTS WHEN WORKING IN THE BUILDING SERVICES INDUSTRY

Vital as it is to our working lives, electricity has its own unique hazards. These include:

- shock – caused by direct contact with a live part such as an exposed conductor or terminal. Shock can result in a burn and, in some cases, cause unconsciousness and even heart failure
- fire – caused by:
  - o an overheated cable
  - o arcing caused by loose connections
  - o short circuit, which is an explosive increase in current caused by a direct contact between line and neutral, or line and earth.

Case study

A small manufacturer of agricultural equipment had set up its operation in a former warehouse. Although production had started, there were still alterations to be completed in order to convert the premises into a fully functioning unit.

A plumber was installing pipework at high level above a three-phase distribution board located in the plant room. As he climbed the stepladder he was using, he reached out to grab the top of the distribution board to steady himself. It was an instinctive action and would not usually be dangerous.

What he did not know was that someone had removed a number of circuits and, instead of disconnecting the cables from the distribution board and pulling them out, had simply snipped the cables and left them protruding from the top of the board, unseen from ground level. The plumber's hand came to rest on these cables.

The 230 V shock he received was painful, potentially lethal and made him feel ill for several hours afterwards.

1   Who (if anyone) was to blame in this situation?

2   Is the incident reportable? If so, who should it be reported to?

3   How could this incident have been avoided?

Common electrical dangers to be aware of on site

Construction sites are harsh environments and, because of this, electrical supplies, equipment and tools should be designed to survive the rough treatment they are likely to receive. Some common electrical hazards in the workplace are covered below.

Faulty and damaged equipment

All electrical work equipment should be regularly inspected and tested to make sure that there is no damage to:

- the casing
- the lead
- the plug.

It should also be tested to make sure that it works correctly and that there are no hidden electrical faults such as loose connections and poor earth **connectivity**. If any equipment appears to be damaged, you must not use it and it must be replaced or sent away for repair.

The care and use of electrical work equipment and tools is covered by PUWER (see page 7) and the testing of electrical equipment is described in the IEE Code of Practice for In-service Inspection and Testing of Electrical Equipment.

Classes of electrical tools and equipment

The classes of electrical equipment are among the key information contained in the code of practice.

- Class I – has an earth conductor connected to the frame of the equipment.
- Class II – relies on extra or toughened insulation to protect the user from an electric shock. There must be **no** earth connection.
- Class III – protects the user from electrical shock by transforming the voltage down from the mains voltage of 230 V to an extra low voltage of, for example, 24 V. This extra low voltage cannot harm a human being, so even if there was a fault and the user received a shock, it would not cause them injury. There must be **no** earth connection.

Test intervals for electrical tools and equipment

The IEE Code of Practice for In-service Inspection and Testing of Electrical Equipment also contains a table with recommended intervals between testing. This depends on:

- the type of electrical equipment, for example hand-held or portable
- the environment the equipment is used in, for example electrical equipment used on a construction site will need to be tested more often than that used in a shop.

Types of electrical tools and equipment

The equipment testing code of practice classifies electrical equipment into a number of main types. These are described in Table 1.07.

Back-up supply and back-feed hazards

Some items of electrical equipment can pose an electric shock hazard even when the supply is switched off.

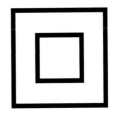

Class II

Class III

Figure 1.11: Class II and Class III equipment labels

Electrical equipment	Description
Hand-held appliance or equipment	Held in the hand during its normal use, for example an electric drill or a soldering iron
Moveable equipment	Weighs up to 18 kg and can be unplugged and moved, for example a welding set
Portable appliance and equipment	Plugged in and can be moved while it is being used, for example a kettle
Stationary appliance or equipment	Fixed in place or weighing more than 18 kg and not intended to be moved regularly, for example a wall-mounted heater or a washing machine

Table 1.07: Main types of electrical tools and equipment

Key term

Mechanical damage – physical damage to cables or electrical equipment. Sheaths, enclosures and containment, cable armouring and toughened casings are all called mechanical protection.

 Safe working

When using extension leads:

- Always use them fully extended because they will become hot if left wound on the reel.
- Do not plug one extension lead into another. Get a longer extension lead instead.
- Always use a three-core extension lead.
- Fit an RCD plug if the lead is longer than 3 m.

Figure 1.12: Cable cover for leads and cables laid across a road or walkway

Faulty or damaged cables

There are a lot of electrical cables on a construction site – equipment and tool leads, extension leads, as well as wiring for temporary lighting and supplies to site huts and offices.

All leads, whether connected directly into electrical equipment or used as extension leads, must be regularly inspected for damage. The main hazards posed by cables are electric shock and fire.

Electric shock

The cable used must be of a standard and type designed to withstand the **mechanical damage** likely on a construction site. If the sheath and insulation are damaged, the conductors will be exposed, posing a shock hazard. Also, cables need to be connected securely so that live conductors do not slip out of the terminal and become exposed.

Fire caused by short circuit or overheating

A cable of the correct size must be used because, if the conductor cross-sectional area is too small for the current it carries, it will heat up and pose a fire risk. Likewise, a loose connection will arc and become hot.

Tripping hazard

Extension leads and equipment leads tend to be strewn about the site and can be a tripping hazard. If a lead has to be run across a walkway or road, it should be covered by a cable protector of some sort (see Figure 1.12). If the cable is to be in place for some time, it is a good idea to secure it with cable ties.

Proximity of cables

When working, you must always be aware of any nearby electrical cables. It is easy to cut or damage a cable.

Buried and hidden cables

When digging trenches or drilling or chopping into walls, you must be aware that there might be cables hidden under the surface. All underground cables should be marked on a layout drawing so that they are not damaged when excavation takes place.

For cables buried in walls, check before drilling that there isn't a switch or socket or other item of electrical equipment directly below. If there is, be aware that it is probably near a cable passing through the area you are about to work on. It is not always possible to be sure where a cable runs, so you made need to use a cable detector to locate any hidden cables.

Sources of electrical supply for tools and equipment

One way of reducing the risk of electric shock from power tools is to supply them at a reduced voltage. There are three classes of voltage:

- extra low voltage – 0 V to 50 V (not considered harmful)
- low voltage – 51 V to 999 V (including our 230 V mains supply, which is considered harmful and potentially fatal)
- high voltage – 1000 V and higher (extremely dangerous and harmful).

By using an extra low voltage, power tools are made safe because, even if there is a fault, the voltage will be too low to cause injury. The two main methods of reducing power tool voltage are described below.

Battery supplies

The simplest method is to supply the tools with a battery. These are usually rechargeable. Most battery-powered power tools are supplied with two batteries and a charging unit. Having two batteries means that one can be kept on charge while the other is in use.

Apart from safety, the other advantage of battery-powered tools is that there is no power lead, which can become damaged, or constantly plugged in and unplugged as you move around the work area causing a tripping hazard.

110 V supplies

An alternative method of supplying power tools at a reduced voltage is to transform the supply down to 55 V. This is achieved by plugging the power tool into a 230 V/110 V centre-tapped transformer. This means that:

- the supply is 230 V
- the output to the power tool is 55 V.

The 110 V transformer produces an output of 55 V (half of 110 V) because it is centre-tapped.

110 V centre-tapped transformers, their sockets and associated plugs are coloured yellow to distinguish them from other voltages (see Figure 1.14).

 Safe working

Only good-quality cable detectors should be used for finding hidden cables. Ask an electrical wholesaler for advice. Look for a BS, EN or ISO number on the detector.

Figure 1.13: A battery-powered drill

 Safe working

- Always use the correct charger for power tool batteries.
- Always use the correct battery with each tool.

 Safe working

Never try to recharge non-rechargeable batteries as they could heat up and catch fire.

Figure 1.14: A centre-tapped transformer

The advantage of this type of voltage reduction method is that it is constant and does not run out like a battery. The main disadvantages are:

- although there are portable versions available, the transformers are heavy
- transformers need power leads
- 230 V supply is present, both in the transformer and in the transformer's supply lead.

Generators

A generator is a mechanical machine that produces electricity. Smaller, portable generators are available for supplying power to construction sites. You may find yourself in an old building or remote site that has no electricity supply. In this case, you will need a portable generator.

These machines are usually powered by petrol or diesel and can supply electricity at various voltages. Because fuel is needed, there is a risk of fire so the generator fuel must be stored and handled carefully. Remember that, although a generator is a petrol-driven machine, it is producing electricity and electricity is always dangerous! Another hazard is fumes from the generator exhaust, so make sure it is placed in a well-ventilated area.

Safe isolation procedure

The correct procedure for safe isolation of electrical supplies is covered in more detail in Chapter 2: Electrical principles and process for building services engineering, pages 72–75.

Back-up supply hazards

Most electrical equipment and components are safe when isolated from the supply. However, some components are dangerous even when the power is switched off.

Many organisations are equipped with back-up power so that vital systems continue to run if there is a mains power failure. Typical back-up systems are:

- generators
- uninterrupted power supplies (UPS).

Generators

Generators are often installed to start up automatically and provide back-up power in the event of a power failure. When shutting down and working on electrical circuits in generator-supported areas, you must take care to isolate the generator from the installation. If not, the generator might start and cause circuits to become live.

Uninterrupted power supplies (UPS)

Even if there is a generator installed as a back-up for mains power, there is often a delay of several seconds before the machine starts and emergency power is fully restored. The UPS is designed to supply emergency power during the intervening time. It consists of a bank of batteries that are under constant charge while the mains power is available. When mains power fails, the batteries are switched in.

A UPS is not intended to supply power for long periods of time but to fill in during the generator start-up. Because it is essentially battery power, a UPS supply has to be converted to AC using inverters.

As with generators, you must take care when working on UPS-protected circuits because the UPS will automatically switch in once mains power is lost.

Batteries

Batteries cannot be switched off and, as a result, can be dangerous. While charged, a battery can deliver an electric shock at any time.

Banks of batteries must be well ventilated because some types give off toxic fumes and contain harmful chemicals. Occasionally, batteries will explode. You must wear appropriate PPE at all times when working with batteries. This PPE consists of:

- goggles
- gloves
- a protective overall or apron.

Back feed hazards

Photovoltaic arrays

More and more houses and other premises are being supplied using photovoltaic cells. These are banks of cells, usually mounted in the roof, that convert sunlight to electricity. Like the power supplied from a UPS, this electricity is created as DC and then converted to AC.

Photovoltaic arrays are always producing electricity as long as there is light, so without the correct equipment installed, or if handling the panels directly, there is a risk of electric shock.

Capacitors

Capacitors are designed to charge while the power is on. Once the power is switched off, they will discharge and sometimes this discharge can be high voltage. Banks of capacitors are used for power factor correction in large industrial installations and as part of electronic equipment. They are also used for smoothing voltage when it is converted from AC to DC by use of rectifiers.

Transformers

Transformers consist of conductors wound around an iron core and this can cause **back EMF**. Back EMF is produced by the following process.

1 When the supply is switched on, the iron core becomes an electromagnet.

2 When the supply is switched off, the current in the windings drops away almost instantly.

3 The magnetic field in the core, however, takes longer to fade.

4 As the core magnetic field collapses, it induces a current back into the windings, which could cause a shock to anyone in contact with them.

> **Key term**
>
> *Back EMF (electromotive force)* – a voltage produced within electromagnetic windings (such as those in motors and transformers). As well as producing the voltage required, the magnetic field in the iron core produces this second, opposing voltage. The current from back EMF flows against the supply current. This opposition is called reactance, which reduces the amount of power produced. This inefficiency causes a low power factor.

Electric shock procedure

If a colleague receives an electric shock, you should follow this procedure.

1 Do not touch the victim – or you will receive an electric shock as well.

2 Switch off the power – when someone is hurt, shutting down an area or operation no longer matters so, if necessary, the main isolator can be operated.

3 If safe to do so, remove the victim from the supply. You will have to use a non-conducting object to do this such as a wooden broom handle or piece of wood.

4 Call for help – nearly everyone carries a mobile phone so dial 999 for an ambulance as quickly as possible. Also send someone, or go yourself, for a first aider.

Cardiopulmonary resuscitation (CPR)

One of the results of a severe electric shock is stoppage or interference with the heartbeat. Because of this, an electric shock victim may be unconscious with no pulse or breathing. If this is the case, you can apply CPR while you are waiting for an ambulance to arrive. CPR consists of chest compressions and rescue breaths, which keep blood and oxygen circulating through the body.

Before beginning CPR, you must do the following:

- Open the airway. An unconscious person's airway may be narrowed or blocked. The main reason for this is that muscular control in the throat is lost, allowing the tongue to sag back and block the throat. Lift the chin and tilt the head back to lift the tongue away from the entrance to the air passage.
- Remove any obvious obstructions from the mouth. Place two fingers under the point of the casualty's chin and lift the jaw. At the same time, place your other hand on the casualty's forehead and tilt the head well back (see Figure 1.15).

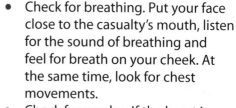

- Check for breathing. Put your face close to the casualty's mouth, listen for the sound of breathing and feel for breath on your cheek. At the same time, look for chest movements.
- Check for a pulse. If the heart is beating adequately, it will generate a pulse in the neck. To check for the pulse, ensure the head is tilted back and feel for the Adam's apple with two fingers. Slide your fingers back into the gap between the Adam's apple and the strap muscle that runs from behind the ear across the neck to the top of the breastbone. Feel for 10 seconds before deciding there is no pulse.

Figure 1.15: Opening the airway

CPR (based on NHS guidelines)

1 Place the heel of your hand on the breastbone at the centre of the person's chest. Place your other hand on top of your first hand and interlock your fingers.

2 Using your body weight (not just your arms), press straight down by 5–6 cm on their chest at a rate of 100 chest compressions a minute (see Figure 1.16).

3 After every 30 chest compressions, give two breaths. Tilt the casualty's head gently and lift the chin up with two fingers. Pinch the person's nose. Seal your mouth over their mouth and blow steadily and firmly into their mouth. Check that their chest rises. Give two rescue breaths, each over one second.

4 Repeat this until an ambulance arrives.

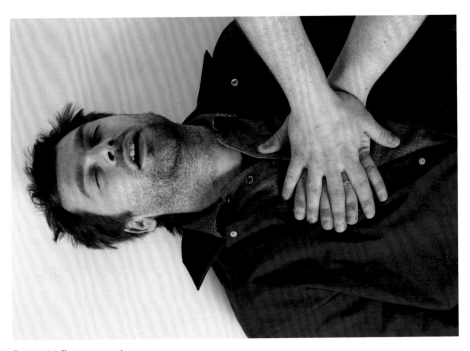

Figure 1.16: Chest compression

When you call for an ambulance, telephone systems exist that can give basic life-saving instructions, including advice on CPR. These are now common and are easily accessible with mobile phones.

Treatment of minor burns

Contact with a live part can cause burns to the skin. This can be from either prolonged contact or as a result of an electrical flash or explosion. Serious burns must be treated in hospital, but you can apply the following first aid to minor burns.

1 Immediately get the person away from the heat source to stop the burning.

2 Cool the burn with cool or lukewarm water for 10–30 minutes. Do not use ice, iced water or any creams or greasy substances, such as butter.

3 Remove any clothing or jewellery that is near the burnt area of skin, but do not move anything that is stuck to the skin.

Chapter 1

Progress check 1.06

1 What are the two main hazards associated with electricity?

2 What do you check when you carry out a visual inspection of an electric drill?

3 What should you do and not do when using an extension lead?

4 Make sure the person keeps warm – for example by using a blanket – but take care not to rub it against the burnt area.

5 Cover the burn by placing a layer of cling film over it.

6 Use painkillers, such as paracetamol or ibuprofen, to treat any pain.

Reporting to supervisors

Electric shock must be reported to your supervisor using the RIDDOR procedure.

KNOW THE SAFETY REQUIREMENTS FOR WORKING WITH GASES AND HEAT PRODUCING EQUIPMENT

One of the most immediate sources of fire is gas used for operations such as welding and brazing. This type of equipment must only be used by trained, skilled and competent people and always used correctly and following safe procedures.

Types of gas used on site

Gases fall into two main categories:

- flammable – gases that burn if subjected to heat or high pressure
- inert – gases that do not burn and can sometimes be used to put out fires, for example carbon dioxide. Inert gases are not necessarily safe. They can affect breathing or cause ice burns.

The main types of gas used on site are described in Table 1.08.

Gas	Uses	Hazards
Propane	• Barbecues • Portable stoves • Oxy-gas welding • Fuel for engines • Residential central heating	• Explosion • Fire
Butane	• Used as a propellant to drive other gases at pressure • Cigarette lighter fuel • Bottled fuel used for gas barbecues and cooking stoves • Propellant in aerosol sprays such as deodorants • Refrigerants – chemicals that convert heat to cold • Cordless hair irons (powered by butane cartridges)	• Explosion • Fire • Temporary memory loss • Frostbite if skin is exposed to butane at high pressure • Asphyxiation (suffocation) • Ventricular fibrillation – irregular heartbeat that can stop the heart and cause death
Oxygen	• To increase flame temperature in oxy-gas welding and cutting	• Explosion • Fire
Acetylene	• Used as a fuel gas in oxy-gas welding and cutting • Can burn at about 3500°C	• Explosion • Fire
Nitrogen	• Propellant • Coolant • Fertiliser (in solid state)	• Asphyxiation • Freeze burns (caused by liquid nitrogen)

Table 1.08: Gases used on construction sites

Safely transporting, storing and using bottled gases and equipment

Because most gases used in the building engineering services environment are hazardous in some way, they must be handled carefully. A damaged gas cylinder, for example, can cause explosion, fire or release of poisonous fumes into the air.

Gas cylinders

Gas is normally contained in cylinders, which have a valve for filling and releasing the contents. This valve must be in good working order and checked regularly.

The following PPE must be worn when cylinders are filled or refilled:

- eye protection
- protective overalls
- gloves
- ear protectors.

Figure 1.17: Valve on a gas bottle

All gas cylinders and associated equipment such as valves and regulators and safety equipment must be given a visual inspection before use. The things to look for are:

- bulges in the cylinder
- scorch marks from fire damage
- deep scratches or other damage.

Only cylinders made by approved companies can be used. All new cylinders must also be inspected by an approved inspection body. The inspection body stamp must be visible on the container. Once in use, all cylinders and their valves and other equipment must be tested regularly.

Transporting

Gas cylinders have to be moved around and delivered to site, but this is a very dangerous operation. It is vital that the following precautions are followed.

- Clearly mark the cylinder to show what it contains.
- Only use suitable lifting equipment such as cradles and slings.
- Never lift a cylinder by holding its valve.
- Secure cylinders to stop them falling or moving; this will usually be upright although some cylinders can be stacked horizontally.
- All regulators and hoses must be disconnected and protective valve caps fitted.
- The cylinders must not stick out from the sides of the vehicle that is carrying them.
- The vehicle must be marked to show what sort of gas it is carrying.
- The driver of the vehicle must be trained.
- There must be documentation in the vehicle stating the types of gas being carried.

Storage

Great care must also be taken when storing gas cylinders. If they are kept in damp or hot conditions, some gases will explode or change from a relatively safe substance to a more dangerous one.

- Store the cylinders in an area designed for the purpose. It should be flat and well ventilated with no chance of damage.
- If necessary, secure the cylinders in place so that they do not fall over.
- Make sure the cylinders are clearly marked to show what is inside.
- Do not store gas cylinders for too long. This means that you should only buy the amount of gas that you need for each job.
- When organising the storage area, keep the oldest cylinders at the front and use these first.
- Keep the cylinders away from extreme heat or naked flame.
- Do not store the cylinders where they will be standing in water.
- Keep the valves closed.

Figure 1.18: Correctly stored gas cylinders

Figure 1.19: A fire triangle

How combustion takes place

Fire and combustion need three elements, as shown in Figure 1.19. If one of these elements is not present, the fire will not burn.

The dangers of working with heat producing equipment

The main types of heat producing equipment used in the building services engineering environment are:

- welding sets
- cutting equipment
- gas torches
- soldering irons
- electric paint removers.

When using this type of equipment, remember to follow these key safety points.

- Do not use near flammable materials.
- Make sure the work area is well ventilated.
- Firefighting equipment such as extinguishers should be available in case of an accident.
- Wear the correct PPE, including gloves, protective overalls and eye protection.
- You should be trained to use the equipment. For some equipment, you may just need to be shown how to use it and supervised for a short while. For other types of equipment, such as welding and cutting, you have to be fully qualified.

Welding and cutting equipment

There are complete sets of regulations and guidelines for using and handling welding and cutting equipment. There are three types of welding and cutting equipment:

- oxy-gas – a combination of pure oxygen mixed with a flammable gas such as acetylene (see Figure 1.20)
- electric arc – the heat is created using a high electric current between the welding handle and the workpiece
- combination electric and gas – examples of this are MiG and TiG.

You must observe the main safety points for heat producing equipment, plus the following points:

- You must be fully trained and qualified to carry out these operations.
- PPE must include a metal face protector, protective overalls and gloves.
- Other people must be protected from the heat and, in the case of arc welding, the bright light from the welding process.
- You must give all equipment a visual inspection before use. This includes inspecting the valves, cables and hoses.
- All equipment must be tested for leaks using an approved leak detection chemical.
- You must carry out the procedures for starting the welding process and closing it down afterwards in the correct order.
- Only use approved hoses and other equipment. You must never swap around oxygen and fuel gas hoses.

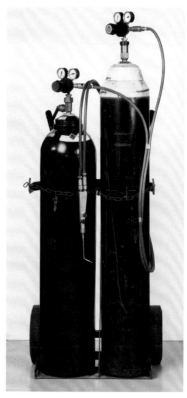

Figure 1.20: An oxy-gas welding set

Procedures on discovery of fires on site

Good housekeeping is key to fire prevention. Many flammable materials have to be on site so careful storage and clear work areas are vital to prevent a blaze. However, if a fire does occur, it is important to follow the correct procedures in order to ensure your own and others' safety.

Many companies carry out fire alarm exercises so that the workforce can practise evacuating a building in the event of a fire. It is important to know what to do if the alarm does sound and to familiarise yourself with fire exits and escape routes. The main steps in evacuating a building during a fire are described on the following page.

1 If you discover a fire, sound the alarm. You can do this by smashing the nearest 'break glass' fire alarm button. You should also shout a warning to anyone in the area and, if you have a mobile phone, dial 999 and ask for the fire service as soon as you are able.

2 If you feel confident enough, and it is safe to do so, attempt to fight the fire. It is important that you use the correct extinguisher. Using the wrong one can result in injury or even make the fire worse.

3 On hearing the alarm, leave whatever you are doing immediately.

4 If it is possible and safe to do so, shut all the windows. This will help limit the amount of oxygen available for the fire.

5 Leave the area as quickly as you can. Do not try to gather up your belongings. Things can be replaced, but human lives cannot.

6 Head for your assigned assembly point. Your fire warden (see page 10) will meet you there and take a register of all the people from your area. This is to make sure that everyone is accounted for.

7 Do not go back into the building under any circumstances.

Activity 1.2

Walk round the department at your college and locate all the fire alarm pushes and exits. Write down the location of each. You could obtain a copy of the layout drawings for the building or mark them on a sketch. Is there a pattern? What reason can you think of for the positions chosen for the exits and alarm buttons?

Classes of fire and types of fire extinguisher

The class given to a fire depends on its fuel. It is important to recognise the class of a fire so that you use the correct extinguisher and the fire is dealt with efficiently and as safely as possible. Table 1.09 lists the classes of fire and the types of extinguisher that should be used to tackle them.

Class of fire	Description	Fire extinguisher	
		Type	**Colour**
A	wood, paper, textiles, etc.	water	red label
B	flammable liquid, petrol, oil, etc.	foam	cream label
		dry powder	blue label
		carbon dioxide	black label
C	flammable gases	dry powder (multi-purpose)	blue label
D	burning metals, e.g. magnesium	dry powder (special purpose)	blue label
E	fire caused by electrical faults and where there is still an electrical supply present	carbon dioxide	black label
F	cooking oils and fats	wet chemical	yellow label

Table 1.09: Classes of fire

Figure 1.21: Classes of fire extinguishers

Progress check 1.07

1 What are the three elements needed for combustion?

2 What class of fire is a petrol fire?

3 Which extinguisher would you use to tackle a paper or wood fire?

4 What are the considerations when transporting a gas cylinder?

5 What should you do with a gas welding set before using it?

KNOW THE SAFETY REQUIREMENTS FOR USING ACCESS EQUIPMENT IN THE BUILDING SERVICES INDUSTRY

A considerable amount of building services engineering work is carried out at height. This means that you will need to use access equipment. This can be anything from a step-up to a full set of scaffolding. Because of the risk of collapse or falling, the equipment must be right for the job, in good condition and erected correctly by people who have been trained to do so. The Work at Height Regulations 2005 are the legislation covering this type of work.

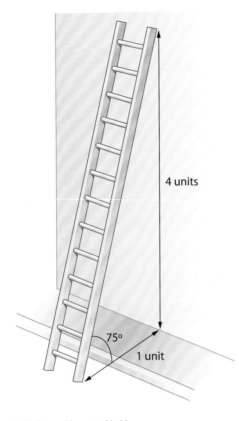

Figure 1.22: A correctly erected ladder

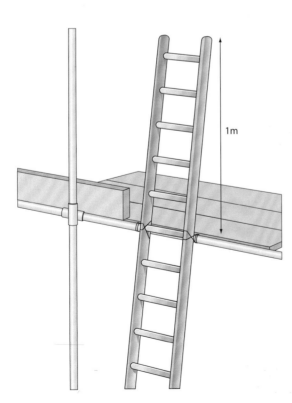

Figure 1.23: A ladder secured to scaffolding

Types of access equipment and their safe use

There are several types of access equipment, each designed for a different type of job. The main types are outlined below.

Ladders

Ladders enable you to climb to the work area but limit you to the direct area of the job. The HSE recommends that you only work at the top of a ladder in one position for a maximum of 30 minutes. If you have to reach across from the top of a ladder to continue working, move the ladder because stretching sideways can cause the ladder to fall. Apply the belt buckle rule. If your belt buckle hangs over either side of the ladder when you are trying to carry out a job, move the ladder. The HSE also recommends 10 kg as the maximum weight you can carry up a ladder without a manual handling assessment.

Ladder safety is covered by the HSE guide *Safe use of ladders and stepladders*. Below are the main points to remember when erecting a ladder.

- Ladders must be leant against solid walls or structures. The ideal angle is 75°.
- Ladders must be placed on an even, solid surface.
- If possible, a ladder should be secured at the top. This cannot be done if, for example, the ladder is leaning against a wall. If this is the case, you might be able to tie the ladder at some other point, for example a window halfway between the top and bottom of the ladder.
- Do not use a metal ladder near electrical equipment.
- Only one person should be on a ladder at any time.
- Only carry equipment or materials up a ladder if you can carry it in one hand. Use your free hand to help you to climb.
- Do not climb a ladder if your footwear is slippery, for example if it is covered with mud.
- Only use a ladder if the weather is suitable. For example, do not use a ladder outside if there is a strong wind.

Activity 1.3

Working in pairs, transport and erect a ladder. If possible, select a structure that allows you to tie the ladder at the top. Make sure the ladder is at the correct angle. Take turns to climb the ladder to tie it off while the other foots the ladder. When both of you have completed the exercise, take the ladder down and take it back to the starting point for the next pair.

Extension ladders

Extension, or telescopic, ladders are designed to be fully adjustable. The ladder must be of a high quality because there is a lot of strain on the ladder when extended. Here are a few points to remember when using an extension ladder.

- The two parts of the ladder must not move against each other when it is extended.
- The two sections must be secured while in the extended position so that the top part does not slide back down.
- The point at which the two sections meet is called the overlap. Extension ladders should have a minimum overlap label on the ladder. The general rule, however, is to allow at least 1 m overlap.

Stepladders

A stepladder is a short ladder that can stand on its own. There is usually a small platform at the top. This platform is not designed for you to stand on, but forms part of the structure of the stepladder and provides a place to rest tools and materials. When using a stepladder, you must remember the following points.

- Do not use the top two steps of a stepladder. If it isn't high enough to carry out the work, use a taller one or consider a different type of access equipment.
- Do not place a stepladder on a moveable base such as a pallet or mobile elevating platform.

Figure 1.24: Take appropriate precautions when using stepladders

Roof access equipment

Working on a roof has its own special hazards. A roof may look solid, but many are constructed from materials that cannot take your weight and will give way if you try to stand or crawl on them. The first thing to consider is: can the work be done without going onto the roof? If not, you must use the correct equipment and techniques (see Figure 1.25).

Roof ladder

The top end of a roof ladder is formed into a hook, which should be laid over the ridge and against the opposite side of the roof. Once securely in place, the ladder provides a set of steps that you can use to climb up and down the roof.

Crawling boards

Crawling boards can be laid across the roof, each end located over a joist. The board will then spread your weight. Crawling boards must be secured to stop them sliding down the slope of the roof.

Edge protection

If you are working on a roof and are within 2 m of the edge, you must erect edge protection. This prevents both you and your equipment and materials falling and injuring yourself or someone below.

Edge protection can consist of boards erected around the perimeter of the roof or on the edge of scaffolding. Edge protection boards should be supported from ground level and not fixed to the roof itself.

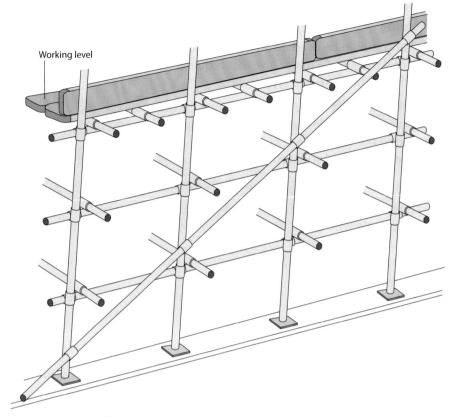

Working level

Figure 1.27: Fixed scaffolding

Mobile elevated work platforms

Mobile elevated work platforms (MEWP) are mechanical machines that allow you to work at height, known as scissor lifts and cherry pickers. They are usually towed by, or mounted on, the back of a vehicle and consist of a platform and lifting mechanism. You will often see them being used for work on street lamps and overhead wiring.

The main safety points are as follows.

- Make sure the ground is stable and solid enough to take the weight of the MEWP without tipping over or sinking.
- The MEWP should be fitted with outriggers for stability.
- The platform must be fitted with guard rails and edge protection.
- Place the MEWP close enough to the work so that you don't have to lean out.
- Cordon off the area around the machine using barriers or cones.
- Never use a MEWP on your own. There must be another person on the ground to go for help if the elevator mechanism fails or there is some other problem.

Figure 1.28: A cherry picker

Progress check 1.08

1 How can you secure a ladder if you cannot tie it off at the top?

2 What is edge protection?

3 Who needs to be involved if scaffolding needs to be more than 38 m high?

KNOW THE SAFETY REQUIREMENTS FOR WORKING SAFELY IN EXCAVATIONS AND CONFINED SPACES IN THE BUILDING SERVICES INDUSTRY

Although building services engineering does not normally require us to carry out major excavation work, there are times when you may have to work in trenches, cable jointing or installing water or gas services. An excavation can be a hazardous environment with dangers such as collapse or falling. Remember, one cubic metre of soil can weigh as much as one tonne.

On the other hand, as a building services engineer you will certainly be required to work in confined spaces. This type of work is physically and mentally demanding, so good health, preparation and robust health and safety procedures are needed before you start.

How excavations should be prepared for safe working

The main hazards when working in excavations are:

- collapse of the excavation, burial or injury
- equipment, people or material falling into the excavation
- the presence of electrical or gas services that could cause fire, explosion or shock.

If possible, it is always better to carry out work without using a trench. However, if trenches have to be dug, preparations should be made before any actual digging work takes place. These are described below.

Locate any underground services in the area

Obtain drawings showing any pipe and cable runs in the area. This could prevent a breakage or power cut caused by digging equipment. If drawings are not available, use locators to trace any services. Mark the ground accordingly. Look around for obvious signs of underground services. These are valve or manhole covers, electrical and telecommunications boxes and patching of the road surface.

Training

Make sure that all the people involved in the work are familiar with safe digging practices and emergency procedures as well as safe access into the excavation.

Trench support

Decide what temporary support you will need for the trench. Make sure this equipment is ready and on site before work starts. Examples of support equipment include props, trench sheets and edge protection.

Props

The sides of the trench are covered with boards or metal sheets. These are held in place by adjustable steel poles called props. This stops the sides of the trench collapsing.

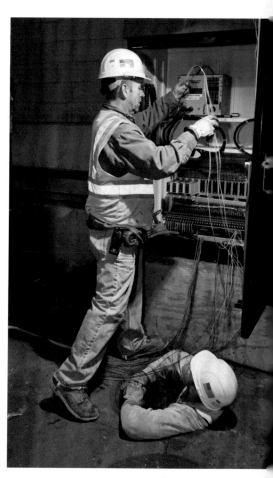

Figure 1.29: People working in an excavation

Hazard	Causes and effects
Fire or explosion	• Presence of flammable substances • Excess oxygen (e.g. from oxy-gas welding)
Toxic gas, fumes or vapour	• Fumes from previous contents of the area • Fumes from outside (e.g. vehicle exhaust)
Lack of oxygen	• Insufficient air supply
Ingress of liquid or other material	• A leak may allow water or other materials to enter and fill the space
Excessive heat	• A confined space can become very hot, especially if there is nowhere for the heat to escape to. • This may cause heat stroke or unconsciousness.

Table 1.10: Potential hazards of working in a confined space

The main points to consider when working in a confined space are:

- freedom of movement so that the work can be completed safely
- ventilation so that there is a constant air supply and any gases resulting from the work can escape
- escape routes for the people working in the space.

Ventilation

A good oxygen supply is needed. This can be achieved by making sure that the space is kept open. If necessary, oxygen can be pumped into and out of the space. Breathing apparatus might be needed in some cases.

Lighting

These areas can be very dark, so good lighting must be provided. This serves two purposes, as it:

- illuminates the work area itself so you can carry out your work safely
- illuminates escape routes so you can get out quickly and safely if something goes wrong.

PPE

As with all work in the building services engineering sector, everyone involved with the excavation work must be issued with, and wear, the correct PPE. This includes:

- hard hat
- hi-vis jacket or waistcoat
- protective footwear
- gloves
- eye protection
- ear protection when using power tools such as pneumatic drills to break hard surfaces.

Depending on the work, you may need ear defenders, dust masks or even respirators. The PPE you need to wear will be specified in the risk assessment (see pages 13–14).

Evacuation procedures

There must be a set procedure for evacuating the space. Before you start work, you must know what this is. There must be clear and unobstructed escape routes. Someone must be posted outside the space to assist the person to get out in the event of an emergency.

Medical conditions

Working in a confined space can be stressful. It is difficult to move and can be hot, frustrating and claustrophobic. Anyone with a medical condition such as heart disease or a respiratory problem should not work in a confined space.

Lone working

You should not work in a confined space alone. There are, of course, spaces where only one person can enter and carry out the job, but there must be someone else present. The permit to work will notify your supervisor:

- that you are working in a confined space
- the start time
- the finish time
- the work you are carrying out.

> **Progress check 1.09**
>
> 1 What are the signs that there may be underground services?
> 2 What is a prop?
> 3 What are the basic precautions needed when working in a confined space?

Figure 1.31: Working in a confined space

Chapter
1

BE ABLE TO APPLY SAFE WORKING PRACTICE

It is not just lifting heavy items that can cause injury. Lifting light but awkward objects can strain your back and arms. When it comes to heavy items, you must remember that your body is not a lifting machine and can be damaged by attempting to lift heavy weights.

There is no shame in asking for help! Even better, why not use lifting equipment, such as a simple sack barrow, a trolley or wheelbarrow? If you do have to lift an object of any sort, remember to use your legs to lift and not your back. Page 45 shows the correct method for lifting. The statutory document dealing with safe manual handling is the Manual Handling Operations Regulations 1992.

The main points for manual handling

If at all possible, avoid hazardous manual handling operations. This can be done by:

- rethinking the job or by using mechanical lifting equipment
- carrying out a risk assessment. This does not have to be a written assessment – just look at the job and try to think about the risks to yourself and others, and how they can be avoided.

Team handling

Some loads need to be carried by a team of people. The team should work together. There are a number of points to remember when lifting and carrying as a pair or team:

- Make sure there is enough space for the team to work.
- All team members should be able to take hold of the load. Lifting and carrying equipment should be used if the load is small or difficult to hold.
- One person should take charge of the operation. This person must make sure that the group work as a coordinated team.
- Good communication between team members is needed.
- Ideally, the team members should be of similar build and strength.
- If the weight of the load is distributed unevenly, it is better for the heavier part to be lifted by the strongest members of the team.

Progress check 1.10

1   What sort of injuries can be caused by manual lifting?

2   How can you reduce the risk of injury when lifting a heavy object?

3   How do you ensure that a lifting team works in a coordinated way?

① Always check the nature of the object to be lifted before going straight in to lift it. Objects can be heavier than they appear. If you think that lifting aids and/ or assistance might be required, do not attempt to lift unaided.

② Clear your path to where the object is to be moved to and make sure that you have adequate space to set it down. Adopt a secure stance with your legs shoulder-width apart either side of the object.

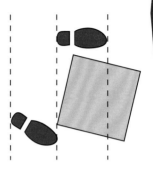

③ Keeping your back straight, bend your legs and get a solid grip on the object with arms straight out.

④ Begin to lift the object by extending your legs straight and keeping your back straight.

⑤ Move smoothly to your destination, keeping the object close to your body at waist height. Avoid jerky movements and never run.

⑥ When putting the object down, do so steadily, bending your legs rather than flexing your back (as with lifting). Remember that putting down can be as hazardous as lifting!

Figure 1.32: Step-by-step guide to manual lifting

Knowledge check

1 The Health and Safety at Work Act places responsibility for health and safety on:

   a employers and shareholders
   b employers and employees
   c employers and the government
   d employers and the local council

2 The Electricity at Work Regulations are intended to ensure that:

   a electricians are the only people allowed to use an electrical system
   b electrical installations are in the control of a technician
   c tools and equipment are inspected and tested regularly
   d the electrical system in a workplace is safe for everyone who works there

3 A water-filled fire extinguisher is suitable to extinguish fires involving which of the following?

   a electrical appliances
   b flammable gases
   c flammable liquids
   d wood, paper, straw, textiles, etc.

4 Construction site power tools should be powered using:

   a a UPS system
   b a three-core lead
   c a 110 V transformer
   d a capacitor

5 When transporting gas cylinders, the vehicle should:

   a be registered with the Gas Licensing Board
   b marked to show what sort of gas it is carrying.
   c driven by a gas engineer
   d have a sealed trailer for the cylinders

6 Stored gas cylinders should:

   a be laid down and stacked as tightly as possible
   b have their valves removed
   c be kept for as long as possible
   d be stored with the oldest at the front

7 The type of ladder used for accessing a roof is a:

   a roof ladder
   b extension ladder
   c stepladder
   d ladder rack

8 The lining placed into a trench is called:

   a trench plating
   b battering
   c trench sheeting
   d baulk

9 When should an excavation be inspected?

   a every six weeks during the excavation work
   b once a month during the excavation work
   c at the start of each day of excavation work
   d only when an incident or accident occurs

10 What hazard can fumes present in a confined space?

   a delayed working
   b asphyxiation or explosion
   c heat exhaustion or heat stroke
   d stress and restricted movement

Electrical principles and processes for building services engineering

This chapter will cover the following learning outcomes:

- **Understand electrical supplies used in domestic plumbing systems**

- **Know the components used in electrical installations**

- **Understand the procedure for safely isolating supplies**

- **Understand how to identify safety critical faults on electrical components and systems**

- **Understand how to undertake basic electrical tasks**

Introduction

At Plumbing Level 2, electrical work is restricted to basic practical skills on components that are not connected to the supply, which means you will be doing non-live work. On completion of Level 2 you must not regard yourself as competent to work on electrical systems. You will cover electrical work in much more detail at Level 3.

The job of a plumber involves installing and maintaining a wide range of appliances and components, many of which are powered by electricity. You need to be able to work safely and competently on the electrical supply to items such as heating controls, immersion heaters, showers, etc. This will include being able to inspect and test electrical systems and to connect wiring from outlets to appliances or components.

UNDERSTAND ELECTRICAL SUPPLIES USED IN DOMESTIC PLUMBING SYSTEMS

Identify documents required to design electrical systems

There are many British Standards (BS) and approved codes of practice that apply to electrical installation. You can find out more about these by visiting the website of the Health and Safety Executive (HSE) and searching for electrical standards and approved codes of practice. Some of the better known regulations that apply to electrical work you might carry out are:

- BS 7671 (17th Edition of the Wiring Regulations) amended January 2012
- Electricity at Work Regulations 1989
- Health and Safety at Work etc. Act 1974 (HASAWA)
- The Building Regulations Part P.

In addition, you should also consult any manufacturers' instructions relevant to the equipment, machinery and components you are working with.

Identify the different types of supplies used in domestic plumbing systems

In this section you will look at how electricity is supplied to a property from a power station and explore the electrical installation of plumbing components inside a property.

Electrical principles

Electricity is the movement (flow) of electrons through a **conductor**. We need to measure how much electricity will flow through a given **circuit** in a given time, and to control how much electricity is flowing (see Figure 2.01).

A single electron is too small to measure practically. Instead, we measure coulombs of electricity, where 1 coulomb is many millions of electrons. The flow of electricity through a conductor is given in terms of coulombs

Key terms

Conductor – an electrical conductor is a material that allows electricity to flow through it, such as metal. (Thermal conductors allow heat to pass through them.)

Circuit – a network of components (e.g. battery, wire, switch) joined in a loop to allow the flow of electricity.

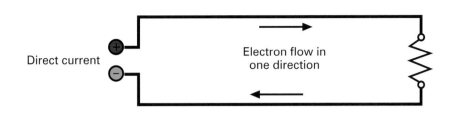

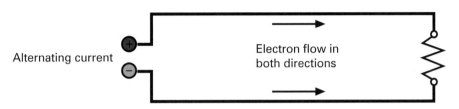

Figure 2.01: Flow of electrons in electrical circuits

per second. This is the electrical current, measured in **amperes** – usually abbreviated to amps (A) and given the symbol I. There are two types of electrical supply:

- alternating current (AC)
- direct current (DC).

These topics are covered in more detail later in this chapter and in Chapter 3: *Scientific principles for domestic, industrial and commercial plumbing.*

Generation of electricity

Almost all electricity used in domestic and industrial premises is generated at power stations. All power stations use the principle of a turbine turning a three-phase alternator, which is the actual machine that produces the electricity.

Power stations are either **thermal** or **kinetic** depending on the fuel or process used to turn the turbine that works the alternator. The term 'generator' is sometimes used instead of alternator. Generator is an old term and refers to a machine that produces direct current (DC). All modern power stations produce alternating current (AC) and use alternators.

Most commercial power stations today are of the thermal type, using fuel such as gas, oil, coal or nuclear fusion to turn water into high-pressure steam. Figure 2.03 (on page 50) shows a thermal power station and its basic components.

Key terms
Ampere (amp/A) – the measure of electrical current; the flow of 1 coulomb in 1 second.
Thermal – a type of power station where water is heated so it turns to steam (heated gas turns water to steam in nuclear power stations), which then drives a turbine.
Kinetic – energy produced by motion or movement, as in wind- or water-driven power stations.

Figure 2.02: A typical UK power station, showing cooling towers in the foreground

Chapter
2

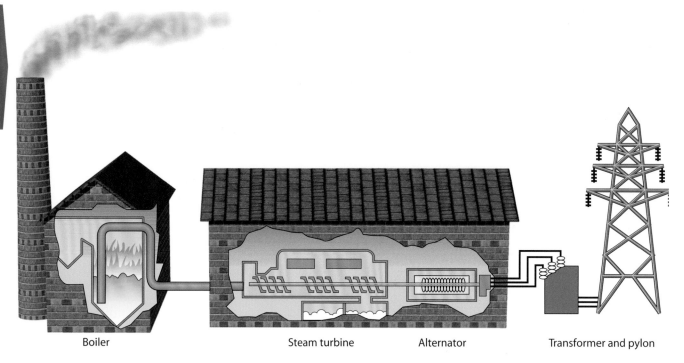

Boiler   Steam turbine   Alternator   Transformer and pylon

Figure 2.03: Simple thermal power station layout

Key terms

Renewable energy source – an energy source that will not be used up, such as wind power.

Finite – available in limited amounts, and will therefore run out eventually.

As fossil fuels such as oil and gas have begun to run out, there has been increased research into the use of kinetic power stations. One type that is becoming more common is the wind turbine, where wind power – a **renewable energy source** – turns giant propellers, which then turn an alternator to produce electricity (see Figure 2.04).

Hydroelectric stations harness the power of water when it falls from a great height, such as down a mountain or from a dam. Experimental stations are also being developed, which use wave power from the sea to generate electricity.

Activity 2.1

Fossil fuels currently used to produce electricity are **finite** resources and are running out. Huge investment is being made to find alternative sources of energy to produce electricity.

Working with a partner or in a group, discuss the energy sources mentioned in this section and identify at least three advantages and three disadvantages of each one.

Alternating current (AC)

Most plumbing installations and systems are supplied with alternating current. Electricity is invisible to the naked eye but an oscilloscope allows us to 'see' the flow of electricity.

Figure 2.04: Wind farms use kinetic energy to produce electricity

If we look at alternating current on an oscilloscope, we see a shape called a sine wave (see Figure 2.05). This represents one complete cycle of alternating electromotive force (emf) which is induced into a coil of copper wire when it is rotated within a magnetic field. During the top half of the cycle the current will flow in one direction and during the bottom half of the cycle the current will flow the other way – hence the term 'alternating'. In the UK, the electricity supply has 50 of these single cycles per second. This is known as the frequency of the supply. The unit used for frequency is the hertz, abbreviated to Hz. The frequency is therefore stated as 50 Hz.

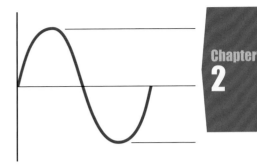

Figure 2.05: An AC sine wave as seen on an oscilloscope

Direct current (DC)

In a direct current electrical circuit the electron flow is in the same direction all the time. One example is from the anode (positive electrode) to the cathode (negative electrode) of a battery around a simple circuit (see Figure 2.06).

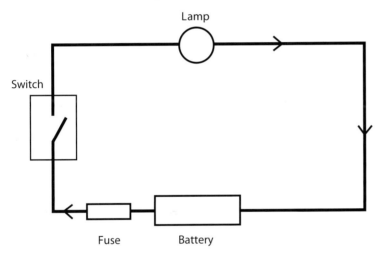

Figure 2.06: A simple circuit

A simple electrical circuit can be created using a power source such as a battery connected to a light bulb or lamp using some wire between the battery and the bulb. This allows electron drift between the two ends of the battery. As the electrons flow through the wires and bulb, the bulb will light up.

Direct current from batteries is produced by a chemical reaction where plates containing dissimilar metals are placed in a solution known as an electrolyte. When the battery is charged and a load is connected across it, current flows. Batteries and direct current supplies are generally only used where very small amounts of electricity are required – for example, in torches, light current back-up power supplies and battery-operated tools.

As a plumber, you will probably only use direct current when using small batteries, or perhaps in heating or boiler control circuits. In this case, the direct current is produced from an alternating current source using:

- a transformer to reduce the **voltage** (covered in more detail later in the chapter)
- a rectifier (a solid-state electronic device) to convert AC to DC.

Direct current on an oscilloscope appears as a straight line where the electricity flows in one direction only.

> **Key term**
>
> *Voltage* – the pressure that pushes electricity around a circuit, known as electromotive force (emf). Voltage may be supplied by a battery or a mains supply.

Key terms

Joule – International System of Units (SI) unit of energy.

National Grid – a network of nearly 8,000 km (5,000 miles) of overhead and underground power lines that link power stations together throughout the UK.

Ring circuit – ensures continuity of the electricity supply to all consumers, even at peak demand times. As one area requires more energy, another area will probably require less.

Identify the common voltages used in domestic plumbing systems

Voltage is measured in terms of the number of **joules** of work required to push 1 coulomb of electrons along a circuit. It is measured in joules per coulomb, more commonly referred to as volts (V). Voltage can come from a battery (DC) or a mains supply (AC).

The alternator in a power station produces electricity at high voltage which is then fed through devices called transformers. These increase the voltage to even higher levels before passing the electricity into the **National Grid**. Electricity is transported through the National Grid at voltages of between 130 and 400 kV. The voltage is raised to these levels so that losses are reduced when the electricity is distributed.

Electricity is 'taken' from the National Grid through a series of distribution stations. These transform the grid supply back down to 11 kV and distribute electricity at this level to a series of local substations. Substations can be seen dotted throughout cities and are normally connected together on a **ring circuit** basis.

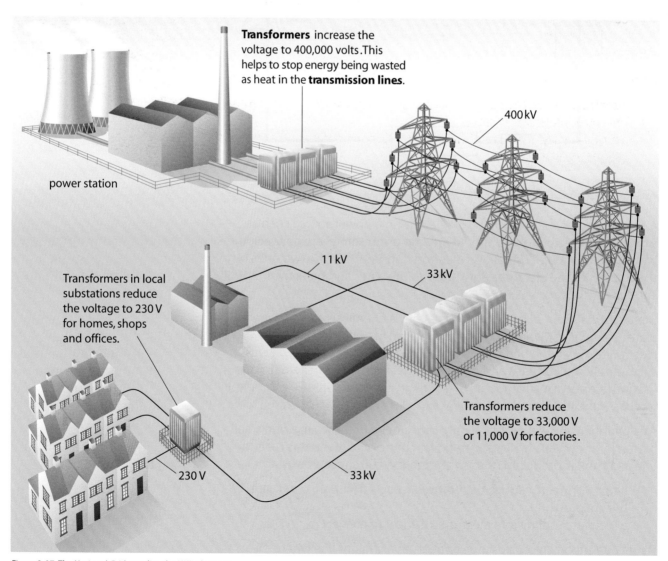

Transformers increase the voltage to 400,000 volts. This helps to stop energy being wasted as heat in the **transmission lines**.

400 kV

power station

11 kV

33 kV

Transformers in local substations reduce the voltage to 230 V for homes, shops and offices.

Transformers reduce the voltage to 33,000 V or 11,000 V for factories.

230 V

33 kV

33 kV

Figure 2.07: The National Grid supplies the UK's electricity

At substations the 11 kV supply is transformed down to 400 V and then distributed through a network of underground **radial circuits** to customers. In rural areas, this distribution sometimes takes place using overhead lines. The electricity is generated and distributed in the form of a 'three-phase' supply, which is used without modification by some large factories and industrial sites (see Figure 2.07).

Different consumers such as factories, offices or schools will need different voltages. The supply to homes in the UK is a single-phase 230 V, 50 Hz supply with phase and neutral conductors and a main earth point provided.

The common voltages you will come across include:

- 230 V – the usual mains electricity supply to homes in the UK
- 110 V for all on-site power tools – achieved using a portable transformer (large building sites may have fixed transformers)
- 12 V, 18 V and 24 V for portable battery-operated power tools such as drills and screwdrivers
- even lower voltages in everyday items such as clocks, measuring equipment, etc.

Describe the layouts of electrical supplies and connections

Generally, at a domestic property main intake point you will find the following:

- a sealed overcurrent device – **fuse** or **circuit breaker** – which protects the supply company's cables
- a metering system to determine the customer's (consumer's) electricity usage (see Figure 2.08).

Key terms

Radial circuit – a single cable is taken from the electricity mains into the consumer's property and terminates at the consumer's meter.

Fuse – a conductor inside a cartridge or holder, designed to melt and break when its rated current is exceeded.

Circuit breaker – an automatic switch that opens when its rated current is exceeded.

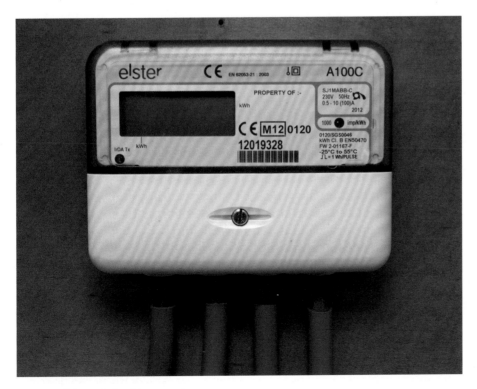

Figure 2.08: A domestic electricity meter

Did you know?

Under the new requirements of BS 7671, the individual circuits in the consumer unit will need to include a residual current device (RCD), as most domestic circuits include embedded cables. These are covered later in the chapter.

The consumer's installation must be controlled by a main switch, which should always be as close as possible to the supply company's equipment. In the average domestic installation, this device is merged with the means of distributing and protecting the final circuits in what is known as the consumer unit or distribution board (formerly called a fuse box). Protective devices are covered in more detail later on pages 63–66.

Figure 2.09 illustrates the final part of the electrical supply journey.

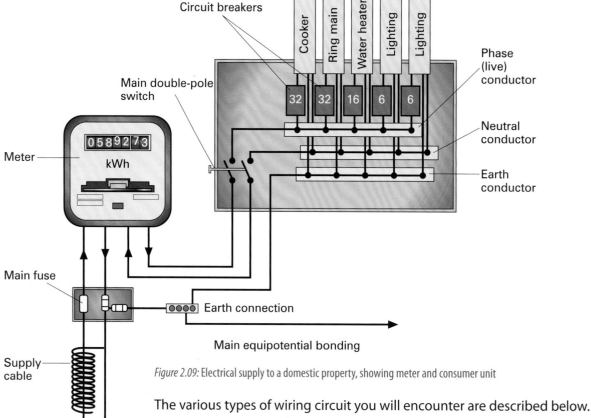

Figure 2.09: Electrical supply to a domestic property, showing meter and consumer unit

The various types of wiring circuit you will encounter are described below.

Lighting circuit

This is a radial circuit that feeds each overhead light or wall light in turn. To allow the lights to be turned off and on, the live or phase wire passes through a wall-mounted switch. Two-way switches are also used (at the top and bottom of stairs or in long passageways) and these require special switches.

The lighting circuit is usually fed by a 1.5 mm^2 twin and earth PVC-insulated cable and is protected by a 6 amp circuit protection at the consumer unit. The lighting in domestic houses is often split into two separate circuits, one for upstairs and one for downstairs.

Ring main circuit – 13 amp socket outlets

The sockets in domestic properties feeding appliances such as televisions and music systems are normally 13 amp sockets fed from a continuous ring circuit. As with the lighting circuit, cables circulate from the consumer unit round each socket. The difference is that they return to the consumer unit, forming a continuous loop – hence the term 'ring main'. The ring main permits the cables to be kept to an optimum size, as electricity can flow in two directions to reach the socket.

The ring main circuit is fed by a 2.5 mm^2 twin and earthed PVC cable and is protected by a 32 amp circuit protection device.

Spur outlets

Spur outlets are usually used to connect into a ring main circuit on an existing system where it is inconvenient to place a socket from the ring main using the conventional two cables. The spur is connected to the ring main through a junction box or is wired directly from the back of an existing socket. Spurs can be either fused or non-fused. You would not usually encounter spurs on new installations. Figure 2.10 shows examples of all the features described above.

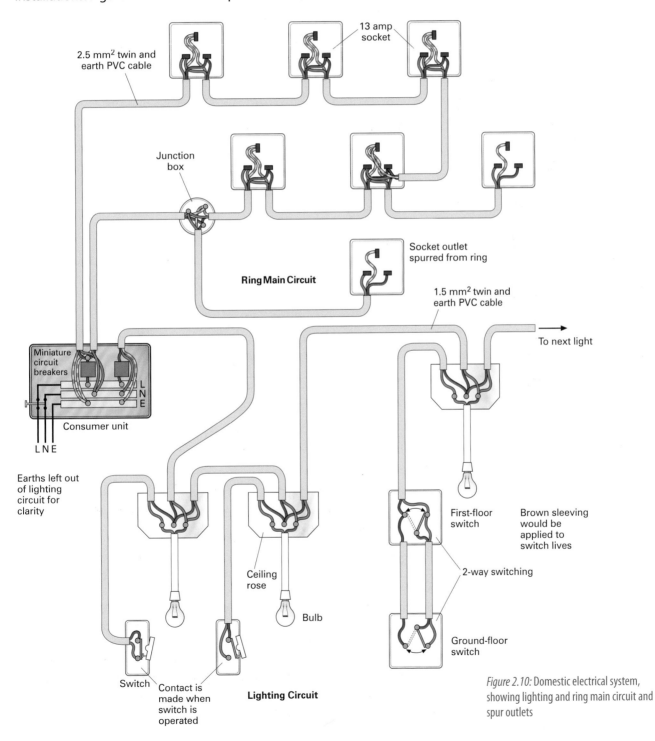

Figure 2.10: Domestic electrical system, showing lighting and ring main circuit and spur outlets

Key term

Thermostat – device which senses the temperature in the system and opens or closes the respective zone valve.

Heating controls

The most common heating controls are based around a concept introduced about 30 years ago by Honeywell called the 'Sundial' system. The two most common variants are the S and Y plan systems.

S plan

The S plan uses two two-port valves (one for heating and one for hot water) to provide independent temperature control of heating and hot water circuits in fully pumped central heating installations. Each valve is controlled by a **thermostat**, and time control must be provided by a programmer.

When either thermostat demands heat, the respective zone valve will open. Just before the valve reaches its fully open position, the auxiliary switch will close, switching on the pump and boiler. When both thermostats are satisfied, the valves close and the pump and boiler switch off (see Figure 2.11).

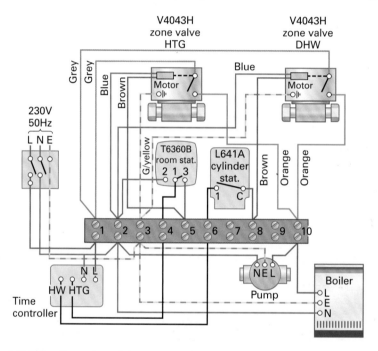

Figure 2.11: Wiring diagram for an S plan system

Y plan

This is similar to the S plan, but uses one three-port mid-position valve (which handles both hot water and heating) to provide independent temperature control of heating and hot water circuits:

- **Hot water only** – When the cylinder thermostat demands heat, the valve remains open to domestic hot water only and the pump and boiler switch on.
- **Heating only** – When the room thermostat demands heat, the valve remains open to central heating only and the pump and boiler switch on.
- **Heating and hot water** – When both thermostats demand heat, the valve plug allows both ports to open and the pump and boiler are switched on. When neither thermostat is demanding heat, the pump and boiler are switched off. The valve remains in the last position of operation while the time control is in the on position (see Figure 2.12).

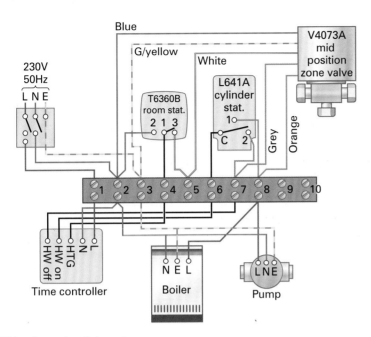

Figure 2.12: Wiring diagram for a Y plan system

Identify the different types of earthing system

In order for electricity to flow, there must be:

- a complete circuit of conductors
- a voltage source to create a potential difference across the ends of the circuit.

For the circuit to be of practical use, there must also be:

- a known value of circuit resistance to limit the amount of current flowing.

In the case of a **short circuit**, when the known circuit resistance is bypassed or breaks down, very high currents can flow. This causes serious fire risks due to the heating effect of the current and danger to life from the risk of electric shock. Therefore, to protect people from electric shocks, all electrical equipment and installations must be **earthed**.

You can think of the earth as a very large conductor at zero potential (voltage). By connecting together all the metalwork of an installation (other than the metalwork such as copper circuit conductors, which are designed to carry current), dangerous potential differences cannot exist between metal parts of the installation and the earth (ground). If they did, anyone touching the 'live' metalwork could get a serious electric shock or even be killed.

One simple way to remember what earthing is all about is to think of electricity as being 'lazy'. It always flows around a circuit through a load with a known resistance to control the current, and it flows from a high potential to a low potential. As it is lazy, it will always try to find the path of least resistance to get to the zero potential of earth. If it can, it will do that through you (you have a low resistance) when you touch live unearthed metalwork instead of working its way around the circuit through the high resistance.

> **Key terms**
>
> *Short circuit* – occurs when electricity flows along an unintended path.
>
> *Earth/earthing* – connecting exposed conductive parts of an installation to its main earthing terminal.

Earthing provides a very low-resistance path to earth through the exposed installation metalwork so that, in the event of a fault, a very high fault current will flow. This will almost instantaneously trigger circuit protective devices such as fuses and circuit breakers. Figure 2.13 illustrates the concept of earthing.

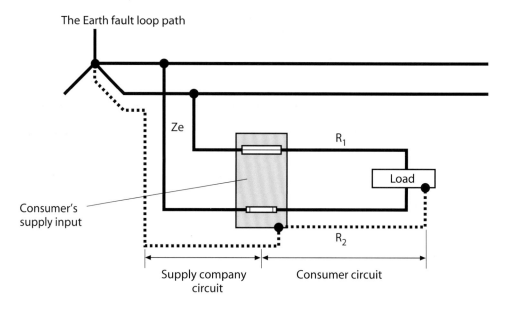

Figure 2.13: Earth fault loop path

Figure 2.13 shows a three-phase supply (the star-shaped part at the top left of the diagram) which the local electricity authority supplies to a substation. One phase conductor (single 230 V) supply is taken into the premises to the consumer's supply input, and a second 'neutral' conductor is also provided. This is connected at the substation to the centre of the three-phase supply, which is called the 'star point'. This star point is connected to earth at the substation.

You will also notice a dotted line (the earth conductor) which is taken, again from the star point, into the consumer's supply input. At the consumer's end, the earth conductor is connected to a separate terminal called the main earthing terminal. It is important to notice at this point that the neutral and earth conductors are connected together at the same point at the electricity authority's supply, but are connected to separate terminals when they arrive inside the premises. The supply authority must maintain the value of the star point as near to zero (earth) potential as possible so that a good low-resistance earth path is always available.

Inside the premises, the consumer must ensure that all the earthing conductors connect to the main earthing terminal to maintain the low-resistance path. The total earth path, all around the installation and back to the star point of the transformer, is called the 'earth fault loop' and is given the symbol Ze.

During normal operation, the current will flow through the fuse from the consumer's supply along the phase conductor R_1, through the load (which has a separate earth connection), and will return via the neutral.

We know that the neutral is connected to earth back at the substation, so the only high potential difference that exists is between the phase conductor R_1 and earth.

Imagine now that the load (an electrical appliance such as a kettle or toaster) has not been earthed and an internal fault develops that allows electricity to flow in the casing of the appliance. If the consumer were to touch it, the result could be death.

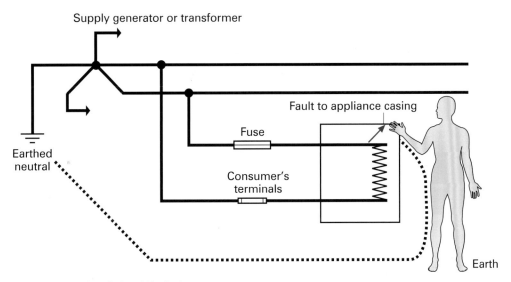

Figure 2.14: An earthing fault could be fatal

Because the appliance has not been earthed, the fault current flows through the consumer to the low-resistance earth return path (see Figure 2.14).

TN-S system

The type of installation discussed in this section is a TN-S system and is very common in the UK (see Figure 2.15), but it is being overtaken by others. You do not need an in-depth knowledge of different types of installation earthing arrangements, but a brief explanation may be useful for reference purposes.

- T is for TERRA (Latin for earth).
- N means that the exposed metalwork is connected to the main earthing terminal.
- S means that a separate earth conductor is used throughout the installation.

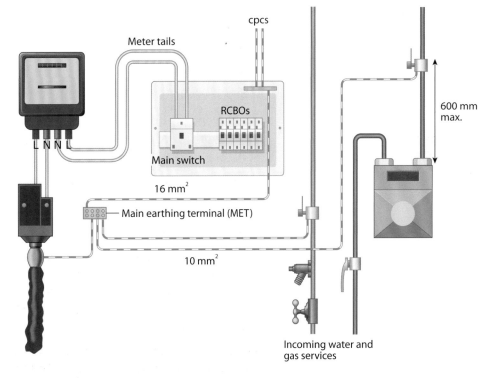

Figure 2.15: Customer's intake position for a TN–S system

The most common types of installation system are outlined below.

TT system

The first T again stands for TERRA. The second T means that the exposed metalwork is connected directly to earth by a separate earth electrode (usually a copper rod driven into the ground). This system is most common in rural areas, where the electricity is often fed by overhead lines (see Figure 2.16).

TN-C-S system

This is sometimes called protective multiple earthing (PME):

- The T is for TERRA.
- N means that the exposed metalwork is connected to the main earthing terminal.
- C means that, for some part of the system (usually in the supply section), the function of the neutral conductor and earth conductor are combined in a single common conductor.
- S means that, for some parts of the system (usually in the consumer's part), the functions of neutral and earth are performed by separate conductors.

This type of system is the most commonly installed in new-build environments. While it is very effective, this system has a few potentially very serious risks if it is not installed or maintained properly. Therefore, certain rules are laid down for TN-C-S systems. The main ones are:

- The neutral conductor must be earthed at a number of points.
- Neutral and phase conductors must be made of the same material and be of the same cross-sectional area.
- The neutral conductor must not be fitted with any neutral link or device that can break the neutral path.

You are unlikely to remember all this, but the information is provided for reference in case you need it.

Describe protective equipotential bonding

To prevent any build-up of dangerous voltages in the event of a fault, all metalwork in an installation should be connected using conductors and earth clamps, or clips, in a process called earth **bonding**.

The bonding of all exposed metal components in a dwelling that are not part of the electrical installation is called equipotential bonding. This is because all the metalwork is kept at equal potential so that dangerous potentials cannot exist. The equipotential bonding conductor should be close to the consumer unit.

As you have learned, earthing provides a very low-resistance path to earth through exposed installation metalwork so, in the event of a fault, an extremely high current will flow. This will then trigger circuit protective devices such as fuses and circuit breakers.

When considering this low-resistance path to earth, you must also think about the metalwork inside a modern house that could conduct electricity if a fault occurs. Figure 2.18 on page 62 lists several conductors that are not specifically designed to carry current.

> **Key term**
>
> **Bonding** – connecting non-current carrying metalwork (metallic devices) together and into the earthing system to protect from electric shock.

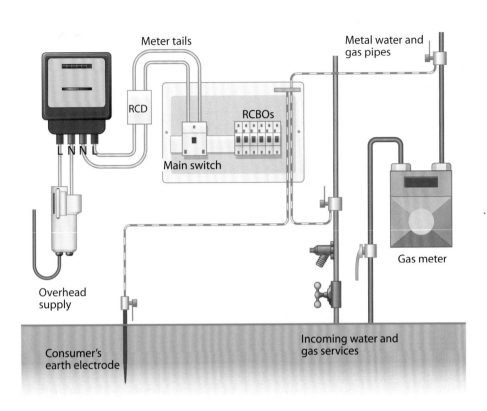

Figure 2.16: Customer's intake position for a TT system

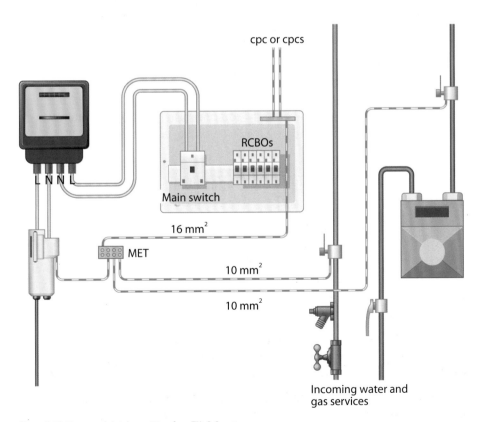

Figure 2.17: Customer's intake position for a TN-C-S system

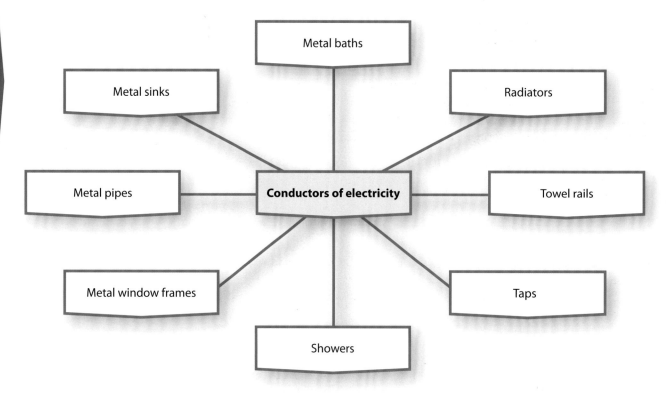

Figure 2.18: Conductors of electricity in a modern house

The use of plastic and non-metallic plumbing fittings and PTFE tape (thread seal tape) makes it even more important to connect all these devices together with low-resistance copper wire. This is because we cannot safely assume that the total metallic pipework system provides a low-resistance return path for fault currents.

Bonding is very important, as most of the systems that plumbers install require some form of bonding protection. You should be able to tell if the earthing and bonding have been installed correctly. You may also have to carry out work on these systems, but you should do so only if you are competent.

Main equipotential bonding

Bonding pipework, as shown in Figure 2.19, will provide a safe route to earth and is described as main bonding. In the diagram, note the size of the bonding conductors and the need for the clamp to be labelled. The bonding conductor from the main terminal to the earthing clamp is 10 mm^2. The bonding to gas, water or other services should be as close as possible to the point of entry. For the gas supply, it should be within 600 mm of the meter.

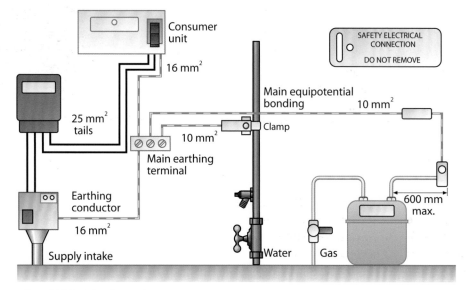

Figure 2.19: Earth bonding conductors

The method explained here of protecting against indirect contact using equipotential bonding and automatic disconnection of supply meets the requirements of the Wiring Regulations (BS 7671).

Consider, for example, a hot water radiator central heating system in a house. You could naturally assume that, as the whole system is metal and a good conductor of electricity, the total resistance must be low. However, if you think about it more deeply, you will realise the following.

- Each of the valves connected into the system is likely to have PTFE tape around the thread to improve the water seal. PTFE tape is an excellent insulator.

- The system may contain one or more plastic fittings on the cold water feed side. Plastic is also a good insulator.

Electrical Wiring Regulations (BS 7671) require all exposed metalwork in a building to be bonded together and connected to the earthing block within the consumer unit. The following bullet points briefly describe the steps you must take to ensure the safe earthing of all metallic materials within domestic properties.

- Gas, oil and water pipes can provide a path for stray electrical current. This could lead to corrosion of the pipework and also the possibility of electric shock for anyone touching or removing a section of the pipework.
- Certain areas of a domestic property may need supplementary (extra) bonding to link sections of central heating or cold or hot water pipework where the metal pipework has been separated by a plastic fitting or length of pipe (see Figure 2.20). This will ensure **earth continuity** throughout the property.

Describe supplementary earthing bonding

You need to be able to tell if a plumbing installation is correctly bonded. You can see in Figure 2.19 on page 62 that the equipotential bonding connects to only one point of the pipework.

There are other exposed metal parts within the domestic hot water, cold water and central heating systems that may not be protected because they have been isolated from the earth by plastic fittings, cisterns, etc. in the system.

It is important that, on completion of a job, the services shown in Figure 2.19 (gas and water pipes, electrical supply cable) – including any supplementary bonding requirements – are installed correctly. This will normally be done by a qualified electrician at the same time as the plumbing systems installation if necessary.

Describe the protection methods used on electrical systems

The Electricity at Work Regulations 1989 require that an electrical current is disconnected automatically if a greater current flows through a conductor (cable) than the circuit was designed for. This should happen within 0.4 seconds for anything with a socket outlet and 5 seconds for a fixed appliance such as an immersion heater.

> **Key term**
>
> *Earth continuity* – all exposed metalwork in a building is bonded together and connected to the earthing block in the consumer unit.

Figure 2.20: Earth bonding to a pipe

Faults in circuits may be caused by:

- overloading – when the load is too big for its supply cable
- short circuits – when line and neutral accidentally make direct contact
- earth faults.

Fuses and circuit breakers protect installations and individual circuits from overcurrent. No protective device operates instantaneously; these devices will always take more than their rated value before operating. For most devices, this is 1.4 times their rating. BS 7671:2008 states that all protective devices must be able to withstand the full **prospective fault current (PFC)**. Faults cause large amounts of energy to be released in the protective devices, so they must be able to contain this energy.

Fuses

Fusing is a safety measure to prevent high electrical current passing through wires that are not designed to carry such large charges. This is important because when a current flows it generates heat. If a current is too high for the wire it is passing through, there is a risk of overheating and fire.

Figure 2.21 shows different types of fuse – rewirable, cartridge, high breaking capacity (HBC) – while Figure 2.22 illustrates how a fuse works.

> **Key term**
>
> *Prospective fault current (PFC)* – a large fault current that will flow from the source of supply to the fault itself when an electrical fault occurs.

Figure 2.21: Fuses can be rewirable, cartridge or high breaking capacity (HBC)

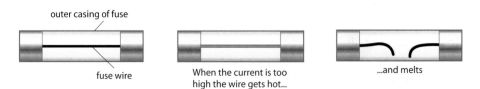

outer casing of fuse

fuse wire

When the current is too high the wire gets hot...

...and melts

Figure 2.22: How a fuse breaks under fault

High rupture capacity (HRC) fuses are designed to take extra current for several seconds and are used in circuits where the load draws a very high current on start-up before settling down to its normal value. An example of this is an electric motor, which draws a high starting current, especially if it is connected to a mechanical load.

Miniature circuit breakers (MCBs)

An MCB is a device that will trip a switch to break the electrical circuit if an excessively high current is detected. When electricity flows, it produces heat and generates a magnetic field around the conductors. The MCB uses a combination of the heating and magnetic effects to break the circuit under fault conditions.

MCBs are more expensive than fuses but they are also more accurate and can be reset. They are now found in all electrical consumer units in newer domestic properties (see Figure 2.23).

Residual current devices (RCDs)

These are highly sensitive devices that provide excellent protection to high-risk parts of electrical systems (such as plug socket outlets and electric showers).

An RCD measures the difference between the current in the electrical conductors in the system (e.g. live and neutral) and measures changes in the electrical current. If a small change occurs, the system is automatically disconnected. A typical operating current for an RCD in the event of a fault is 30 mA (30 1000ths of an amp). This may seem very low but it has been calculated that the average person can 'feel' electricity at current levels of around 1 mA.

Residual current circuit breaker with integral overload protection (RCBO)

This is a new type of miniature circuit breaker that incorporates earth fault protection. It is generally a combination of a thermal-magnetic type MCB and an integrated RCD and provides both overcurrent protection and earth fault current protection in a single unit (see Figure 2.24).

This restricts earth fault protection to a single circuit, so only the circuit with the fault is interrupted. A distribution board/consumer unit is often protected by an RCD, in which case all circuits will be disconnected in the event of a fault.

Explain the relationship between the size of fuses and the current in the system

Fuses come in different sizes (ratings) to protect against different levels of current. Fuse ratings are measured in amps (A).

Most electrical equipment – lamps, televisions, washing machines, music systems, vacuum cleaners – is rated in **watts**. Fuses in domestic environments are often 'overrated' – this means that a fuse with too high a rating is used for safety reasons.

You must always make sure you use the appropriate size of fuse. You can work out the fuse rating using this simple formula:

$$\text{amps} = \frac{\text{watts}}{\text{volts}}$$

For example, if a lamp uses a 100 watt bulb, the fuse rating is then calculated as:

100 ÷ 230 (the voltage of the domestic mains supply) = 0.435 amps

Manufacturers do not produce fuses rated at 0.435 amps, so the next practical size up is used – in this case, a 3 amp fuse. In this example, the bulb may be used in a stand-alone table lamp fitted with a cord and a 13 A plug top, which should be fitted with a 3 A fuse. Plugs are usually

Figure 2.23: An MCB now forms an essential part of most installations

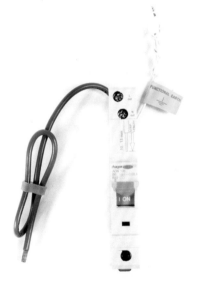

Figure 2.24: An RCBO is an MCB that incorporates earth fault protection

Key term

Watt (W) – the SI unit that indicates the amount of power an electrical item consumes.

supplied with a 13 A fuse, but this is unnecessarily high in this example and does not provide enough protection for the user. Technically, this set-up would be classified as unsafe.

However, if the lamp was one of several connected to a fixed lighting circuit, the entire circuit would usually be fitted with a 6 A fuse or circuit breaker.

There is potential for confusion here because 1 amp of current is more than enough to kill at 230 V. As already mentioned, the average person can 'feel' 1 mA (1000th of an amp).

Progress check 2.01

1   Which part of the current Building Regulations deals with electrical safety?

2   There are three types of earthing arrangements and supply systems. Which one is currently the most common and which type is gradually replacing it?

3   What does the term 'supplementary bonding' mean?

4   Produce a list of appliances and any parts of a building that supplementary bonding should be applied to.

5   What formula would you use to work out the fuse rating needed for an electrical appliance?

KNOW THE COMPONENTS USED IN ELECTRICAL INSTALLATIONS

Identify incoming electrical systems in domestic dwellings

This topic has been covered earlier in the chapter, so refer back to pages 53–57.

Identify the types of wiring used in electrical systems

As a plumber, you are likely to encounter various types of wiring, cabling and flex. Some of them are discussed in more detail below.

Cables

Cables carry current from the main supply to the sockets, switches, lights and appliances. There are many types of cable, each designed to do a certain type of job. Figure 2.25 shows how a typical cable is made up.

Did you know?

In March 2006, the colour coding of the live and neutral wires in cables was changed from red for live and black for neutral to brown for live and blue for neutral. However, you may come across the old colours when working on existing installations.

Conductor – usually copper and the part that actually carries the current. Note: some conductors are solid, some are stranded

Circuit protective conductor (cpc) – the conductor that carries the earth current if there is a fault. This is not normally live and is not insulated. Must be sleeved when cable is connected.

Figure 2.25: A twin and earth cable

PVC insulation – colour-coded and prevents short circuit and electric shock

PVC-insulated and sheathed flat-wiring cable

Polyvinyl chloride (PVC) insulated and sheathed cables are now the most popular type used for domestic and industrial wiring. PVC is versatile, tough, cheap, and easy to work with and install.

For these reasons it is generally the most economical material for this type of work. However, the level of insulation it provides is limited in conditions of excessive heat and cold, although this is not usually a problem in domestic properties. It can also suffer mechanical damage unless you apply additional mechanical protection in certain situations.

It is suitable for service wiring where there is little risk of mechanical damage. It is available as two- or three-core. Two or three plain copper, solid or stranded conductors are individually insulated with PVC and sheathed overall with PVC. An uninsulated plain copper circuit protective conductor (cpc) lies between the cores.

Two-core cables are coloured brown and blue plus earth. Three-core cables are brown, black and grey plus earth. The sheaths are normally grey or white.

Multistrand PVC-insulated and sheathed cable

This is suitable for surface wiring where there is little risk of mechanical damage. It is normally used for 'meter tails' to connect the consumer unit/distribution board to the public electricity supply (PES) meter and as single-core for **conduit** and trunking runs where conditions are difficult.

This cable comprises PVC-insulated and PVC-sheathed solid or stranded plain copper conductors. The old core colours are normally black and red; the new colours are blue and brown. Sheaths are normally black, red or grey, although other colours are available.

Single-core PVC-insulated and sheathed cable

This is used for domestic and general wiring. It comprises a PVC-insulated copper conductor with a PVC sheath. The old core colours are red or black (see Figure 2.26); the new colours are brown or blue. The sheath is normally white or grey.

Heat-resisting PVC-insulated and sheathed flexible cords (flex)

These cords are suitable for use in ambient temperatures up to 85°C. They are not suitable for use with heating appliances. They comprise plain copper flexible conductors insulated with heat-resisting (HR) PVC and are HR PVC-sheathed.

The core colours for single- and twin-core flex are brown or blue; for three-core, they are brown, blue and green/yellow; and for four-core, they are blue, brown, black and green/yellow (see Figure 2.27). The sheath is white.

> **Key term**
>
> **Conduit** – a pipework system designed to carry wiring and protect it from mechanical damage.

Figure 2.26: Single-core cable showing the old colour

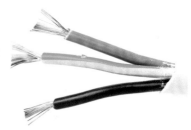

Figure 2.27: Heat-resisting PVC-insulated flex

Figure 2.28: A selection of metal conduit fixings

Figure 2.29: Examples of tee, terminal and 90° conduit boxes

Figure 2.30: Metal trunking must be joined securely

Identify the types of wire protection

Conduit

Conduit is typically used for electrical installations in factories, workshops and farms. It is normally used to protect single cables, which have only one conductor and one layer of insulation. There are two main types of conduit: metal and plastic (PVC).

Conduit sizes are measured using the outside diameter of the pipe. The *IET On-Site Guide* provides information about a number of conduit sizes, from 16 mm to 75 mm. However, the most commonly used sizes are 20 mm and 25 mm.

Conduit does not carry gas or liquid, so it does not have to be airtight or watertight. However, all conduit joints must be secure so that there is no movement that might damage the cables inside.

Metal conduit

Metal conduit is used as a heavy-duty way of enclosing electrical cables. Metal conduit is exposed metalwork and must be at the same electrical potential along its entire length. The conduit itself is made of steel and the lengths are joined by screwed threads (see Figure 2.28). Metal conduit is used in industrial installations and workshop type areas. It can be cut to length, rethreaded and formed into bends using conduit bending machines. It may also incorporate tees, terminal and conduit boxes (see Figure 2.29).

Plastic conduit

This is used where extra mechanical protection is needed. For example, it might be used in schools or colleges as a means of running cables in areas where they cannot be clipped directly to the surface or hidden in cavity walls. Plastic conduit is supplied as heavy and light gauge. The gauge is given by the thickness of the conduit wall. Common sizes are 20 mm and 25 mm.

Plastic cable ducting is also available. This is large piping that can be installed underground as a route for supply and data cables.

Trunking

Trunking is used to hold large amounts of cable. It is also used where it is not possible to clip cables directly to a surface or hide them by chasing them into walls or running them through ceiling and floor voids.

Metal trunking

This is usually made from sheet steel and is available in 2 m lengths which are then riveted or bolted together. The joints need to be mechanically secure to protect the cables inside from damage and to provide a good earth connection (see Figure 2.30). Small copper earth bars must be fitted across every joint.

Plastic trunking

Like plastic conduit, plastic trunking is used in lighter commercial installations. It is used where some mechanical protection is needed or where cables cannot be clipped to the surface or chased into walls. It is installed in a similar way to metal trunking, but the joints can be glued. The lid is usually a snap-on type.

Mini-trunking

Mini-trunking ranges in size from 16×8 mm to 60×60 mm. It provides minimal mechanical protection and is not acceptable as a wiring enclosure for single insulated cables. Typically, mini-trunking is run in domestic installations in situations where the cable cannot be clipped directly to the surface, chased into walls or run through floor and ceiling voids. It can be installed using standard fixings such as screws, cavity fixings and rawl plugs. Some versions also have double-sided tape on the back for quick and easy fitting.

State the relationship between the size of wire and the voltage carried

Some of the following areas have been covered earlier in this chapter and are also discussed in Chapter 3, pages 118–119. Refer to the relevant pages as necessary.

For practical purposes, three things must be present to create an electrical circuit: current, voltage and resistance.

Resistance

We can compare the flow of electricity to water being supplied to a bath via a cistern using gravity. The head of water in the cistern will provide the pressure (or voltage). With the tap open, the water will flow into the bath at a fixed rate per second (current). In order to flow from the cistern to the bath, the water must pass through a pipe (cable).

If we increase the diameter of the pipe, more water will be able to pass through it each second – there is less resistance to the flow of the water. If we decrease the diameter of the pipe, there will be more resistance and less water will pass through the pipe each second.

Resistance to current flow in an electrical circuit is measured in **ohms**. The bigger the conductor, the lower the resistance to current flow.

Current, voltage and resistance are interrelated. If we know the value of any two of them, we can calculate the third using Ohm's Law:

voltage (V) = current (I) $\times$ resistance (R)

See Chapter 3, page 119 for more on Ohm's Law.

Insulation

An insulator is a material that does not conduct electricity. Electrical conductors must be insulated to protect against short circuits. For example, the conductors in a cable or flex are protected with a plastic coating that insulates the live parts of the system that carry the current. Components such as pumps and valves using electric motors are also insulated internally.

Cross-sectional area of a conductor

The correct choice of cable size for a circuit depends on how much current will be passing through it. The higher the current, the larger the cable required. Cable size is based on the **cross-sectional area** of its conductor.

Key terms
Ohm – the SI unit of resistance, denoted by the Greek letter omega (Ω).
Cross-sectional area – the area of a cable conductor's face. Look at a conductor end-on and that is the area to be measured.

BS 7671 contains tables of how much current various cables can carry. The amount of current a cable can carry is also affected by:

- surrounding temperature – the higher the temperature, the less current it can carry
- length of cable run – the longer the run, the less current it can carry
- installation method – how a cable is installed. Wiring methods include clipping a cable to the surface of a wall, running it in conduit or trunking and running it on cable tray.

As a general rule:

- A lighting circuit will use a 1.5 mm^2 twin and earth PVC-insulated cable and be protected by a 6 A circuit protection device at the consumer unit.
- A ring main circuit will use a 2.5 mm^2 twin and earth PVC cable and be protected by a 32 A circuit protection device.

Identify components on electrical systems

Table 2.01 outlines some of the most common electrical components you are likely to encounter as a plumber.

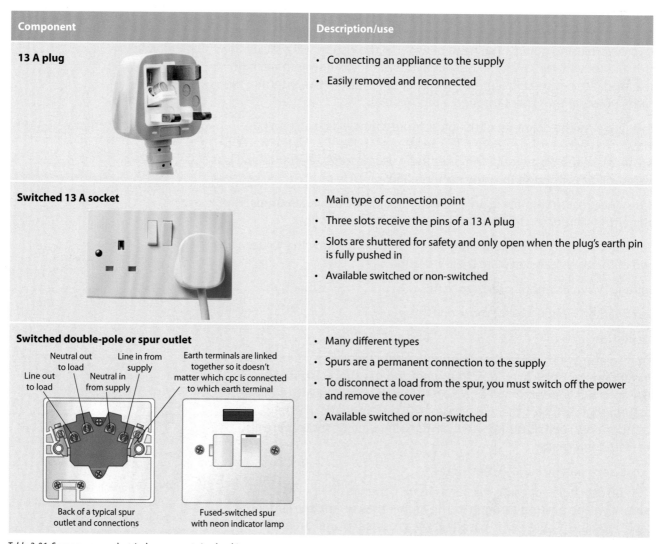

Component	Description/use
13 A plug	• Connecting an appliance to the supply • Easily removed and reconnected
Switched 13 A socket	• Main type of connection point • Three slots receive the pins of a 13 A plug • Slots are shuttered for safety and only open when the plug's earth pin is fully pushed in • Available switched or non-switched
Switched double-pole or spur outlet Neutral out to load · Line in from supply Line out to load · Neutral in from supply Earth terminals are linked together so it doesn't matter which cpc is connected to which earth terminal Back of a typical spur outlet and connections · Fused-switched spur with neon indicator lamp	• Many different types • Spurs are a permanent connection to the supply • To disconnect a load from the spur, you must switch off the power and remove the cover • Available switched or non-switched

Table 2.01: Some common electrical components in plumbing

Continued ▼

Light switch 	• One-way, two-way or intermediate light switches available in many finishes and types • More than one switch can be fitted into a switch plate to operate separate lights • Light switches should only switch the line conductor. They must never switch the neutral
Junction box 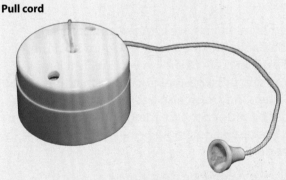	• A connection point for twin and earth cables • Should always be accessible, even if hidden under floors or in roof space • Available for different current ratings • Available in three, four and six terminal versions
Pull cord 	• Electrical switch for lighting in areas where there is a risk of electric shock due to wet or damp conditions (e.g. bathrooms) • User is insulated from the supply by the cord
Isolator	• This type of isolator is used to switch off the supply to fixed machinery • It has the option of a lockable feature using a padlock to prevent unauthorised use or to enable maintenance work to be carried out safely
Electrical timer 	• Used to electronically control equipment during a 24-hour period • Has an override facility to enable immediate use outside planned periods • More complex versions allow settings to be varied over a seven-day period • May be used for central heating purposes, but now largely replaced by programmers, which give greater flexibility and control

Table 2.01: Some common electrical components in plumbing (continued)

UNDERSTAND THE PROCEDURES FOR SAFELY ISOLATING SUPPLIES

Most fatal accidents involving electricity occur at the isolation stage. This is when you must be really careful and aware of what you are doing, as you may have no idea of the type of supply you are working with. Do not take any risks and ask for help if you are unsure of anything.

The electrical contracting industry has drawn up standardised procedures, which are also used as the standard for safe working in the plumbing industry.

Identify the test equipment required to carry out the safe isolation of an electrical supply

Before working on any electrical supply, you must make sure that it is completely dead and cannot be switched on accidentally without you knowing. Not only is this a requirement of the Electricity at Work Regulations 1989, but it is also essential for your personal safety and that of your customers and co-workers.

The proper way to test whether a circuit is live is to use an approved voltage indicating device, similar to the one shown in Figure 2.31. These use an illuminated lamp or a meter scale to show a voltage. Test lamps normally have a 15 watt lamp and must be constructed so that they are not dangerous if the lamp is broken. They must also be fitted with protection against excess current, either by a fuse not exceeding 500 mA or a current-limiting resistor and a fuse. The test leads should be held captive and sealed into the body of the voltage detector.

Test lamps and voltage indicators should be clearly marked with the maximum voltage that the device can test and the maximum voltage that the device will withstand.

In order to safely isolate an electrical supply you will also need a proving device, labels and a locking-off device. These are covered in more detail in the following sections.

Describe how to test voltage indicators on a known source

It is important to test your equipment regularly to make sure it is in good and safe working order. Your test equipment must have a current calibration certificate indicating that it is working properly and providing accurate readings. If it is not calibrated, test results could be inaccurate. If you have any doubts about an instrument or its accuracy, ask for assistance.

Before starting work:

- Check the equipment for any damage. Look to see if the case is cracked or broken, indicating a recent impact, which could result in false readings.
- Check that the insulation on the leads and the probes is not damaged, and that it is complete and secure.

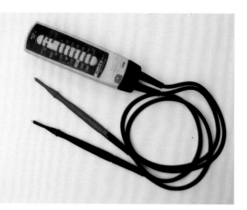

Figure 2.31: A voltage indicating device tests for live circuits

Safe working

You should never carry out live testing of supplies unless absolutely necessary.

It is not acceptable to use a home-made test lamp or neon screwdriver, as these will not show a supply that has low voltage.

- Test the voltage indicator on a proven supply. This will confirm that the kit is working. The best piece of equipment for doing this is a **proving unit**.

Once you have tested your equipment, you can use it to check whether the circuit you are going to work on is dead. You should then recheck the voltage indicator on a known supply to make sure it is still working correctly and has not been damaged during testing.

Identify the correct locations to carry out safe isolation

Identify sources of supply

Before you begin any work on an electrical circuit, you will need to identify the type and source of supply. In domestic dwellings this will be a single-phase 230 V supply.

In most cases there are two types of circuit:

- 32 A ring main for 13 A socket outlets
- 6 A lighting circuit.

You might also encounter:

- 16 A supplies to fixed components such as immersion heaters
- supplies to other fixed components such as cookers (32 A), central heating systems and showers. (The amperage protection will vary, depending on the kW rating of the unit.)

As a plumber, you are unlikely to work on lighting circuits, but you should be able to recognise circuit layouts and cable sizes. In older properties you might also find a 16 A radial circuit. This will supply the socket outlets but, unlike a ring main, the circuit will terminate at the last socket rather than returning to the consumer unit. This type of circuit is less popular than the ring main because the number of sockets that can be installed is limited.

To identify the source of supply:

1 Locate the plug socket nearest to the point where you intend to work on the component or appliance.

2 Make sure the socket is live by testing it – you can do this by plugging in a power tool to check that it works.

3 If the test shows that the socket is live (sound), locate the consumer unit and identify which circuit breaker (or fuse in an older property) will isolate the supply. Each circuit breaker or fuse will be labelled clearly.

Once you have identified the source of supply, you must identify the correct point of isolation. This could be at a consumer unit, a fused spur outlet or an electrical isolator.

Isolate at the consumer unit

How you isolate at the consumer unit depends on which type of fuse or circuit breaker has been used. If it is a cartridge fuse, remove the fuse and keep it with you. If an MCB or RCD has been used, switch it off and lock it off.

Isolate at a fused spur outlet

If the supply is from a fused spur outlet, switch off and remove the cartridge fuse, then lock off the fuse holder.

Safe working

It is vital that you follow correct procedures in order to work safely on electrical systems.

Plumbers in domestic dwellings will only work on individual circuits that supply power to an appliance or a component (boiler, pump, immersion heater, etc.) or on the appliance or component itself.

Isolate at an electrical isolator

Isolating devices (fuses, MCBs, RCDs) must comply with British Standards requirements. The isolating distance between the contacts must comply with BS EN 60947-3 for an isolator. The position of the contacts must be externally visible or clearly, positively and reliably indicated.

Describe the procedure for preventing the supply being turned back on

You must take precautions to prevent the supply being turned on accidently by the customer or a co-worker. These precautions will depend on how the circuit has been isolated.

Circuit breakers

You should lock circuit breakers using an appropriate locking-off clip and padlock that can only be operated with a unique key or combination, which you should keep with you while you are working on the circuit. You should also place a warning notice at the point of isolation, saying, for example, 'Work in progress – system switched off'.

Fuses

If the circuit is isolated by a fuse, you should remove the fuse and keep it with you while you are working on the circuit. You should also fit a lockable fuse insert in the gap where the fuse was. You should lock this using a padlock that can only be operated with a unique key or combination, which you should keep with you at all times. You should also place a warning notice at the point of isolation.

If a lockable fuse insert is not available, consider:

- fitting a 'dummy' fuse (a holder with no fuse in it)
- padlocking the distribution board door (keep the key with you and put up a warning notice)
- disconnecting the circuit (put up a warning notice).

Describe how to check the supply is dead

To check that the system is dead, you will need an approved voltage indicating device (see page 72). Use a proving unit (see page 73) to make sure the voltage indicator is working correctly. Then use the voltage indicator to check:

- phase (live)-to-neutral conductors
- phase-to-earth conductors
- neutral-to-earth condctors.

If all connections are dead, recheck the voltage indicator to make sure it is still working correctly and has not been damaged during testing. Once you have confirmed that it is working correctly, you can consider the circuit to be dead and safe to work on.

Figure 2.32 shows the procedure for isolating an installation, individual circuit or item of fixed equipment.

 Safe working

Do not place insulating tape over a circuit breaker to stop someone else switching it on. This is not a safe means of isolation.

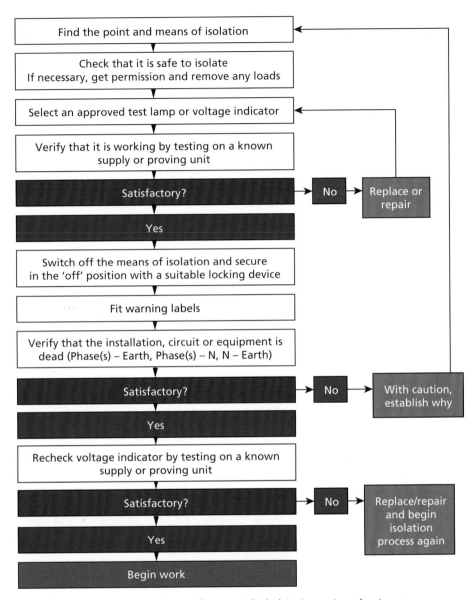

Figure 2.32: Safe isolation for a complete installation, an individual circuit, or an item of equipment

Progress check 2.02

1 Produce a list of equipment and other items that are required to carry out a safe isolation procedure.

2 What steps should you take to prevent the electricity supply being turned back on after safe isolation has been carried out?

3 What is the sequence of checking the conductors against one another when the voltage indicating device has been proven, to confirm the system is dead?

4 List the three steps you should take to identify the source of supply.

5 What does RCD stand for and what is its purpose?

UNDERSTAND HOW TO IDENTIFY SAFETY CRITICAL FAULTS ON ELECTRICAL COMPONENTS AND SYSTEMS

All electrical tools, appliances, components and wiring systems could present a danger to users, installers or anyone within the vicinity. Therefore, it is extremely important to identify, recognise and correct critical faults.

This is why the UK has such strict and robust rules and regulations regarding the type, use and installation of appliances, systems and components. These rules and regulations ensure that fires caused by electrical faults, injuries and deaths are kept to a minimum. Always take care when working with electricity, whether on an installation or using a power tool. Don't become part of the statistics.

Describe safety critical faults on electrical installations and the consequences of the failure to rectify them

Each year about 1000 accidents in the workplace involving electric shock or burns are reported to the Health and Safety Executive. Around 30 of these are fatal. Table 2.02 outlines some common faults and the consequences of not rectifying them.

Critical fault	Electric shock	Burns	Death	Fire	Component failure	Overheating	Reduced lifecycle of component
Damaged cables and flex	✓	✓	✓	✓	✓	✓	✓
Loose wires	✓	✓	✓	✓	✓	✓	✓
Incorrect cables (size/type)	✓	✓	✓	✓	✓	✓	✓
Broken junction box	✓	✓	✓	✓	possible	possible	possible
Missing earth bonding	✓	✓	✓	✓	possible	✓	✓
Incorrect size fuse	✓	✓	✓	✓	✓	✓	✓

Table 2.02: Safety critical faults and the consequences of not rectifying them

Identify responsible persons to be informed of any electrical faults

Table 2.03 lists who you should contact in the event of faults. The ticks indicate which reasons are relevant for each person.

Responsible person	Reasons why					
	Safety	Conform with warning notices and notification	Cost of rectification	Identification of similar faults on site	Interruption of work processes	Insurance issues
Home owner	✓	✓	✓			✓
Tenant	✓	✓				
Landlord	✓	✓	✓	✓	✓	✓
Site manager	✓	✓		✓	✓	✓
Supervisor	✓	✓	✓	✓	✓	✓
Co-contractor	✓	✓		✓	✓	✓
Site agent	✓	✓	✓	✓	✓	✓
Caretaker	✓	✓		✓	✓	✓
Managing agent	✓	✓	✓	✓	✓	✓

Table 2.03: Who should be informed of faults and why

Explain the actions to be taken when finding a fault on an electrical installation

Use the following scenario to consider what you should do if you find a fault.

Working practice

A large modernisation project on an occupied tenanted housing site is being carried out. It involves (among other upgrades) the installation of a new central heating system. A plumbing and heating engineer has identified that the earth connection has not been made into a junction box located beneath the floor. Additionally, the junction box is not secured to a joist.

The junction box is to give a fused spur outlet for the new boiler and heating system controls and the plumbing and heating engineer must make a connection into it. He checks with a colleague who is carrying out another installation in the adjoining property to see if the same error applies to any other houses. The same issue is found.

- Who would be responsible for leaving an electrical installation in this condition?
- Which safety check was obviously not carried out at the time of the installation, which would have identified the fault?
- Who are the responsible people to whom this fault should be reported?
- If the plumbing and heating engineer supervisor instructed the operative to correct the fault and continue, list the procedure that should be adopted.

Refer back through this chapter for help with the answers if necessary.

Remember
Plumbers are in constant contact with metal pipework and metal surfaces when carrying out their work. Customers can be too.

Safe working
Remember: when you remove a section of pipework, you are removing a section of the system from the main bonding conductor. You may, therefore, be exposing yourself to real danger – this one can kill!

UNDERSTAND HOW TO UNDERTAKE BASIC ELECTRICAL TASKS

In this section you will look at other tests that confirm the safety of an electrical installation. You will need to test for earth continuity and the polarity of wires at an appliance or a component. You should also carry out an insulation-resistance test, which will tell you if the cabling used is in good condition.

The systems you will deal with include:

- continuity of earthing conductors
- polarity
- insulation resistance.

Whatever the test, you must use the correct equipment. For electrical work associated with plumbing, the testing and equipment required includes:

- continuity: low-reading ohmmeter (an instrument to measure resistance in circuits)
- polarity: low-reading ohmmeter or insulation and continuity tester
- insulation resistance: insulation resistance tester.

Explain the importance of electrical temporary continuity bonds

As described on pages 60–63, testing for earth continuity involves making sure that all exposed metalwork in a building is bonded together and connected to the earthing block in the consumer unit. This means that if there is a fault, the current will be led to earth and the supply will be disconnected automatically.

Describe the procedure for applying temporary electrical continuity bonds

When working on a repair or maintenance job, it is important to check that any metal pipework is bonded correctly.

- It is essential that earth continuity is maintained before any cutting, disconnection or removal of pipework or components takes place. This is done by 'bridging' the gap exposed by the removed section of pipe or component with a temporary bonding wire.
- The temporary bond must be securely fixed in place before the length of pipe is removed.
- Typical kit consists of crocodile clips (clamps) and a 10 mm conductor with a minimum 250 V rating. It is suitable for up to 28 mm metal pipe. You should use earth clips when connecting bonding wire to pipework. These will indicate clearly the importance of the connection and show that it ensures a safe electrical connection.
- Correctly identify the pipe or component to be removed/taken apart. If the pipe has been painted, you must remove the paint around the area where the clamps will be positioned to ensure good continuity.

- Position the clips so that they bridge the gap of the pipe, fitting or component that is going to be removed. Work can then safely take place.
- Only remove the clamps once the job is complete.

Figure 2.33 shows a typical bonding arrangement for domestic premises. All of the incoming services to the property, such as gas, water and electricity, are bonded close to the point of entry to the building and are connected to the main earthing terminal.

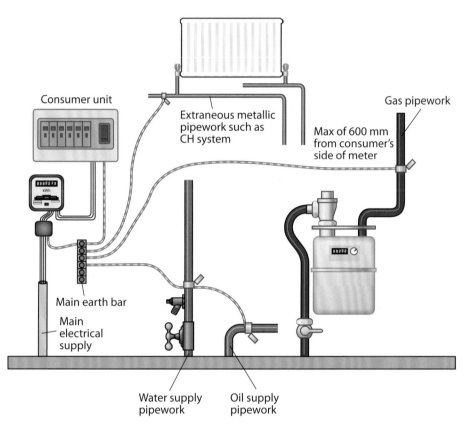

Consumer unit

Extraneous metallic
pipework such as
CH system

Gas pipework

Max of 600 mm
from consumer's
side of meter

Main earth bar

Main
electrical
supply

Water supply
pipework

Oil supply
pipework

Figure 2.33: Example of the main bonding system in a domestic property

Describe the process for wiring a three-pin plug

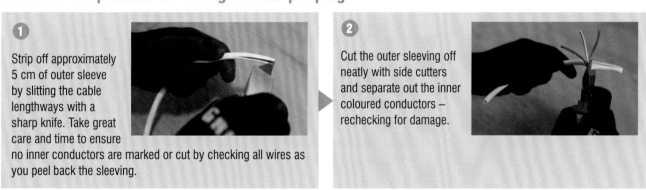

① Strip off approximately 5 cm of outer sleeve by slitting the cable lengthways with a sharp knife. Take great care and time to ensure no inner conductors are marked or cut by checking all wires as you peel back the sleeving.

② Cut the outer sleeving off neatly with side cutters and separate out the inner coloured conductors – rechecking for damage.

3

Cut each conductor to length according to the manufacturer's instruction card for that plug. If there are no dimensions available, there needs to be enough slack to make a sound termination but not too much to cause bunching or over-crowding/crushing when putting the lid back. Remember, if the cable is pulled out by accident when in use, the earth must be slightly longer so it breaks last for safety reasons.

4

Remove 5–6 mm of the coloured sleeving from each conductor with correctly set wire strippers and check that no strands have been damaged or cut as you twist the strands tightly on each individual conductor.

5

Loosen the cord grip/clamp and the terminal screws. Leave the earth terminal tight if the appliances is double insulated and has no earth. Place the twisted wires individually into the correct terminal before re-tightening. Follow the colour code: brown (old colour, red) to live; blue (old colour, black) to neutral; yellow/green (old colour, green) to earth.

6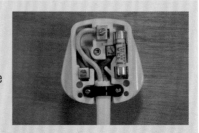

Tighten down the cord clamp over the outer flex sleeve, making sure not to clamp over the coloured conductors and then check that:

a the fuse is correct for the appliance:
   <700 W = 3 A, 700 W–1200 W = 5 A, >1200 W = 13 A)
b the wires and sleeves are undamaged
c no copper is showing in the terminals.

Finally, replace the top of the plug, making sure no wires are trapped between the two parts of the plug, and replace the screw.

Identify the tools required to cut and join cable

Refer to Table 4.01 on pages 124–135 of Chapter 4 for common plumbing tools and equipment. The tools you will need for this task are:

- side cutting pliers/electrician's pliers
- side cutters
- insulation stripper
- a selection of insulated screwdrivers.

Describe the process for attaching a cable to a junction box

The junction box has no specific line, neutral or earth terminals. When connecting cable to a junction box (see Figure 2.34), make sure:

- the cable sheath is taken into the junction box
- no conductor is showing outside the terminal
- the conductors are laid neatly in the box.

Figure 2.34: Attaching a cable to a junction box

Describe basic safety electrical checks

Earth continuity

For this you need an ohmmeter. The ohmmeter leads should be connected between the points being tested, for example:

- between simultaneously accessible extraneous (external) conductive parts such as pipework, sinks, etc.
- between simultaneously accessible extraneous conductive parts and exposed conductive parts (metal parts of the installation).

To check that the conductor is sound, move one probe into the main earth terminal and the other to the metalwork to be protected. It doesn't matter which probe goes to which location; the test will work either way.

You can also use this method to test the main equipotential bonding conductors. The ohmmeter should show a low resistance reading (see Figure 2.35).

Did you know?

If you do not carry out these tests using the correct procedure, you will be breaking the law.

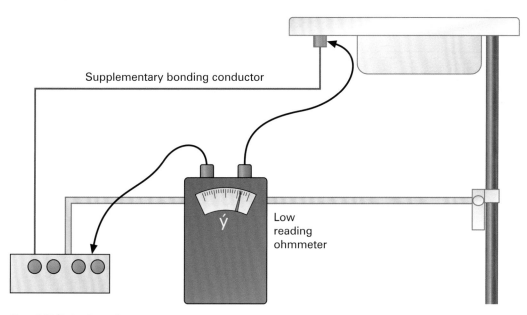

Supplementary bonding conductor

Low reading ohmmeter

Figure 2.35: Testing for earth continuity

Short circuit

A short circuit occurs when live and neutral come into direct contact. A large fault current (sometimes thousands of amps) will flow, resulting in a violent explosion and heat. This can cause serious damage (see Figure 2.36).

You can check for short circuits using a multimeter. For instance, on a boiler:

- Turn off the power to the boiler at the fused spur and remove any other electric plugs or wires.
- A reading of over 100 ohms across live and neutral is needed for the test to be satisfactory – no short circuit present.

Figure 2.36: Short circuit damage to a component

Safe working

Always ensure that testing equipment is calibrated on an annual basis or as directed by the manufacturer.

Alternatively, to test for a suspected short circuit on an appliance:

- Safely remove every appliance from the supply.
- Replug each appliance back into the supply until either a trip is activated or a fuse is blown.

Resistance to earth

Use the same methods as when checking for a short circuit, but with a multimeter to check for a fault between live and earth.

Polarity

Testing for polarity makes sure that phased conductors are not crossed somewhere – neutral from mains connected to live and vice versa (reversed polarity). In this situation, the system might still function as expected. However, when isolated from a switch, the system will be in a dangerous condition.

- Close lighting switches before carrying out a polarity test.
- You should test across the conductors to make sure that no wires have been crossed (see Figure 2.37).

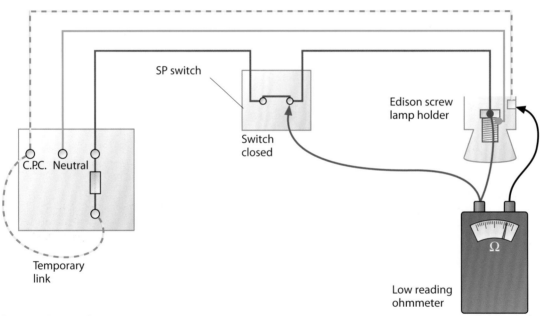

Figure 2.37: Testing polarity

Insulation resistance

Insulation resistance tests (see Figure 2.38 on page 83) make sure that:

- the insulation of conductors, electrical appliances and components is satisfactory
- electrical conductors and protective conductors are not short-circuited or do not show a low insulation resistance (which would indicate a defective insulation).

Before testing, ensure that:

1 pilot or indicator lamps and capacitors are disconnected from circuits to avoid an incorrect value being obtained

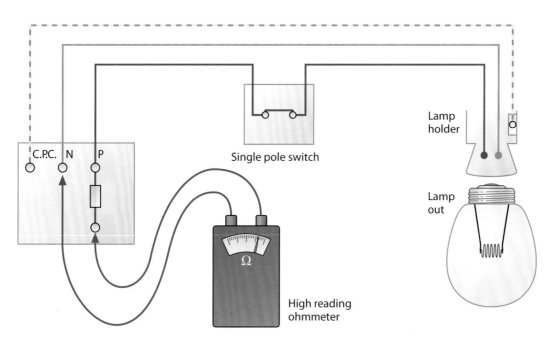

Figure 2.38: Insulation resistance check

2 voltage-sensitive electronic equipment – such as dimmer switches, delay timers, power controllers, electronic starters for fluorescent lamps, emergency lighting, RCDs, etc. – are disconnected so that they are not subjected to the test voltage

3 there is no electrical connection between any phase or neutral conductor (e.g. lamps left in).

An insulation resistance of no less than 0.5 M ohms should be achieved.

Socket tester

A socket tester plugs into a mains outlet socket to check that the socket is wired correctly and is grounded to earth. Although there are unlikely to be any problems with this, it is still worth checking a socket before using it, particularly in older properties or if you have any doubts about the installation conforming to standards.

Sources of additional information on this chapter include:

- Building Regulations Approved Document P:
  www.planningportal.gov.uk
- HSE website dealing with Electrical Safety:
  www.hse.gov.uk/electricity

Safe working

Never carry out any form of electrical testing unless you have been trained and are competent to do so.

Progress check 2.03

1. Name the three conductors found in a consumer unit. Which one of these will the fuses or circuit breakers be attached to?
2. Name four circuits that could be connected inside a domestic consumer unit.
3. Describe how an immersion heater is connected to the electricity supply of a building. Your answer should mention connection, fuse protection size, termination and size and type of cable used.
4. What is the new colour coding for the live, neutral and earth wires in a three-pin appliance plug?
5. What do the terms 'conduit' and 'trunking' refer to when talking about electrical wiring?

Chapter 3

This chapter will cover the following learning outcomes:

- **Understand the properties of common plumbing materials**
- **Understand the scientific properties and principles of water**
- **Understand the pressure, force and flow of water**
- **Understand the principles of heat in relation to plumbing systems**
- **Know the principles of combustion and heating gases**
- **Know the basic principles of electricity**

Key terms

Corrosion – the gradual damage or destruction of a hard material through chemical action.

Compound – a mixture of two or more elements.

Introduction

A plumber's job involves the installation, maintenance and servicing of many different systems using a large range of materials for pipework, fittings and components. Therefore, you must have a good understanding of the properties of these materials and their suitability for the job they will be doing. This chapter outlines the science of plumbing and introduces the basic materials, theories and concepts you will encounter and work with.

UNDERSTAND THE PROPERTIES OF COMMON PLUMBING MATERIALS

Identify different uses of materials in plumbing

Several materials, which have a range of different uses, are commonly used in the building services engineering industry. There is no perfect pipework material that is suitable for all applications. Different materials perform according to different factors and conditions, such as pressure, type of water, cost, bending and jointing method, resistance to **corrosion**, expansion and appearance. The most common materials are described on pages 86–90.

Metal

Metals rarely occur in their pure form. More often they occur as ores, which are **compounds** of the metal and have unwanted impurities. To produce the required metal, a process of smelting is necessary. The most common method of producing metals is by removing the oxygen from the ore in a process known as reduction. Metals commonly used in the building services engineering industry include iron, copper, lead, tin, zinc and aluminium.

The industrial production of iron and steel (smelting)

1 Iron ore (haematite – iron oxide) is loaded into a blast furnace along with coke and limestone.

2 Hot air is blasted into the base of the furnace and carbon from the coke reacts with oxygen in the air to form carbon monoxide.

3 Carbon monoxide reacts with oxygen from the haematite to form carbon dioxide and iron.

4 Limestone combines with impurities in the ore (mainly silicates) to form slag.

5 The molten iron is tapped from the base of the furnace and solidifies into billets known as 'pigs' – hence the term 'pig iron'.

At this point the iron is impure. To form steel, which is an alloy of iron and carbon, the iron is heated to get rid of impurities, and then up to 1.5 per cent carbon is added. Other metals can give the steel particular properties, for example the addition of chromium will produce stainless steel. Alloys can be produced by mixing different metals or by mixing metals with non-metallic elements such as carbon.

Alloys

An alloy is a type of metal made from two or more other metals. Table 3.01 describes some common alloys and their uses.

Alloy	Elements	Common uses
Brass	Copper and zinc	Electrical contacts and corrosion-resistant fixings (e.g. screws, bolts) and pipe fittings
Bronze	Copper and tin	Decorative or artistic purposes and corrosion-resistant pumps
Solder	Lead and tin or tin and copper	Electrical connections and as a jointing material
Duralumin	Aluminium, magnesium, copper and manganese	Aircraft production
Gunmetal	Copper, tin and zinc	Underground corrosion-resistant fittings

Table 3.01: Some commonly used alloys

Copper

Copper has been used for pipework for over 100 years. It is a **malleable** and **ductile** material, which you will often use in your plumbing career.

The plumbing industry uses four main types of copper tube:

- R250 half-hard is widely used for domestic installations but should not be used underground. Pipe diameters range from 12–54 mm (diameter is always specified as an external measurement), with wall thickness of 0.7 mm. Tubes are usually available in 6 m lengths, though most merchants will also supply in 1, 2 and 3 m lengths. R250 half-hard is also available in chromium plate, used where pipework is exposed to the eye and an attractive finish is required.
- R290 hard is not as popular as R250 half-hard, mainly because it is unbendable. The wall is 0.2 mm thinner than R250, so the pipe is hardened during the manufacturing process to make it stronger. R290 hard is available in 1, 2, 3 and 6 m lengths, with pipe diameters of 12–54 mm. It is available in chromium plate, most often found in exposed pipe runs to instantaneous showers in a bath or shower room (particularly where the appliance is a new addition to an existing suite).
- R220 soft coils is used for external underground installations. This is the thickest grade, with a wall thickness of 1 mm. R220 soft coils is classified as fully annealed, which means it is soft. It is available in diameters of 15, 22 and 28 mm, in coils of 10–50 m. It may be supplied in a blue (water service) or yellow (gas) plastic coating.
- R220 soft coils micro-bore is used for micro-bore heating and is fully annealed. It is available in diameters of 6, 8 and 10 mm, in coils of 10–50 m. It may be supplied in a yellow (gas) or white (heating) plastic coating. It is also available with air channels in the plastic coating to improve thermal insulation; this is used where pipework is installed internally in solid floors.

> **Key terms**
>
> *Malleable* – able to be hammered into shape.
>
> *Ductile* – able to be drawn out into wire.

Figure 3.01: Copper is used extensively in plumbing

Steel

Low carbon steel

Low carbon steel (LCS) or mild steel is an alloy of iron and carbon. It is often used in the plumbing and heating industry and is manufactured to BS 1387. LCS tube comes in three weight grades: light, medium and heavy (see Table 3.02). As with copper tube, the outer diameters are similar but the internal bores and wall thicknesses vary.

Grade	Wall thickness	Bore	Colour code
Light	Thin	Larger	Brown
Medium	Medium	Medium	Blue
Heavy	Thick	Smaller	Red

Table 3.02: Grades of LCS (also called mild steel)

Light LCS is usually used for conduits. You will come across it occasionally, but medium and heavy LCS are more common. Medium and heavy LCS tubes are used for water supply services as they can sustain the pressures involved. LCS tubes used for domestic water supplies must be **galvanised**. (See page 101 for more on galvanised coatings.)

Stainless steel

This is the most recently developed pipe material used for water services. It is a complex alloy made up of several elements, as shown in Table 3.03.

Stainless steel is shiny because of its chromium and nickel content and is protected from corrosion by a microscopic layer of chromium oxide. This type of tube is produced with bores of 6–35 mm and has an average wall thickness of 0.7 mm. The outside diameters are similar to those of R250 copper tubes.

Stainless steel is commonly used where exposed pipework and sanitary appliances are needed, as it is very strong (much stronger than copper) and is easy to clean. Stainless steel is commonly used for:

- sink units
- urinal units and supply pipework
- commercial kitchen discharge pipework.

Lead

Lead is a very heavy, valuable metal that requires specialist handling. It is one of the oldest known metals and is highly ductile, malleable and corrosion-resistant. Lead in sheet form is still used for weathering on buildings. The Model Water Byelaws 1986 prohibited the use of lead for mains, sanitary and rainwater pipework because of the risk of lead poisoning. Lead has now been replaced by materials such as plastics.

Cast iron

Cast iron is an alloy of iron and is approximately three per cent carbon. It is very heavy but quite brittle, although it can withstand years of wear and

Key term

Galvanise – to coat iron or steel with a layer of zinc to prevent oxidisation (rusting). The term comes from the 18th-century Italian scientist Luigi Galvani.

Element	% of element (approx.)
Iron	70
Chromium	18
Nickel	10
Manganese	1.25
Silicon	0.6
Carbon	0.08
Sulfur	Trace
Phosphorous	Trace

Table 3.03: Composition of stainless steel

tear. It has been used in the plumbing industry for many years for sanitary pipework both above and below ground, mainly guttering, soil stacks and baths. You will probably come across it in older domestic properties and new industrial or commercial properties.

Plastics commonly used in plumbing

Plastics are products of the oil industry. Ethene, a product of crude oil, is a building block of plastics. It is made up of carbon, hydrogen and oxygen atoms. Molecules of ethene (monomers) can link together into long chains (polymers) to make polythene (poly + ethene) when they are heated under pressure with a **catalyst**. If the ethene monomer is modified by replacing one of the hydrogen atoms with another atom or molecule, further monomers result, producing other plastics. This process is called polymerisation.

The plumbing industry uses two main categories of plastics:

- thermosetting plastics
- thermoplastics.

Thermosetting plastics

These are generally used for mouldings. They soften when first heated, which enables them to be moulded, but when they cool they set hard and their shape is fixed; it cannot be altered by further heating. WC cisterns can be made of thermosetting plastic.

Thermoplastics

Thermoplastics can be resoftened by heating. Most of the pipework materials you will encounter fall into this category. The different types of thermoplastics (see Table 3.04) share many of the same characteristics:

- strong resistance to acids and alkalis
- low specific heat (i.e. they do not absorb as much heat as metallic materials)
- poor conductors of heat
- affected by sunlight, which makes the plastic brittle (also called degradation).

> **Key term**
>
> *Catalyst* – a substance that increases the rate of a chemical reaction while itself remaining unchanged.

Thermoplastic	Max. usage temp. (°C)	Main plumbing industry purpose
Polythene – low density	80	Flexible pipe material used to channel chemical waste
Polythene – high density	104	More rigid; used for chemical or laboratory waste
Polypropylene	120	Tough plastic with relatively high melting temperature; used to channel boiling water for short periods of time (e.g. traps)
Polyvinyl chloride (PVC)	40–65	One of the most common pipework materials; used for discharge and drainage pipework
Unplasticised polyvinyl chloride (uPVC)	65	More rigid than PVC; used for cold water supply pipework
Acrylonitrile butadiene styrene (ABS)	90	Able to withstand higher temperatures than PVC; used for small diameter waste, discharge and overflow pipework

Table 3.04: Types of thermoplastics

Other materials used in the plumbing industry

Ceramics include products that are made by baking or firing mixtures of clay, sand and other minerals. They include bricks, tiles, earthenware, pottery and china. The kiln firing process fuses the individual ingredients of the product into a tough matrix. The main constituent of all these products is the element silicon (Si). Clay is aluminium silicate and sand is silicon dioxide (silica). Ceramics also include products made by 'curing' mixtures of sand, gravel and water with a setting agent (usually cement) to form concrete, or a mixture of sand, water and cement to form mortar. You may come across ceramics in the form of discs in modern taps as an alternative to rubber washers.

Glass is also produced by melting minerals together. The basic ingredients are sand (silicon dioxide), calcium carbonate ($CaCO_3$) and sodium carbonate ($NaCO_3$). The resulting mixture of calcium and sodium silicates cools to form glass. Again, additives can change the character of the product: boron will produce heat-resistant Pyrex®-type glass, and lead will produce hard 'crystal' glass. Glass is still used in laboratories because of its resistance to chemical attack. Joints between glass materials are made using specialised gaskets and O rings.

Describe the properties of materials used in plumbing

Materials are classified according to a variety of properties and characteristics. The following section outlines the main properties before looking at common plumbing materials in more detail. Properties can be measured by the way materials react to a variety of influences (see Figure 3.02).

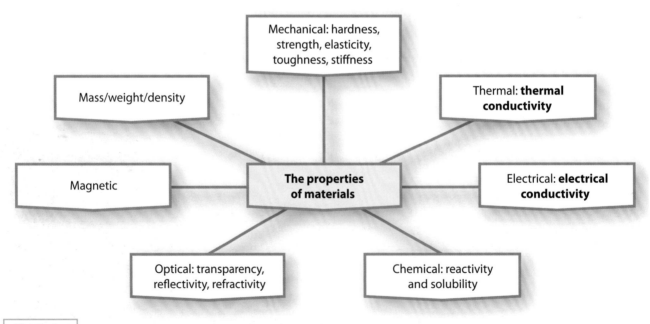

Figure 3.02: The properties of materials

Key terms

Thermal conductivity – how well or poorly a material conducts heat.

Electrical conductivity – how well or poorly a material conducts electricity.

Activity 3.1

Find out about the basic properties of solid materials.

Mass and weight

Mass and weight are described in detail on page 106.

Density

Solids

Solid materials of the same size and shape often have a completely different mass. This is because of the relative lightness or heaviness of the material, which is referred to as density. In practical terms, the density of an object or material is a measure of its mass (for example, grams) compared to its volume (for example, m^3). You can work out the density of a material using the following formula:

$$density = \frac{mass}{volume}$$

You can find the densities of the common materials used in plumbing in many reference sources.

Activity 3.2

Find a list of comparative densities. You could use your learning centre or local library. Make a list of common materials used in plumbing (see pages 86–90) and put them in order of density.

Liquids and gases

Liquids and gases also have different densities depending on the number of molecules that are present within a particular volume of the substance.

Relative density – sometimes known as specific gravity (SG) – is an effective way of measuring the density of a substance or object by comparing its weight per volume to an equal volume of water or air.

The density of solids and liquids is usually described by the relative density of the substance compared to that of water. Water is given a relative density of 1.0 and other materials have a relative density that is smaller or greater than 1.0, depending on whether the substance is lighter or heavier than water.

The density of gases is described in a similar way, by comparing it to the density of air. When comparing gases, the relative density of air is given as 1.0. For example, think about helium-filled balloons. Do you think helium is lighter or heavier than air? Will its relative density be smaller or greater than 1.0?

Malleability

Malleability is a property of materials such as metals. A malleable material is one that can be shaped and deformed by a compressive strength without fracturing, cracking or breaking. Sheet lead is a good example of a malleable material, as it can easily be worked and shaped into quite complicated patterns (see Figure 3.03).

Figure 3.03: Sheet lead is highly malleable, allowing it to be shaped easily

Figure 3.04: This machine draws out (pulls) copper into thin wire using the metal's property of ductility

Scale (1 is softest, 10 is hardest)	Material
1	Talc
2	Gypsum
3	Calcite
4	Fluorite
5	Apatite
6	Feldspar
7	Quartz
8	Topaz
9	Corundum
10	Diamond

Table 3.05: The Mohs hardness scale

Ductility

A metal is ductile if, in a tensile strength test, it can be stretched and elongated without breaking. An example of this is wire, which starts out as a rod of metal. It is then cold drawn through a series of progressively narrowing dies until the desired gauge (thickness) of wire is reached. This process of producing wire takes advantage of the metal's ductility (see Figure 3.04).

Hardness

Hardness is a measure of a material's resistance to permanent or plastic deformation by scratching or indentation. It is an important property in materials that have to resist wear or abrasion such as the teeth on wrenches. A material's hardness must often be considered along with its strength. Some tools (such as chisels) require a hard point and a tough head to prevent shattering when struck with a hammer. Hardness is measured on a scale of 1–10 based on the hardness of ten naturally occurring minerals (see Table 3.05).

Strength

The strength of a material is the extent to which it can withstand an applied force or load (stress) without breaking. The load is expressed in terms of force per unit area (newtons per square metre, N/m^2) and can be in the form of:

- **compression force**, as applied to the piers of a bridge or a roof support
- **tensile or stretching force**, as applied to a guitar string, tow rope or crane cable
- **shear force**, as applied by scissors or when materials are torn (see Figure 3.05).

Materials can therefore have compressive, tensile or shear strength.

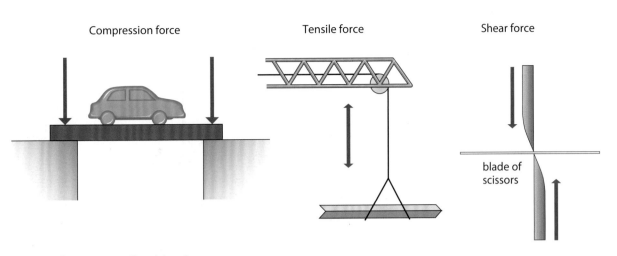

Figure 3.05: Compression, tensile and shear forces

Materials that can withstand a high compression loading include cast iron, stone and brick – hence their common use in building. However, these materials are brittle and will break if subjected to high tension. If a building needs to resist tensile strain – in an earthquake-prone area, for example – steel is a more suitable building material as it has high tensile strength.

Tensile strength

The tensile strength of a material is simply its resistance to being pulled apart. Materials such as brick or concrete have low tensile strength as they snap easily, whereas a ductile material such as steel or copper rod can be elongated before it finally snaps. This breaking point is known as the material's ultimate tensile strength (UTS). A good example of tensile strength can be seen in suspension bridges, where the road deck is held up by lots of cables (see Figure 3.06).

Figure 3.06: A suspension bridge has great tensile strength

Elasticity and deformation

Almost all materials will stretch to some extent when a tensile force is applied to them. The increase in length on loading compared to the original length of the material is known as strain. As loading continues, a point is reached when the material will no longer return to its original shape and size when the load is removed. The material has been stretched irreversibly and permanent deformation has occurred. The material is then said to have exceeded its elastic limit, or yield stress, and is suffering 'plastic deformation'.

Eventually, at maximum stress, the material reaches its breaking point – its UTS – and failure or fracture will follow rapidly. Figure 3.07 illustrates this sequence for four materials:

- **Mild steel** has little elasticity but has the highest yield stress of the samples. It is fairly ductile – i.e. it can sustain plastic deformation over a large range – and it has the highest UTS.
- **Cast iron** is brittle. It has the least elasticity of the samples and no ability to sustain plastic deformation, but has higher tensile strength than concrete.
- **Copper** has little elasticity but is the most ductile of the samples. Its UTS is less than half that of mild steel.
- **Concrete** has little elasticity and the lowest tensile strength of the samples.

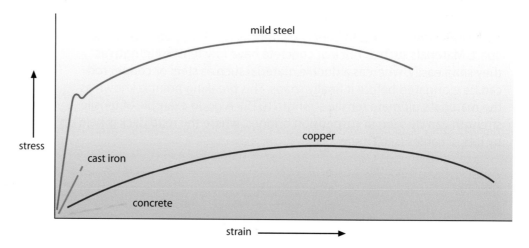

Figure 3.07: The effects of stress and strain on materials

Describe specific heat capacity

To size various plumbing components, such as boilers and radiators, you need to be able to understand the concept of heat. Heat is different from temperature and is a measure of the amount of energy in a substance. The standard unit of measurement of heat is the joule.

To work out the amount of heat required to heat a substance, you must measure the amount of heat required over time or the power required. This is a measure of the energy divided by the time taken to heat the substance and is expressed in kilowatt hours (kWh).

For plumbing calculations involving heat, you usually need to work out the amount of energy required to raise a quantity of a substance such as water from one temperature to another. To do this you need to know the specific heat capacity of the substance.

Specific heat capacity is the amount of heat required to raise the temperature of 1 kg of a substance by 1°C. The specific heat capacity of water is 4.186 kJ/kg°C. Table 3.06 shows the specific heat values for a range of materials.

Material	kJ/kg °C
Water	4.186
Aluminium	0.887
Cast iron	0.554
Zinc	0.397
Copper	0.385
Lead	0.125
Mercury	0.125

Table 3.06: Specific heat values

Example

Calculate the heat energy and power required to raise 200 litres of water from 10°C to 60°C, assuming 1 litre of water weighs 1 kg.

Heat energy = 200 litres × 4.186 kJ/kg°C × (60°C − 10°C) = 48,860 J

Power required to heat the water in 1 hour (3600 s) assuming no energy is lost:

$$\frac{41,860}{3,600} = 11.63 \text{ kW}$$

The power calculation is essential for determining factors such as the amount of energy needed to heat a hot water storage cylinder within a specific period of time. For example, if the 200 litres of water needs to be heated in 30 minutes instead of 1 hour, then the power required will be doubled, as the heating period has been halved (from 3600 seconds to 1800 seconds).

Describe the coefficient of linear expansion

Most materials expand when heated. This is because all substances are made up of molecules (groups of atoms) which move about more vigorously when heated. This causes the molecules to move further apart from each other – resulting in the material taking up more volume.

As the material cools, the molecules slow down and move closer together, so the material gets smaller (contracts). The amount that the material expands in length when heated can be calculated using the following formula:

length (m) × temperature rise (°C) × coefficient of linear expansion

Material	Coefficient (°C)
Plastic	0.00018
Zinc	0.000029
Lead	0.000029
Aluminium	0.000026
Tin	0.000021
Copper	0.000016
Cast iron	0.000011
Mild steel	0.000011
Invar	0.0000009

Table 3.07: Coefficients of linear expansion

Activity 3.3

What happens when materials are cooled down? How will this have an impact on your work as a plumber?

Table 3.07 shows the coefficient values of some of the most common materials used in the plumbing industry.

Example

Calculate the increase in length of a 6 m long plastic discharge stack that is subject to a temperature rise of 19°C. Use the following formula:

length (m) × temperature rise (°C) × coefficient of linear expansion

6 × 19 × 0.00018 = 0.02052 m or 20.52 mm

 Safe working

Plastics tend to expand the most. Therefore, you often need to leave an expansion gap in the pipework system when working with plastics to prevent the material from failing.

You need to take expansion and contraction into account in plumbing systems that are constantly being heated and cooled. Otherwise, the system or component may break down, causing leakage or other problems.

Describe heat conductivity

Conduction/conductivity is the transfer of heat energy through a material. It takes place as a result of the increased vibration of molecules, which occurs when materials are heated. The vibrations from the heated material are passed to the adjoining material, which then heats up in turn. Some materials are better at conducting heat than others. For example:

- metals tend to be good conductors of heat
- wood is a poor conductor of heat.

Gases and liquids also conduct heat, but poorly. In the plumbing industry, conduction is principally between solids. Of the metals commonly used, copper has higher conductivity than steel, iron or lead. Wooden, ceramic and plastic materials, which are poorer conductors of heat, are known as 'thermal insulators'. Pipework and cylinders must be insulated:

- to prevent heat loss
- for efficiency
- for environmental reasons.

Activity 3.4

Find out how good the following materials are as heat conductors: plastic, rubber, ceramic, wool and carbon. How would you use these materials in plumbing?

Conduction also refers to a material's ability to transfer electricity. Most metals are good conductors of electricity, but plastics are not.

Explain the concept of capillarity in liquids and describe its effects

Capillarity, or capillary action, is the process by which a liquid is drawn or hauled up through a small gap between the surfaces of two materials. This phenomenon is especially important for plumbers. In a positive way, it helps when soldering, but it can also affect how water can get into buildings, which can be damaging.

Forces of attraction

Surface tension

Surface tension describes the way in which water molecules 'cling' together to form what is effectively a very thin 'skin'. This can be demonstrated by filling a glass beaker right up to the top and examining the top of the glass. The water will appear above the upper limit of the glass. Why does the water not spill over the side? The answer is because of cohesion (molecules sticking together) as a result of surface tension (see Figure 3.08).

Adhesion

Adhesion is the force of attraction between water molecules and the sides of the vessel containing the water. This leads to a 'meniscus', which is the slightly curved 'skin' that appears when water is held in a vessel (see Figure 3.09). The processes of adhesion and cohesion cause capillary action.

Practical examples of capillary action

You must consider the possible effects of capillary action when planning lead roofing work. Water can find its way into a building's lapped roof if materials are close together. Figure 3.10 shows how this occurs and how it can be remedied.

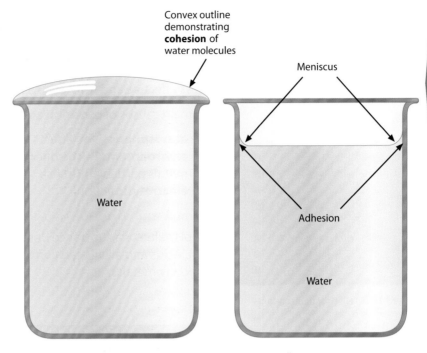

Figure 3.08: Surface tension and cohesion

Figure 3.09: Adhesion

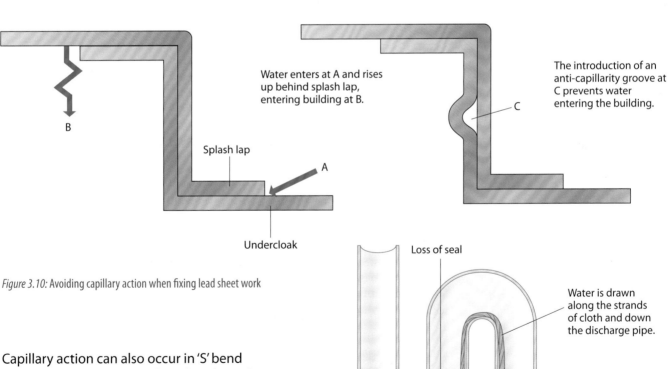

Water enters at A and rises up behind splash lap, entering building at B.

The introduction of an anti-capillarity groove at C prevents water entering the building.

Figure 3.10: Avoiding capillary action when fixing lead sheet work

Capillary action can also occur in 'S' bend drainage traps, which are found under sinks. If waste material, such as a piece of dishcloth, becomes lodged in the 'S' bend, capillary action could take place (see Figure 3.11). This can lead to the loss of the seal at the bottom of the trap, allowing bad smells from the drainage pipe to filter back into the home.

Loss of seal

Water is drawn along the strands of cloth and down the discharge pipe.

Figure 3.11: Capillary action can cause trap seal loss

Figure 3.12: The effects of corrosion on brass

Explain the causes of corrosion in plumbing system materials

If materials are not kept under the correct conditions, they can decay and break down. Once these materials break down, they can become unusable or less effective and will need to be replaced.

The main causes of corrosion are:

- **atmospheric** – the effects of air and water
- **environmental** – the direct effects of acids, alkalis and chemicals, with metals particularly at risk
- **electrolytic** action.

pH value and corrosion

The pH value of a substance describes how acid or alkaline it is (see Figure 3.13). Acids and alkalis can corrode and damage plumbing materials, especially metals.

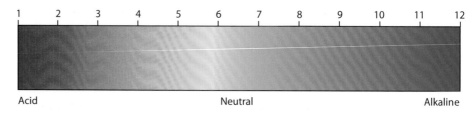

| 1 | 2 | 3 | 4 | 5 | 6 | 7 | 8 | 9 | 10 | 11 | 12 |

Acid   Neutral   Alkaline

Figure 3.13: The pH scale measures acidity and alkalinity

Atmospheric corrosion

Pure air and pure water have little corrosive effect, but together – in the form of moist air (oxygen + water vapour) – they can attack **ferrous** metals such as steel and iron very quickly to form iron oxide (rust). The corrosive effects of rusting can completely destroy metal. This process is called oxidisation.

Various other gases (such as carbon dioxide, sulfur dioxide and sulfur trioxide) that are present in the atmosphere can also increase the corrosive effect of air on particular metals – especially iron, steel and zinc. These gases tend to be more abundant in industrial areas, as they are often waste products from various industrial processes.

Coastal areas also suffer from increased atmospheric corrosion because of the amount of sodium chloride (salt) from the sea, which dissolves into the local atmosphere.

Non-ferrous metals, such as copper, aluminium and lead, have significant protection against atmospheric corrosion. Protective barriers (usually sulfates) form on these metals to prevent further corrosion. This protection is also known as 'patina'.

Corrosion by water

Ferrous metals are particularly vulnerable to corrosion caused by water. This is common in central heating systems in the form of black ferrous oxide and red rust in radiators. A by-product of this process is hydrogen gas, which builds up in the radiator (see Figure 3.14), leading to the need for radiator to be 'bled'. See Chapter 8 for how to bleed a radiator.

> **Key term**
>
> *Ferrous* – metals that contain iron.

> **Did you know?**
>
> Sulfur/sulfuric/sulfide (rather than sulphur/sulphuric/sulphide) are now the standard spellings, as adopted by the International Union of Pure and Applied Chemistry (IUPAC) and later by the Royal Society of Chemistry.

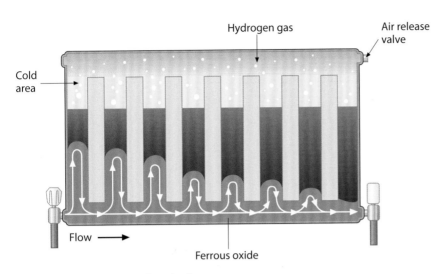

Figure 3.14: Corrosion effects in steel panel radiators

In certain areas of highly acidic water, copper may become slightly discoloured – but this does not affect the quality or safety of the drinking water. However, where lead pipework is still in use, very soft (acidic) water may dissolve minute quantities of lead and thus contaminate the water. This can have potentially toxic effects, especially for children.

Corrosion effects of building materials and underground conditions

Some types of wood (such as oak) have a corrosive effect on lead. Latex cement and foamed concrete will adversely affect copper. Certain types of soil can damage underground pipework. Heavy clay soils may contain sulfates, which can corrode lead, steel and copper. Ground containing ash and cinders is also highly corrosive as these are strongly alkaline. Pipes being laid in such ground should be wrapped in protective material.

Electrolytic action

Electrolytic action (electrolysis) describes a flow of electrically charged ions from an **anode** to a **cathode** through a medium called electrolyte (usually water), as shown in Figure 3.15. In chemistry, an electrolyte is any substance containing free ions that make the substance electrically conductive. The most typical electrolyte is an ionic solution, but molten and solid electrolytes are also possible.

Electrolytic corrosion takes place when electrolysis leads to the destruction of the anode. How long it takes for the anode to be destroyed will depend on:

- the properties of the water that acts as the electrolyte: if the water is hot or acidic the rate of corrosion will increase
- the position of the metals that make up the anode and cathode in the electromotive series.

Key terms

Anode – the negative electrode of an electrochemical current source while being discharged.

Cathode – the positive electrode of an electrochemical current source while being discharged.

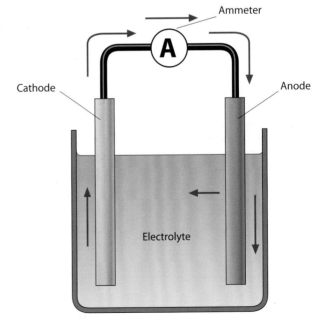

Figure 3.15: The process of electrolysis, as used by a simple battery to produce electricity

The electromotive series

This is a list of the common elements used in the plumbing industry.

- copper (cathodic or more noble)
- tin
- lead
- nickel
- cadmium
- iron
- chromium
- zinc
- aluminium
- magnesium (anodic or less noble).

The order in which they appear indicates their electromotive properties. The elements higher up in the list will destroy those lower down through the process of electrolytic corrosion. The further apart in the list the materials appear, the faster the corrosion will take place. For example, copper will destroy magnesium faster than lead will destroy chromium.

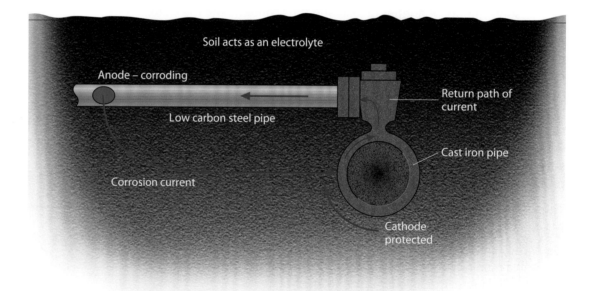

Figure 3.16: Example of possible electrolytic corrosion in pipework systems

You need to be aware of the potential for electrolytic corrosion when two very dissimilar metals, such as a galvanised tube and a copper fitting, are in direct contact. If these metallic elements are then surrounded by water (of a certain type) or by damp ground, a basic electrical cell is effectively created and electrolytic corrosion can take place.

Activity 3.5

Try to think of some practical examples of electrolytic corrosion. What precautions can you take to help prevent electrolytic corrosion in plumbing installations?

Identify methods of corrosion prevention in plumbing materials

The type of corrosion protection needed will depend on the type of material, its use and its location. This section outlines various ways of preventing corrosion.

- Plastic pipes are generally not affected by soil and water.
- Brass fittings used underground need to be wrapped to prevent them corroding.
- Low carbon steel that is not galvanised is not suitable for hot or cold water supplies and, if left unpainted, will corrode both inside and out. To prevent this, you can use galvanised pipe. Low carbon steel pipes underground should be protected from the damp and soil by wrapping them with a non-corrosive covering such as petrolatum tape.
- Copper pipes and fittings need to be installed so they do not come into contact with mortar. The use of a sleeve where pipes pass through holes in brickwork is recommended.

Enamel

Enamel is a type of coating applied to a surface to prevent or delay corrosion. The main types are vitreous enamel and paint enamel. Vitreous enamel is probably used most in plumbing as it can be applied to many different surfaces, usually metals. It is applied as a powder, which is then heated until it melts. Once cooled it forms a very hard (5–6 on the Mohs scale), durable, glass-like coating, which is scratch resistant and also resistant to chemicals. Various colours can be achieved by changing the colour of the powder used. Vitreous enamel is used for coating steel and cast iron baths, kitchen appliances, saucepans and jewellery.

Painted coatings

These can be applied to many different surfaces, but in particular, to plastics to prevent them deteriorating due the effects of ultraviolet light from the sun. Before using the paint, you should check that it will not damage the surface you will be applying to. Enamel is a type of painted coating.

Galvanised coatings

Galvanising iron or steel involves covering the surface with a thin layer of zinc. Water pipes are coated inside and out, usually by hot dipping them in a bath of molten zinc.

Inhibitors

An inhibitor is something that tries to block, restrain or suppress. In plumbing it refers to a chemical inhibitor which is added to a water system during commissioning to prevent corrosion of components – in particular the radiators, which are normally made of steel. It stops the build-up of ferrous oxide, which can prevent efficient flow of the water through the radiator.

> **Did you know?**
>
> The word 'enamel' comes from the High German *smelzan* and later from the Old French *esmail*.

Figure 3.17: Magnesium sacrificial anodes protect well against electrolysis

Sacrificial anodes

A sacrificial anode is attached to another metal object which would be subjected to electrolytic action (see page 99). The anode will then decompose instead of the metal it is attached to – hence the term 'sacrificial'.

Replacing anodes is routine maintenance on some types of expensive plumbing equipment, such as commercial direct water heaters, to ensure their long working life. Magnesium is commonly used as a sacrificial anode as it is lower on the galvanic scale and offers greater protection (see page 100). A sacrificial anode is shown in Figure 3.17.

UNDERSTAND THE SCIENTIFIC PROPERTIES AND PRINCIPLES OF WATER

Water is a chemical compound of two gases: hydrogen and oxygen (H_2O). It is formed when hydrogen gas is burned.

One of the most important properties of water is its solvent power. This means it can dissolve numerous gases and solids to form solutions. The purest natural water is rainwater collected in the open countryside. It contains dissolved gases such as nitrogen, oxygen and carbon dioxide, but this does not affect its potability (suitability for drinking).

Different types of water have different pH values (see Figure 3.13 on page 98 for the pH scale). Here are some examples:

- Rainwater is naturally slightly acidic because of the small amounts of carbon dioxide and sulfur dioxide dissolved in it.
- The pH of groundwater is affected by the different rock types it passes through. For example, water with dissolved carbonate from chalk or limestone is alkaline.
- Pure water has a neutral pH of around 7.

You need to be aware of how water acidity or alkalinity can affect plumbing materials, appliances and components.

Identify the different states of water

Water can exist naturally in three states:

- solid (ice)
- liquid (water)
- gas (steam).

For water to change from one state to another, it must be either cooled or heated.

Describe the changing state of water in relation to temperature

Molecules making up a substance are in a constant state of rapid motion.

- When the molecules are densely packed together, their movement is restricted and the substance is solid.
- When the molecules are less tightly bound, there is a great deal of free movement and the substance is a liquid.
- When molecule movement is almost unrestricted, the substance can expand and contract in any direction and is a gas.

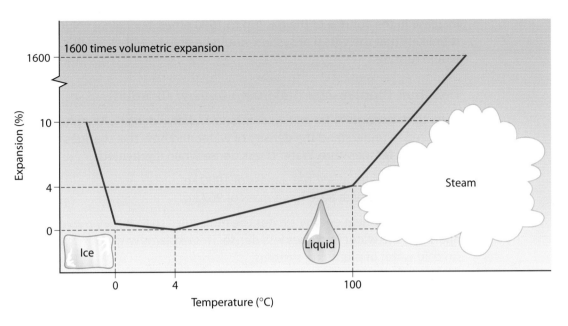

Figure 3.18: How atmospheric pressure affects the state of water

Figure 3.18 shows the behaviour of water at atmospheric pressure. In its natural state between 0°C and 100°C, water is in liquid form. When heated during its liquid state, the water's volume will increase by up to 4 per cent (so it will expand). This increase in volume results in a decrease in density. In other words, water's density depends on its temperature.

When cooled from room temperature, liquid water becomes increasingly dense. However, as it cools further it expands and actually becomes less dense. When frozen, the density of water has decreased by about 10 per cent.

We can demonstrate this very easily, as the solid form of most substances is denser than the liquid phase. For example, a block of most solids will sink in liquid. However, a block of ice floats in liquid water because the ice is less dense.

The freezing point of water is 0°C: at this temperature and below, water changes to ice and expands. This expansion or increase in volume in an enclosed space can result in components rupturing, such as burst pipes.

The boiling point of water is 100°C: at this temperature and above (under atmospheric conditions), water changes to steam. This causes a rapid increase in volume (up to 1600 times), which can have explosive effects if the water is in an enclosed space.

Water stored at above atmospheric pressure

The pressure at which water is stored is also linked to water temperature and volume. If water is stored in an enclosed space (at a constant volume, as in a storage cylinder) at above atmospheric pressure, the temperature at which it boils will rise to above 100°C. For example, the boiling point of water at 1 bar pressure is approximately 120°C.

In this case, why is water stored at above atmospheric pressure and above 100°C so dangerous if it is not boiling? Put simply, if someone opened a tap or a storage cylinder ruptured, the water pressure would rapidly decrease to atmospheric pressure. This would cause the water to boil almost instantly, resulting in a change of state from water to steam and a rapid

Safe working

Hot water temperature must be carefully controlled and maintained below 100°C in order to avoid potentially serious explosions.

increase in the volume of the gas (up to 1600 times its original volume). It is unlikely that the vessel or system would be able to withstand this, so it would rupture.

Describe relative and maximum density of water

Refer back to page 91 for an explanation of relative density. As a plumber, you need to understand the density of water and how this changes with the water's temperature. Water is less dense when it is heated:

- 1 m^3 of water at 4°C has a mass of 1000 kg.
- 1 m^3 of water at 82°C has a mass of 967 kg.

This is because heat energy excites the molecules so that they move further apart and the water becomes less dense. This explains why hot water floats on cold water in plumbing systems.

Water has a relative density (SG) of 1.0. For example:

- 1 m^3 of water has a mass of 1000 kg.
- 1 m^3 of mild steel has a mass of 7700 kg – 7.7 times heavier than water, giving a relative density of 7.7.
- Water is at its maximum density at 4°C.

Explain the concept of latent heat

Latent heat is the amount of energy released by a substance during a change of state. Latent means the energy is 'hidden' in the substance and is only released when heat or pressure causes a change of state.

Sensible heat is the increase in temperature (or energy released) by a substance before it changes state. The rise in temperature is directly related to the amount of heat or pressure the substance is subjected to.

Describe the expansion of water

When water is heated, the molecules of water start to move apart, creating larger gaps between them. This is known as expansion. As the water expands, it becomes less dense and will be pushed upwards by cooler, more dense, molecules moving downwards. Water is unique: it is the only liquid which reaches its maximum density at 4°C: as it is heated above or cooled below 4°C, it becomes less dense. All other liquids will carry on gaining density as they cool, until they become a solid. However, water will start to expand as it drops below 4°C, becoming less dense until it becomes a solid at 0°C. This is why pipework may split in freezing weather. This unique property of water is known as the anomalous (strange) expansion and only applies to water.

Explain how different factors can affect the properties of water

'Hard' water is created when it falls on ground containing calcium carbonates or sulfates (chalk, limestone and gypsum), which it dissolves and takes into solution.

Did you know?

Any substance with a relative density of less than 1.0 will float on water. Substances with a higher relative density will sink.

Water is classified as 'hard' if it is difficult for soap to lather. Conversely, soap lathers easily in 'soft' water because of the lack of dissolved salts such as calcium carbonates and calcium sulfates. However, soft water can corrode plumbing components because it is relatively acidic.

Water hardness can be temporary or permanent:

- Temporary hardness is a result of the amount of carbonate ions in the water. Temporary hardness can be removed from the water by boiling, which results in the carbonate being precipitated out as limescale. This hard scale accumulates inside boilers and circulating pipes, restricting the flow of water, reducing the efficiency of appliances and components and ultimately causing damage and system failure.
- Permanent hardness is a result of ions of nitrates and sulfates, and cannot be removed by boiling the water.

Describe the effects of hard water on plumbing systems and components

Effects of hard water

Hard water itself is not corrosive to plumbing systems, but the type of hard water will determine how it affects the plumbing and heating. Hard water is undesirable for the following reasons:

- It produces limescale in pipework, heating equipment and sanitary appliances (see Figure 3.19).
- Limescale can lead to high maintenance costs and reduced lifespan of components.
- Hard water requires much more soap and detergent for washing, as lathering is more difficult.
- It can affect the performance of appliances and looks unsightly.

Figure 3.19: Effects of limescale produced by hard water

Activity 3.6

Find out about the softness or hardness of water in your home. What effects (if any) has it had on appliances?

Ask your friends or relatives in other areas of the UK whether they have the same problems? If so, why? If not, why not?

Identify methods of water treatment

Water conditioners

One way to prevent hard water damage is to install a water conditioner in the cold water supply. This softens the water before it is heated, which reduces limescale formation. There are three main methods of treating water hardness:

- base exchange softeners
- scale reducers
- magnetic water conditioners.

These methods are described in detail on pages 275–277.

UNDERSTAND THE PRESSURE, FORCE AND FLOW OF WATER

Mass and weight

In its simplest terms, mass is the amount of matter in an object, and is measured in grams (g) or kilograms (kg). Under normal circumstances, and as long as it remains intact, an object should always have the same mass. For example, a 1 kg bag of sugar will have the same mass when on a workbench or on the moon. But will it weigh the same?

The weight of an object is the force exerted by its mass as a result of acceleration due to gravity. On Earth, all objects are being accelerated towards the centre of the planet due to the Earth's gravitational pull. The pull exerted by gravity on the mass of an object is its weight, which is measured in newtons (N). A newton is equivalent to 1 metre per second (m/s) per 1 kg of mass.

The difference between mass and weight

For an example of the difference between mass and weight we must go to the moon.

On Earth:

- the gravitational pull is 9.81 m/s^2
- an object with a mass of 1 kg will weigh 9.81 N.

On the moon:

- the gravitational pull is approximately 1.633 m/s^2
- an object with a mass of 1 kg will weigh 1.633 N.

The mass of the object does not change, whether it is on the Earth or the moon, but the weight of the object changes considerably because of the reduced gravitational pull of the moon.

Expressed as a formula, we can say:

weight = mass × gravitational pull

Back on Earth, if we disregard the effect of height above sea level, the weight acting on 1 kg of mass is equal to 9.81 newtons (N). So, we can say that 1 kg weighs 9.81 N.

Force

Force is a push or pull that acts on an object. If the force is greater than the opposing force, the object will change motion or shape. Obvious examples of forces are gravity and the wind. As mentioned above, force is measured in newtons.

The presence of a force is measured by its effect on a body. For example, a heavy wind can cause a stationary football to start rolling. Equally, gravitational force will cause objects to fall towards the Earth. Therefore, a spring will extend if we attach a weight to it, because gravity is acting on the weight.

As the force of gravity acts on any mass, the mass tends to accelerate and exert a force that depends upon the mass and the acceleration due to gravity. This acceleration due to gravity is agreed worldwide as being 9.81 m/s² at sea level and, therefore, a mass of 1 kg will exert a force of 9.81 N.

Expressed as a formula, this is:

force (N) = mass × acceleration

Pressure

Pressure is force applied per unit area and is measured in newtons per square metre (N/m²), a unit also known as a pascal (Pa).

You will probably come across other terms used to identify pressure, such as the 'bar' or 'pounds per square inch' (lb/in² or psi). These can be expressed as:

- 1 bar = 100,000 N/m²
- 1 lb/in² = 6,894 N/m².

Pressure is therefore a measurement of a concentration of force. You can see the effect of a concentration of pressure if water flowing through a pipe is forced through a smaller gap by reducing the diameter of the pipe. (Think about a hosepipe, and how to maximise the force of the jet of water.)

Pressure can also be lowered by spreading the applied force over a wider area. For example, rescue teams will often spread themselves over fragile roofs or on thin ice to minimise the chance of the surface giving way.

As a plumber, you need a basic understanding of how pressure affects pipes and fittings. What is being transported (water or gas) will affect the pressure inside a pipe or vessel. You must consider this when deciding which material and size of pipe or vessel to use.

Pressure in liquids

The pressure in a liquid increases with depth, so water pressure is higher at the lowest points of a plumbing system and lower at the highest points.

Water pressure is measured using a number of different units:

- metres **head** (m)
- pascal (Pa), also the newton (N) per metre squared (N/m²)
- bar pressure (bar)
- psi (pounds per square inch).

Figure 3.20 illustrates the effects of pressure in solids and liquids.

Key term

Head – the difference in height (vertical distance) between the top and bottom of a water pipe.

Figure 3.20: Solids exert pressure downwards only; liquids exert pressure downwards and outwards.

Describe the relationship between pressure and head of water

An understanding of the head of pressure created in systems is important for determining component sizes (for example, pipe sizes) and confirming that components will be able to withstand the pressure of water created within them.

1 metre head = approx. 10,000 Pa (10 kPa) = approx. 0.1 bar

Describe the procedure for calculating pressures of water

Intensity of pressure is the force created (kPa) by the weight of a given mass of water acting on a unit area (m^2).

Total pressure is the intensity of pressure multiplied by the area acted on.

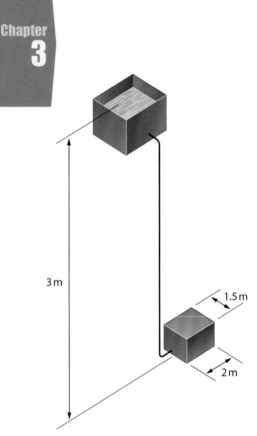

3 m

1.5 m

2 m

Figure 3.21: How to calculate water pressures

Example 1

Calculate the intensity of pressure and total pressure acting on the base of a lower boiler, as shown in Figure 3.21.

Intensity of pressure = head × 9.81 kPa

$$= 3 \times 9.81$$

$$= 29.43 \text{ kPa or approximately 0.3 bar pressure}$$

An alternative method of calculating this is to multiply the head by 0.1 bar (0.1 bar = 1m head):

3 × 0.1 bar = 0.3 bar pressure

Force acting on boiler base = intensity of pressure × area of base

$$= 29.43 \times (2 \times 1.5)$$

$$= 88.29 \text{ kPa}$$

Example 2

If a tap is sited 5 metres below a plumbing cistern feeding it, the pressure created at the tap will be:

5 metres head × 0.1 bar pressure = 0.5 bar

Describe frictional resistance to water flow in pipes and fittings

Flow rate

Flow rate means how much of a substance is passing by a certain point in a given time. Both gas and liquid flow can be measured in volumetric or mass flow rate. There are various measures used to express this. The most common volumetric measures are cubic metres per second (m^3/s), litres per second (l/s) or, for mass, kilograms per second (kg/s).

The rate of water flow through pipework is affected by friction. This is an important consideration when designing pipework installations. To get an idea of how friction works, compare the speed of a car driving freely through a tunnel with the speed of a car that is in contact with the tunnel walls. As you can imagine, the result of rubbing on the wall will slow the car down considerably.

Pipe walls have the same effect on water flow: the water is 'rubbing' on the pipe. This friction is further increased if the inside of the pipe wall is rough. This effect is known as frictional resistance or frictional loss.

Comparison of flow in smooth and rough pipework

- 10 m length of 25 mm diameter copper or plastic pipe (smooth inner walls): water flow flows at approximately 22 litres/minute

- 10 m length of 25 mm diameter galvanised LCS pipe (rougher inner walls): water flows at approximately 18 litres/minute.

Flow rate will drop off with increased pipe length because of frictional resistance to the wall of the pipe. Water travelling down the centre of a pipe will travel faster, as it is not in contact with the pipe wall. The greater the resistance to the flow of water, the greater the loss of flow rate through the pipe.

Pipe size and shape also affect flow rates. Knuckle bends or elbows provide greater resistance than handmade machine bends. Taps and valves in plumbing and heating systems can offer a lot of resistance to water flow. The number of fittings in a pipework run also increases frictional resistance.

When we need to calculate the resistance offered by the system, we say that a single 15 mm elbow fitting has a resistance to the flow of water through it equivalent to 0.37 m of pipe. Therefore, five 15 mm elbows are equivalent to 1.85 m of pipe (see Table 3.08). This clearly shows that taps and valves offer the greatest resistance to the flow of water through them.

Fitting type	Equivalent pipe length		
	15 mm	22 mm	28 mm
Capillary elbow	0.37 m	0.60 m	0.83 m
Capillary tee	1.0 m	1.60 m	2.0 m
Radius bend (machine or spring bend)	0.26 m	0.41 m	0.58 m
Angle pattern radiator valve	2.0 m	4.30 m	6.0 m
Stop tap	4.0 m	7.0 m	10.0 m

Table 3.08: Pipework fittings and equivalent pipe length

Describe principles of velocity

Velocity, pressure and flow rate

Velocity is the speed at which something travels in a particular direction. 'The amount that comes out depends on how hard you push and what gets in the way.' Bear this in mind when considering the following formula describing the flow of a fluid in a pipe:

$Q = A \times v$ where: Q is the flow rate

A is the cross-sectional area (CSA) of the pipe

v is the average fluid velocity in the pipe.

To clarify: a fluid travelling at an average velocity of 1 metre per second (1 m/s) through a pipe with a CSA of 1 m^2 has a flow rate of 1 cubic metre per second (1 m^3/s). Remember that Q gives a flow rate in m^3/s, so it is expressing volumetric flow rate, in other words the volume of fluid passing per second.

Mass flow rate represents the amount of mass that passes per second. We use the following formula to calculate this:

$W = Q \times r$ where: W is the mass flow rate

Q is the volumetric flow rate

r is the liquid density (rho).

If a pipe containing water is lying horizontally on the ground, there is still pressure acting on the water, even though it is not coming out of the pipe. We call this static pressure (no movement).

If we now lift one end of the pipe, there is a difference in height between the top and bottom, known as the head. This can be measured in height or pressure. This is called dynamic pressure (movement) acting on the water.

The head and the pipe's CSA will now determine the velocity of water flow down the pipe. However, the water will be slightly restricted from coming out of the other end by any obstacles or resistance in its way, Obstacles could be fittings on the pipework and a resistance could be the internal surface of the pipe (as discussed on pages 108–109).

We can therefore say that if we increase the pressure, then we are pushing harder and so more water can come out. This is described as being directly proportional, which means that one factor increasing causes the other factor to increase.

However, if we increase the resistance to flow (by having greater friction, more fittings or reducing the pipe size), then less water comes out. This is described as inversely proportional, which means that one factor increasing causes the other to decrease.

An increase in head (lifting one end of the pipe higher) will result in an increase in velocity. In other words, head and velocity are directly proportional.

Work

If an object is moved, then work is said to have been done. The unit of work done is the joule (J). Work done is the relationship between the effort (force) used to move an object and the distance that the object is moved. Expressed as a formula, this is:

work done (J) = force (N) × distance (m)

> **Example**
>
> A combination boiler has a mass of 50 kg. How much work is done to move it 10 m?
>
> $$\text{work} = \text{force} \times \text{distance}$$
> $$= (50 \times 9.81) \times 10$$
> $$= 490.5 \times 10$$
> $$= 4905 \text{ J}$$

Energy

Energy is the ability to do work: to cause something to move or to cause change. It is measured in joules (J). Machines cannot work without energy and we are unable to get more work out of a machine than the energy we put into it. This is due mainly to **friction**. The loss of energy by friction usually ends up as heat, which you can demonstrate by rubbing your hands together. They should get warmer.

The work produced (output) is usually less than the energy used (input). Energy can be transferred from one form to another, but energy cannot be created or destroyed.

There are two types of energy:

- potential – energy of position or stored energy
- kinetic – energy due to the motion of an object.

Both these types of energy can be in many forms, including: solar, electrical, heat, light, chemical, mechanical, wind, water, muscles and nuclear.

Describe the principles of siphonic action

Atmospheric pressure

The pressure exerted by the weight of the Earth's atmosphere pressing down on the ground varies depending on the height above sea level. The pressure at the top of Mount Everest is lower than the pressure at the bottom of a valley below sea level (such as the Great Rift Valley in Africa). The pressure at sea level is 101,325 N/m^2 (approximately 1 bar).

The siphon and how it works

A siphon is simply an inverted tube that has one 'leg' longer than the other (see Figure 3.22). A reduction of pressure in the tube by sucking on it at the

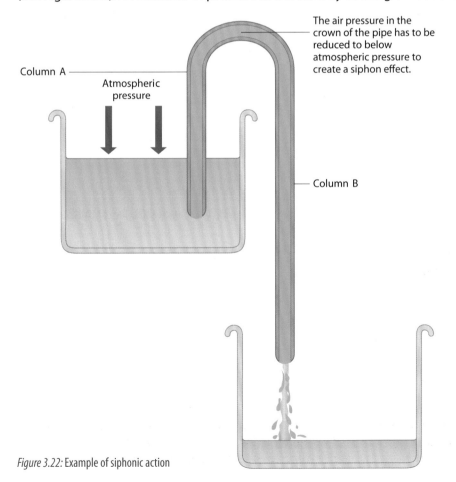

> **Key term**
>
> *Friction* – force that opposes motion, as when two surfaces rub together.

> **Remember**
>
> Plumbers must be aware of the effects of atmospheric pressure to ensure that they avoid creating 'negative' pressure or vacuums within pipework systems. Negative pressure can damage components.

The air pressure in the crown of the pipe has to be reduced to below atmospheric pressure to create a siphon effect.

Column A

Atmospheric pressure

Column B

Figure 3.22: Example of siphonic action

end of the long leg will cause the water to flow from the upper beaker to the lower one. It is because of the reduced pressure that atmospheric pressure pushing down on the water in the higher beaker causes the water to flow. The siphon then works continuously under the fall of the water from the higher beaker to the lower one because of the force of gravity and the cohesiveness and tensile strength of water. Siphonic action will stop when air is let into the siphon, either because the water in the upper vessel has run out or because the siphon tube has been removed from the water.

The principle of siphonic action is used to good effect in plumbing applications such as siphonic WC pans, WC flushing cisterns and automatic flushing urinal cisterns, and when siphoning the contents of a hot water storage cylinder using a hose pipe.

Working practice

A copper pipe carries very hot water from one part of a factory to another. The pipe is well insulated, but there is evidence of damage to the surrounding brickwork where it passes through a wall. The pipe is 50 metres in one straight length; it runs at high level and is adequately clipped throughout. The pipe varies in temperature by as much as 50°C at any time during the process.

Matt is a Level 3 apprentice carrying out some plumbing maintenance at the factory. The factory manager asks him what the problem could be.

- What is causing the damage to the brickwork?
- What should be done to the wall to alleviate the problem in future?
- Calculate how much the pipe will increase in length on each temperature change.
- What steps should Matt take to remedy the problem?
- Is copper pipe the best option for this particular usage? If not, what alternative can you suggest?

Progress check 3.01

1 What units is force measured in?
2 Define the term 'pressure'.
3 In a plumbing system, where is the greatest head of pressure: higher at the lowest points or higher at the highest points?
4 Which actions, components and appliances use the principle of the siphon in plumbing systems?
5 The rate of water flow through pipework is affected by what?
6 Why is it good practice to use manufactured bends in pipework rather than knuckle bends or elbows?

UNDERSTAND THE PRINCIPLES OF HEAT IN RELATION TO PLUMBING SYSTEMS

The main difference between heat and temperature is that heat is a unit of energy, measured in joules (J). Therefore:

- Temperature is the amount of hotness or coldness that the environment or an object or substance has.
- Heat is the amount of heat energy (J) that a substance contains.

For example, imagine an intensely heated short length of wire and a bucket of hot water:

- The wire has a temperature of 350°C.
- The water has a temperature of 70°C.

The wire is far hotter, but contains less heat energy. The water will take longer to heat, although to a lower temperature than the wire. Therefore, more heat energy will be required.

Identify units of measurement for temperature

Heat flow between two objects of differing temperatures is always from the hotter object to the cooler one. This is how a fridge or freezer works: the heat energy in any item placed in the fridge or freezer will flow from the hotter item to the cooler surrounding air, thus cooling the item. This is true in any situation when a hotter item finds itself in a cooler environment.

The temperature of a substance does not depend on its size. For example, a cup of water at 50°C is the same temperature as a pot of water at 50°C, even though the pot contains more water.

Different scales can be used to measure temperature. These scales are:

- degrees Celsius (°C)
- Kelvin (K)
- degrees Fahrenheit (°F).

Celsius is a unit of measurement derived from the seven base units used in the SI system (metric system). It is used to describe how hot something is and is universally accepted. It uses 0°C as the freezing point and 100°C as the boiling point of water at normal atmospheric pressure measured at sea level.

The Kelvin scale is also one of the seven base units of measurement used in the SI system. Kelvin is widely used in science, as it describes the lowest possible temperature that can be reached without using negative numbers. The lowest temperature that can be reached occurs when all molecular activity stops, and thus no heat is being generated. This is known as absolute zero. In °C absolute zero is –273.15°C.

The USA is the only country that now uses the Fahrenheit scale for official temperature readings. In this scale water freezes at 32°F and boils at 212°F at normal atmospheric pressure measured at sea level.

Activity 3.7

Find out about the other base units of measurement used in the SI system.

Explain the procedure for calculating heat capacity

Refer to pages 94–95 for an explanation of heat capacity and an example calculation.

Compare the methods of heat transfer

As a plumber, you will need a good understanding of the methods of heat transfer because you will be dealing with the effects of this process on a daily basis. There are three methods of heat transfer:

- conduction (see page 96)
- convection
- radiation.

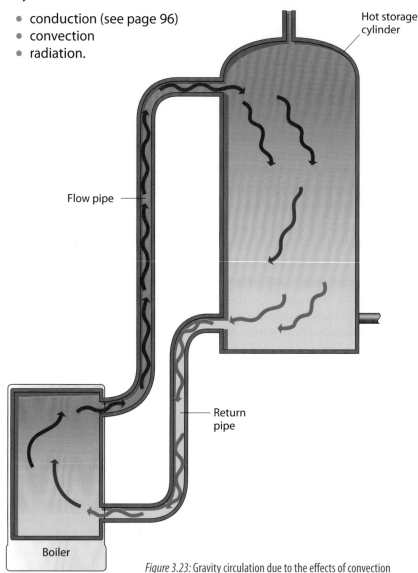

Figure 3.23: Gravity circulation due to the effects of convection

Convection

Convection is the transfer of heat by means of the movement of a locally heated fluid substance (usually air or water). As a fluid is heated, it expands, which lowers its density. The less dense, warm fluid begins to rise and is replaced by cooler, denser fluid from below. Eventually, convection currents are set up, which allow a continuous flow of heat upwards from the source.

Examples of systems that use convection currents for heat transfer are:

- convector panels (radiators), which warm the air at one place in a room – the resulting convection currents transport the heat around the room
- domestic hot water systems, which depend on convection currents to transfer heat from an immersion heater (similar to the element in an electric kettle) to the rest of the water in the hot tank (cylinder).

It is easy to demonstrate the 'updraught' part of a convection current by hanging a piece of light material above a convector heater. The movement of the material will clearly show the presence of rising currents of warm air. Figure 3.23 illustrates the use of convection currents in domestic hot water systems.

Radiation

Radiation is the transfer of heat from a hot body to a cooler one without the presence of a material medium (other than air) by means of 'heat' waves. Heat radiation can be felt as the glow from a fire or the heat from the sun.

Some materials absorb heat radiation better than others, and in these cases colour or texture can be important factors. For example, dull, matt surfaces will absorb radiated heat more efficiently than shiny, polished surfaces.

Most domestic 'radiators', such as those found in central heating systems, warm rooms by convection in addition to radiating heat energy (see Figure 3.24).

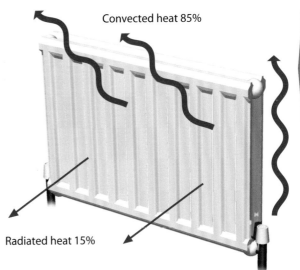

Convected heat 85%

Radiated heat 15%

Figure 3.24: Convector panel radiators warm rooms by convection and radiation.

Describe the effectiveness of different surfaces and finishes in the relationship to heat transfer

Black and dark surfaces absorb heat energy better than white or lighter colours, which reflect the heat energy away (see Table 3.09). When heat energy is absorbed, the electrons in an atom start to vibrate and thus create heat. The more energy that is absorbed, the more heat will be produced.

- Black will absorb the most energy, cause the most vibration and thus generate the most heat.
- White will reflect energy, cause less vibration of the atoms and so feel cooler.
- Shiny surfaces, which are also generally smooth, will reflect the heat energy away rather than absorbing it, so they too will feel cooler to the touch.

Colour	Finish	Ability to emit thermal radiation	Ability to absorb thermal radiation
dark	dull or matt	good	good
light	shiny	poor	poor

Table 3.09: The ability of different surfaces to emit and absorb thermal radiation

Thin and flat objects like radiators in a heating system will radiate heat energy better and faster than fat objects. It would also be better to paint radiators matt black, which would allow them to emit more heat energy, rather than gloss white, which is the most common finish for radiators.

Progress check 3.02

1 What is the specific heat capacity of water?

2 Calculate the increase in length of a 9 m long plastic gutter subject to a temperature rise of 17°C.

3 What are the three methods of heat transfer?

Describe the advantages of insulators used in plumbing systems

Insulation works by keeping heat energy in. Even cold water contains heat energy, otherwise it would not be a liquid at all – it would be ice.

Plumbing and heating systems should be insulated to make them more energy efficient. This means reducing the amount of heat energy lost to the surrounding air. Using less energy to heat properties saves money and reduces the need to burn or generate more energy, thereby conserving what we have.

If the level of insulation thickness is increased beyond what has been calculated to be the effective requirement, then the prevention of heat loss levels off very quickly. Any additional insulation beyond this optimum level will only result in a reducing return for any further heat loss prevention and is not cost-effective.

State the negative aspects of heat transfer

Insulation also helps control heat transfer to other pipes and surfaces. It is not good practice to run cold water pipework alongside heating or hot water pipework under floors or through voids because heat energy can be transferred easily to the cold water pipe and raise its temperature, which can be unpleasant for customers.

The purpose of insulating pipework and components is to prevent heat loss from the water that is circulating around the system. It is a waste of expensive energy to heat water up and then allow some of that energy to warm unused areas such as attics and basements as it circulates around the system to the areas of the dwelling it is required to warm up.

In some instances, condensation can form on the pipework as warm moisture-laden air comes into contact with the pipework, particularly the cold water pipe if it is left uninsulated. This can cause problems such as corrosion of joints and pipework, black mould growth and even wood rot if the condensation is bad enough to drip onto any timber below it.

KNOW THE PRINCIPLES OF COMBUSTION AND HEATING GASES

Describe the requirements for combustion

Combustion (fire or burning) is the rapid combination of a fuel and oxygen (air) at high temperature. A fire can reach a temperature of up to 1000°C within minutes. For a fire to start there are three requirements:

- a **combustible** substance (fuel)
- oxygen (usually as air)
- ignition – a source of heat (spark, friction, match).

When these come together in the correct combination, a fire occurs. See Figure 1.19 on page 30 of Chapter 1, which illustrates the fire triangle.

Key term
Combustible – able to burn or be burned (flammable).

In order for a gas to burn, it has to mix with oxygen (in air) to create the correct mix for combustion to occur. An example is a car engine in which air and fuel are mixed in the carburettor before being injected into the cylinder. The mixture of gas to air is different for each gas and is known as the 'gas to air ratio'. When this ratio is at its lowest possible limit to support a flame, it is known as the 'flammability limit'. Complete combustion will ensure that maximum heat energy is given off for each unit of gas burned. The gas : air ratio for three gases used for heating purposes are:

- natural gas – 1:10
- propane – 1:25
- butane – 1:30.

Identify combustion temperatures and properties of gases used for heating purposes

Table 3.10 outlines the temperature at which different gases will combust, their relative density in air and their **calorific value**.

Fuel type	Relative density	Combustion temperature (°C)	Calorific value (MJ/m3)
Natural gas (in air)	0.60–0.70	1960	38
Propane (in air)	1.5219	1980	95.8
Butane (in air)	2.0061	1970	126

Table 3.10: Combustion of gases

Natural gas

Natural gas (CH_4) is used as a fuel for heating and cooking appliances and is normally supplied through the national network of gas mains. It is made up of 94 per cent methane, 3 per cent ethane and 3 per cent other constituents.

Liquefied petroleum gas (LPG)

Liquefied petroleum gas includes butane and propane. LPG is mainly used for heating equipment in the plumbing industry: hot water and central heating boilers, fires and stoves. Gas for these is normally supplied in large cylinders or storage tanks. LPG is also used in most blowtorches for soldering copper pipe and fittings, in which case it is supplied in small cylinders or canisters.

Butane gas boils at a far higher temperature than propane, which makes it suitable for outdoor use during warm weather. It can also be used indoors for portable gas heaters, and is popular for use in caravans and for barbecues. Butane comes in cylinders like those for propane but in a different colour for easy identification.

Propane is widely used where outside temperatures are too low for butane to operate. For example, the bulk tanks you may see in people's gardens in rural areas will contain propane.

Key term

Calorific value – the amount of heat energy a substance releases when burned.

Did you know?

Natural gas is odourless. so an additive is used to enable us to smell it. This means it is detectable when it leaks, otherwise gas build-up could reach dangerous levels without detection.

KNOW THE BASIC PRINCIPLES OF ELECTRICITY

It is essential for the modern plumber to have an understanding of electricity in order to work effectively and, above all, safely. Electricity is important in powering many plumbing appliances and devices, including control systems for central heating, electric showers and immersion heaters.

To recap learning from Chapter 2:

- A simple electrical circuit can be created using a power source such as a battery connected to a light bulb using some wire between the battery and the bulb.
- This allows electron drift between the two ends of the battery.
- As the electrons flow through the wires and bulb, the light bulb will be illuminated.

Identify units of measurement for electricity

The main units you need to know are:

- current – amps (A)
- voltage – volt (V), or joules/coulombs
- resistance – ohms (Ω).

Many of the following topics are covered in detail in Chapter 2.

Current

Just as a plumber measures water current flow in litres per second, an electrician considers the flow of electricity through a conductor in terms of coulombs per second. The electric current flow (given the symbol I) is measured in amperes (usually abbreviated to amps or A). The amount of current flowing in amps is measured using an ammeter, which is connected into a circuit in series.

Voltage

Voltage can be thought of as the pressure that pushes the electricity around the system. By applying a voltage to the end of a conductor, we provide an electrical pressure that causes a current to flow. The voltage may come from a battery (DC) or a mains supply (AC).

Resistance

To use a comparison with water, think about water being supplied to a bath via a cistern using gravity. The head of water in the cistern will provide the pressure (voltage in electrical terms). With the tap open, the water will flow into the bath at a fixed rate per second (current in electrical terms).

In order to flow from the cistern to the bath, the water passes through a pipe of a certain diameter. Increasing the size of the pipe (and tap) will reduce the natural resistance to the flow of water, and the bath will fill more quickly. If you decrease the size of the pipe, resistance will increase and the bath will take longer to fill.

Resistance to current flow in an electrical circuit is given the symbol R and is measured in ohms (denoted by the Greek letter omega, Ω). In plumbing terms, the bigger the pipe the lower the resistance to water flow; in electrical terms, the bigger the conductor the lower the resistance to current flow.

Remember
A voltmeter is used to measure the voltage of a circuit and is connected to a circuit in parallel.

It is very important to note that ammeters and voltmeters are connected to live electrical circuits to measure current and voltage, and they pick up the electricity that they need to operate from the live circuit.

Gravity drives the water through the system, but for this to happen there must be a difference in level between the supply (the tank) and the load (the bath). This is the equivalent of potential difference. The tap performs the same function as an electric switch (see Figure 3.25).

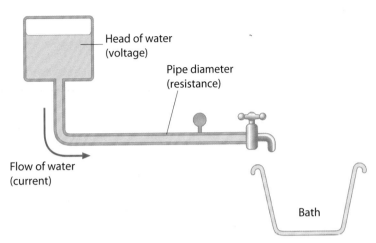

Figure 3.25: Analogy of a water system as an electrical circuit

Explain the procedure for calculating basic electricity relationships

Ohm's Law

The three major components that make up an electric circuit – voltage, current and resistance – are interrelated, and, if we know any two of the quantities, we can calculate the third using a basic rule called Ohm's Law (see also Chapter 2, page 69).

Ohm's Law describes the relationship between electrical quantities: voltage equals current multiplied by resistance. This can be expressed by the equation:

$V = I \times R$ where: V = voltage
                       I = current
                       R = resistance.

This can also be shown in a simple form, often called the Ohm's Law triangle (see Figure 3.26).

Figure 3.26: Ohm's Law triangle

Using the Ohm's Law triangle

If you know any two of the quantities (voltage, current or resistance), you can find the third by covering up the quantity you are looking for and calculating the other two:

- To find the voltage (V) of a circuit if you know the current (I) and resistance (R): cover the V to show that you need to multiply I by R.
- To find the current (I): cover I and you will see that you need to divide V by R because V is above R.

🚫 **Safe working**

The resistance of an electric circuit is measured with an ohmmeter connected across the component whose resistance is being measured.

The ohmmeter has its own internal power supply from a battery. You must never connect it to a live circuit.

Example

A circuit has a resistance of 115 ohms. A current of 2 amps is flowing. What is the voltage applied to the circuit?

$V = I \times R = 2 \times 115 = 230$ volts

Identify how AC and DC currents are generated and describe their differences

These topics are covered in detail in Chapter 2, pages 50–51.

Electromagnetism

Key term

Electromagnetism – the relationship between electric currents or magnetic fields.

All electrical currents produce a magnetic force. This is the basic principle that underpins the creation of almost all the electricity used today, and is known as **electromagnetism**. The application of this fact was demonstrated in principle by Michael Faraday in the 1830s. He discovered that electricity could be generated by moving a magnet in and out of, or around, a coil of wire wound onto a soft iron core. The same effect is created if we move a coil of wire on an iron core within a fixed magnetic field. While direct current is caused as a result of a chemical reaction in a battery, alternating current is produced as a result of electromagnetism.

Knowledge check

1   What term is used to describe a material that can be worked without breaking?

   a   optical
   b   magnetic
   c   malleable
   d   conductive

2   What is the cause of atmospheric corrosion?

   a   oxygen and water
   b   oxygen and carbon dioxide
   c   nitrogen and water
   d   water and carbon dioxide

3   When water freezes, how does it change?

   a   It contracts and becomes denser.
   b   It contracts and its density remains the same.
   c   It expands and becomes less dense.
   d   It expands and its density remains the same.

4   What does temporary hard water contain?

   a   calcium sulfates
   b   calcium bicarbonates
   c   calcium sulfide
   d   calcium carbonates

5   Which of the following statements is true?

   a   Pressure is defined as area plus the head.
   b   Pressure is defined as force minus the area acted on.
   c   Pressure is defined as force applied per unit area.
   d   Pressure is defined as bar.

6   What is the force in newtons of a boiler that weighs 15 kg?

   a   147.15 N
   b   141.26 N
   c   148.60 N
   d   149.30 N

7   How else can 1 bar be described?

   a   20 m/head
   b   15 m/head
   c   10 m/head
   d   5 m/head

8   What is the main method of heat transfer from a steel panel radiator?

   a   radiation
   b   conduction
   c   continuation
   d   convection

9   What is the base SI unit of temperature measurement?

   a   Kelvin
   b   Celsius
   c   Fahrenheit
   d   Centigrade

10   A flammable gas must be mixed with another gas in order to burn clearly and efficiently. Which gas should it be mixed with?

   a   carbon dioxide
   b   oxygen
   c   hydrogen
   d   nitrogen

11   What is the unit used to measure the amount of current that flows through a conductor?

   a   ampere
   b   volt
   c   ohm
   d   watt

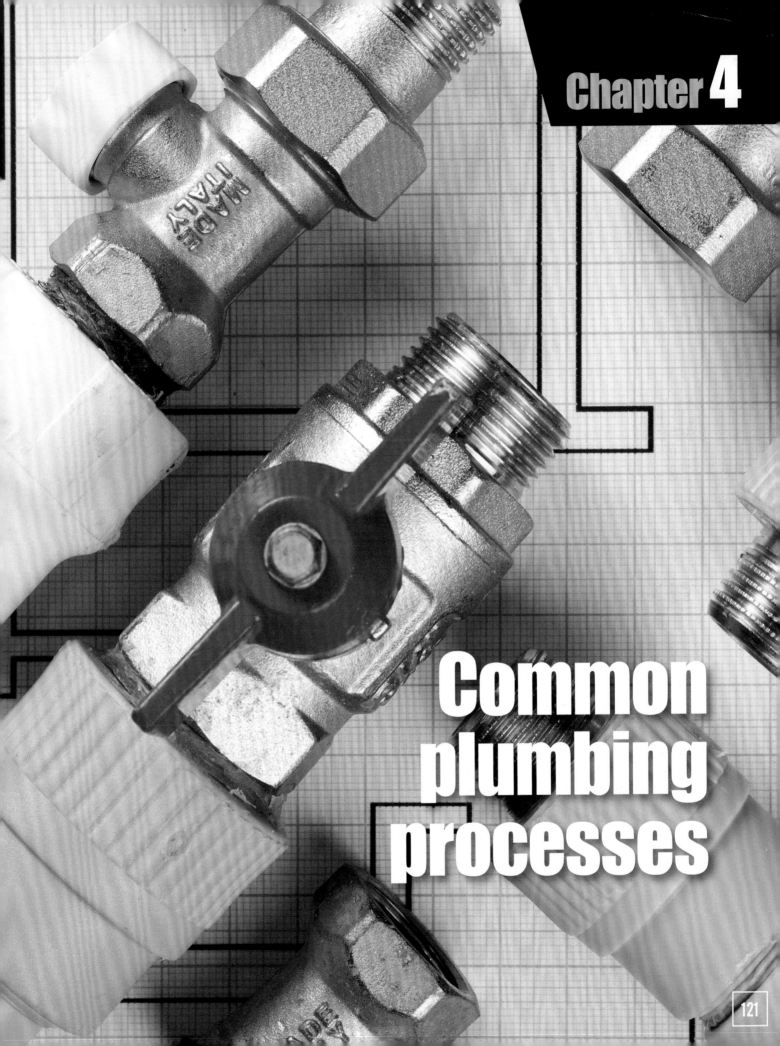

Common plumbing processes

This chapter will cover the following learning outcomes:

- **Know common plumbing hand and power tools**
- **Know fixings and components used in common plumbing processes**
- **Understand the procedures for measuring and bending plumbing tubes**
- **Understand how to joint common plumbing materials**
- **Know common plumbing preparation techniques**
- **Know symbols used for identifying plumbing pipework and fittings**

Remember

Cheap tools are not always a bargain. They are likely to be damaged more easily or wear out simply because they are not up to the job.

Introduction

This chapter will help you gain the knowledge, understanding and skills that you will need to carry out common plumbing processes. This includes an introduction to measuring, bending and jointing tubes and the tools and equipment required. You will also learn how to carry out common basic preparation tasks used in plumbing.

KNOW COMMON PLUMBING HAND AND POWER TOOLS

As a plumber, you need to measure, mark out, cut, fabricate, joint and fix a range of materials. In most cases, this involves using tools. There is a vast range available, but you may only need some for particular jobs. Once you have put together a toolkit, look after your tools to ensure their long life and to avoid the need to keep buying replacements.

When selecting tools, you must consider four main factors:

- quality
- cost
- safety
- probable usage.

Quality

Good-quality tools that can do the job they are designed for are extremely important. Ask other tradespeople what their choice would be before purchasing a particular tool. They may even advise you of the best place to buy them.

Cost

When starting to make up a toolkit, it may be tempting to try to spend as little as possible. While this is understandable, it is not a good idea. Again, seek advice from other tradespeople. You will usually find that they are still using the quality tools they bought when they first started work because they spent that little bit more on them. Good-quality tools of well-known brands are expensive but are a good investment.

Safety

Naturally, safety is of the utmost importance. The safety of a tool depends to a great extent on the individual who is using it. High quality normally includes better safety. Again, it is worth paying a little extra, as cheap tools are generally not as safe as more expensive ones.

Probable usage

Consider what you will use a tool for before you buy it, as there is little point in paying a lot of money for a tool you may seldom or never use.

The best way of dealing with all these considerations is to set a budget for yourself, decide what you require and by when. This will help you to make a calculated purchase of the tools you will require over a period of time.

Identify plumbing hand and power tools

Table 4.01 on pages 124–135 lists an extensive range of the tools and equipment that you may need. It also includes advice on their maintenance, use and recommended personal protective equipment (PPE). The table covers most of the criteria for this learning outcome, so refer to it as necessary.

Checklist

Tool safety and maintenance

- Make sure you clean your tools regularly.
- Lubricate the working parts.
- Once cleaned, *lightly* coat the tools with oil spray to prevent rusting.
- Always use the right tool for the job. Screwdrivers are not chisels.
- Keep file or rasp teeth clean with a wire brush.
- Never use tools with split handles.
- Always replace used, worn or defective hand tools.
- Make sure you use all the necessary personal protective equipment (PPE).

Tool	Care, operation and maintenance	Usage	Recommended minimum PPE
½" hexagon radiator valve spanner	Always: • keep it clean, especially the small opening and slot on the end • dispose of spanners when the hexagon shape is becoming rounded or damaged. Never: • hit the tool with anything to aid tightening/untightening.	• Will fit most common radiator valves, which have a hexagon keyway inside the valve tail, enabling ease of tightening it into the radiator. • Adaptable to tighten most air release valves into the radiator without removing the threaded needle type valve (pip).	• Protective footwear • Protective overalls • Barrier cream
Radiator spanner		• A multipurpose spanner offering a great range of uses and a huge range of universal tasks. • Offers much more flexibility than the hexagon type spanner.	
Radiator air release key (bleed key)	Always: • replace when showing signs of wear • protect the floor area when venting radiators • ensure the circulation pump is inactive before venting radiators • place absorbent material under the key when venting to prevent spillage. Never: • use grips or other levers to aid usage • remove the valve (pip) completely when venting – the water may appear more quickly than you expect.	• Inserts over the head of the air bleed valve head (pip). • Turn anticlockwise to release air (if present) and clockwise to close.	• Protective footwear • Protective overalls • Barrier cream
Blowtorch with disposable cylinder	Always: • check for gas soundness before use • completely turn off when not in use. Never: • leave gas running without being ignited • leave an ignited torch running unattended • run the torch with the storage cylinder laid on its side or upside down.	Used for: • heating pipework for annealing • desoldering/soldering plumbing fittings, either integral solder ring or end feed up to and including 22 mm diameter tubes and fittings. For larger diameters, other torches with a higher temperature burning gas are available. Alternatively, you could use a larger torch connected to a blow lamp and hose (as on page 125).	• Protective footwear • Protective overalls • Barrier cream • Gloves*

Table 4.01: Plumbing hand tools and power tools and their uses (*Possible, depending on method used)

Continued ▶

Tool	Care, operation and maintenance	Usage	Recommended minimum PPE
LPG torch kit (propane)	**Always:** • check for gas soundness when assembling and before use • relieve the regulator when not in use – turn the valve head anticlockwise • check hose regularly for damage • check the manufacture date, which is printed along the hose: it should be replaced every three years. **Never:** • leave gas running without being ignited • leave an ignited torch running unattended • run the torch with the storage cylinder laid on its side or upside down.	Used for: • heating pipework for annealing • desoldering/soldering fittings, either integral solder ring or end feed from small to large diameters by adjusting the regulator and torch. This has more gas storage than the disposable cylinder type, but is much heavier. The torch head naturally becomes very hot, so make sure no one comes into contact with it, even after the flame is extinguished.	• Protective footwear • Protective overalls • Barrier cream • Gloves*
Pipe vice	**Always:** • ensure the tube is sufficiently tightened into the vice, otherwise sharp shards of metal may be formed on the outer walls of the tube • keep the moving parts lightly oiled or greased. **Never:** • use these types of vice to hold copper tube • try to secure LCS tubes in a flat-faced engineer's vice. (There is not enough grip onto the tube.)	• There are basically two types: the yolk and the chain vice. Both can be portable on a tripod or fastened to a workbench. • Be careful with the tripod versions, as they are quite heavy and awkward to handle. Make sure the legs are fully extended outwards before using the equipment.	• Protective footwear • Protective overalls • Barrier cream
LCS pipe cutter	**Always:** • replace damaged or blunt cutter wheels • use the correct blade for the material being cut • make sure that the wheel and rollers are lubricated and move freely • use a reaming tool to deburr the inside of the tube. **Never** • use on any material other than LCS tubes • put too much tension on the wheel or it will shatter.	• Only LCS tubes should be cut with this heavy-duty cutter. • The general purpose type of cutter can cut a range of 3–50 mm diameter, although larger diameters are catered for in larger cutters.	• Protective footwear • Protective overalls • Barrier cream
LCS tube spiral ratchet reaming tool/deburring tool	**Always:** • take care not to drop the tool or you may damage the sharp edges of the spiral • keep the ratchet assembly well lubricated • be careful that the spiral does not wind itself into the tube and form an enlarged external diameter. **Never:** • be tempted to pull the swarf of metal left by the process with your fingers – it is very sharp.	• One size fits all because of its tapered design. • The spiral is offered into the tube end previously cut and the user operates the ratchet. • The swarf from the burr is easily removed by this action. • A similarly designed reaming tool is located on larger electric floor-mounted threading machines.	• Protective footwear • Protective overalls • Barrier cream • Gloves*

Table 4.01: Plumbing hand tools and power tools and their uses (*Possible, depending on method used) (continued)

Tool	Care, operation and maintenance	Usage	Recommended minimum PPE
Large hacksaw	Always: • make sure the teeth are pointing away from you and towards the forward cut • check the blade does not have defective or worn teeth • make sure the blade is tightened correctly when using a large hacksaw • use the right blade type for the job – 32 teeth per inch for light gauge pipe such as copper and plastics, and 24 teeth per inch for heavier gauge LCS tubes • take great care – hacksaw blades are very sharp and may inflict deep cuts to fingers or hands • change the blade when the sharpness is lost. Never: • use for cutting wood or electrical cabling.	Used for cutting: • larger diameter copper tubes • LCS tubes • plastics, including gutter, downpipes, soil and waste pipes.	• Protective footwear • Protective overalls • Barrier cream
Junior hacksaw		Used for cutting: • smaller diameter tubes • smaller diameter plastic tubes • small items to size. Useful for cutting tubes in tight spaces	
Mini tube cutter	Always: • replace damaged or blunt cutter wheels • make sure the wheel and rollers are lubricated and move freely • remove the burr after cutting. Never: • use on any material other than copper tubes.	Used for cutting copper tubes of 6–22 mm diameter (adjustable).	• Protective footwear • Protective overalls • Barrier cream
Pipe slice	Always: • replace damaged or blunt cutter wheels • make sure that the wheel and the rollers are lubricated and move freely • make sure the spring that applies pressure to the wheel when cutting is lubricated • remove the burr after cutting. Never: • use on any material other than copper tubes.	Used for cutting: • 15 mm copper tube (non-adjustable) • 22 mm copper tube (non-adjustable). Tube may be cut in position (clip distance is required).	• Protective footwear • Protective overalls • Barrier cream

Table 4.01: Plumbing hand tools and power tools and their uses (*Possible, depending on method used) (continued)

Continued ▶

Tool	Care, operation and maintenance	Usage	Recommended minimum PPE
Adjustable copper tube cutter	Always: • replace damaged or blunt cutter wheels • make sure that the wheel and the rollers are lubricated and move freely • make sure any deburring blade fitted is lubricated • remove burr after cutting. Never: • use on any material other than copper tubes.	• Used for cutting copper tube of 6–30 mm diameter. • Larger diameter copper tube cutters are available. • Note that all tube cutters are susceptible to 'tracking' (spiral cuts around the tube being cut) if the wheel becomes blunt or the rollers become loose.	• Protective footwear • Protective overalls • Barrier cream
Files	Always: • replace damaged or broken tools • discard files with worn teeth. Never: • use a file with a damaged or missing handle.	• Files have a rough face of hardened metal which is pushed across the surface of a workpiece to remove particles of material. • Used to make an object smoother or smaller, or to change its shape. • Classified by length, cut, grade and shape. • Main shapes: flat, square, half-round, round. • Main cuts: bastard (intermediate), fine, medium. • Must be used with a firmly fitted handle. • Surform files have a perforated blade (like a cheese grater) held in a frame. Can be used on wood, plastic and mild steel. Good for removing material quickly without clogging.	• Protective footwear • Protective overalls • Barrier cream • Eye protection*
Copper tube pipe reaming tool	Always: • replace damaged or broken blades. Never: • operate the tool too quickly, or the blade may lose contact with the tube and cut the hand/wrist holding the tube.	• A useful tool, which removes the burr to the internal bore of cut copper tubes. • Not designed to remove burrs caused by a hacksaw from the external bore of the tube. For this, you need a different type of pipe reamer or a file.	• Protective footwear • Protective overalls • Barrier cream • Eye protection*
Bevelled wood chisels	Always: • keep blades sharp; you can use an oilstone for this • keep the blade protector in place when not in use. Never: • use it on metal objects such as nails or screws, as the blade will easily chip and will need regrinding.	• Used for cutting notches from joists etc. • Limited use for plumbers; carpenters and joiners use these tools more.	• Protective footwear • Protective overalls • Barrier cream • Eye protection*

Table 4.01: Plumbing hand tools and power tools and their uses (*Possible, depending on method used) (continued)

Continued ▶

Tool	Care, operation and maintenance	Usage	Recommended minimum PPE
Typical fixed stem tap spanner	Always: • keep the head lubricated • keep the teeth (serrations) that grip the nut free from jointing compound etc., otherwise the tool could slip and injure the user.	• Allows tightening of taps and tap connectors in awkward situations (e.g. under bath, sink or washbasin). • Head may be changed to allow a smaller or larger jaw to be fitted. • Available in different types, most commonly fixed stem or extending stem jaw.	• Protective footwear • Protective overalls • Barrier cream
Adjustable spanner	Always: • keep the worm screw lubricated • make sure the jaws do not become worn or damaged. Never: • use the tool to knock a chisel etc. – always use the right tool for the job.	• Use on brass or other metal nuts to prevent shards of metal from forming. • Useful to tighten nuts/bolts on some appliances and fixings. • Adjustable, so you do not need to carry a full range of spanners.	• Protective footwear • Protective overalls • Barrier cream
Water pump pliers (pipe grips)	Always: • keep the teeth free from jointing compounds; if clogged, the tool may slip • replace once the teeth become worn • check for wear on the ratchet mechanism, as the pliers may slip under pressure • be careful when loosening a joint or pipe that is difficult to move. It may give suddenly and you could damage your hands or pull a muscle. Never: • misuse the tool (e.g. for knocking chisels etc.) – always use the right tool for the job.	• Probably the most versatile tool in the plumber's toolkit. • Originally used for tightening the large flange nuts found on circulating pumps. • Also used on plumbing fitting nuts and to remove soldered fittings while applying heat. Eye protection is recommended for the latter use.	• Protective footwear • Protective overalls • Barrier cream • Eye protection*
Stillson wrench	Always: • keep teeth free from jointing compounds, otherwise the tool could slip • replace the tool once the teeth become worn • be careful when loosening a joint or pipe that is difficult to move. It might give suddenly and you could damage your hands or pull a muscle. Never: • use on copper tube or fittings • extend the length of the handle by any means. The tool is designed to withstand the leverage gained by the length of the tool handle (see comment in Usage column) • misuse the tool.	• Designed for use on low carbon steel tubes and malleable iron or steel fittings. • Available in a range of sizes to suit various diameters of LCS tube. • Note that the length of handle is reflected in the jaw size, allowing more leverage to be gained on larger diameters.	• Protective footwear • Protective overalls • Barrier cream

Continued ▶

Table 4.01: Plumbing hand tools and power tools and their uses (*Possible, depending on method used) (continued)

Continued ▶

Chapter 4

Tool	Care, operation and maintenance	Usage	Recommended minimum PPE
Spirit level (boat level)	Always: • take great care – spirit levels are probably the most easily damaged tool you will use • select the longest spirit level you can. The longer the level, the more accurate the result will be. Never: • misuse or drop a level • use the level to hit anything or hit the level with anything.	• If you use a level to check mortar, make sure you remove the level before the mortar has a chance to set. • A common misunderstanding is that an item is plumb (vertically) or level (horizontally) when the bubble in the vial is touching one of the graduation lines. This is not true. Instead, the bubble should be located between the two graduation lines.	• Protective footwear • Protective overalls • Barrier cream
Large spirit level			
Pad saw	Always: • check for concealed electrical cabling and/or pipework before starting to cut • check blade condition before use and replace as necessary.	• Sometimes known as a keyhole saw or a dry wall saw. • Has a narrow, tapering blade, which is useful for removing timber flooring materials and cutting out plasterboard neatly and efficiently.	• Protective footwear • Protective overalls • Barrier cream • Eye protection* • Dust mask*
Flooring saw	Always: • make sure the blade is sharp and undamaged before use • check the handle is not cracked or broken • check for concealed electrical cabling and/or pipework before starting to cut.	• Has a rounded nose, with teeth. This allows you to begin cutting in the middle of floorboards. • The blade is short to allow a sawing action in short strokes – usually around 300 mm.	• Protective footwear • Protective overalls • Barrier cream • Eye protection* • Dust mask*
Tenon saw		• Mostly used for cutting and making joints in timber. • Metal strip along the top gives rigidity to the blade. • To start a cut, angle the saw so that it cuts into an edge of wood and then lower the angle for a straighter cut.	

Table 4.01: Plumbing hand tools and power tools and their uses (*Possible, depending on method used) (continued)

Tool	Care, operation and maintenance	Usage	Recommended minimum PPE
Claw hammer	Always: • make sure the head is fitted correctly and securely to the shaft • check that the hammer does not have a defective shaft. Except for the club/lump hammer, the name refers to the shape of the pein (the opposite end to the striking face).	• A very common hammer type, used for all building trades. • Can be used to draw out unwanted nails (using the claw) as well as to knock nails into timber.	• Protective footwear • Protective overalls • Barrier cream • Eye protection • Dust mask* • Gloves*
Ball pein hammer		• The ball is used for bossing or dressing metal. • Widely used by plumbers until only a few years ago, as they would have been using sheet metals such as copper, aluminium and zinc.	
Cross pein hammer		• Can be used in tight spaces. • Has a range of uses, e.g. knocking in panel pins.	
Club hammer or lump hammer		• Used for heavy hammering, e.g. hitting cold chisels, bolster chisels and pointing chisels. • This is the largest hammer that can be used with one hand. • General uses include demolition of concrete, masonry, etc.	
Bolster chisel	Always: • keep the cutting edge sharp using a grinder • keep the striking end of the chisel free from the 'mushrooming effect', using a grinder • discard if the chisel is ground down so much that it is too small to use safely. Both types of chisel are available with a hand guard.	• Sometimes called a brick chisel. • Used to cut masonry (e.g. bricks, blocks) and to cut chases in walls or floors. • Two variations are available: the heavy-duty version (pictured) and a thinner, lighter version, which is used to break the tongue from the groove in tongue and groove flooring, allowing easier removal of boards.	• Protective footwear • Protective overalls • Barrier cream • Eye protection • Dust mask* • Gloves
Cold chisel		• Available in various lengths and a range of thicknesses. • Used to cut through masonry or concrete, or to remove mortar from brickwork etc. May also be used for leverage when removing timber flooring.	

▶ *Continued*

Table 4.01: Plumbing hand tools and power tools and their uses (*Possible, depending on method used) (continued)

Tool	Care, operation and maintenance	Usage	Recommended minimum PPE
(1) Manual pipe threader	Always: • (1) make sure dies are kept free from swarf • (1) make sure dies are sharp • (1) use approved lubricant • (1) ensure handle is not damaged • (1) ensure ratchet type threaders are set for the correct direction	• Sometimes referred to as stocks and dies, as it is essentially made up of two parts: the stocks (the handle and opening which will contain the other part) and the die (which cuts the thread onto the tube). • Must be lubricated during cutting process. Swarf is produced – the cutting waste material from the thread. Swarf is normally very sharp, so never attempt to pull it with bare hands.	• Protective footwear • Protective overalls • Barrier cream • Eye protection
(2) Hand-held electric pipe threader	• (1) clean after use • (1) use cutting solution regularly during use • (2) use pipe clamp assembly when using in situ • (2) apply cutting solution at start of process • (2) inspect electrical lead and plug assembly for damage regularly • (2) keep both hands on the machine at all times during operation • (3) apply cutting solution throughout the process (available through a lubrication arm connected to a reservoir in the machine)	• Used for cutting threads on LCS tubes. Typically, the type shown has interchangeable die heads covering sizes ¼″ ³⁄₈″ ½″ ¾″ 1″ and 1¼″ BSP; for larger sizes, you will need a specialist manual threader. • Deburring is carried out manually. • All tube diameters may be threaded using electrically driven equipment. • A cutting compound must be used: this lubricates the cutting action and also keeps the process cooler. Without cutting compound, the new threads will be stripped out during use.	• Protective footwear • Protective overalls • Barrier cream • Eye protection • Ear protection
(3) Floor-mounted (site) electric threading machine	• (3) service regularly • (3) regularly clean swarf from filter tray. Never: • run your fingers over the threads • (3) stop the machine while threading is taking place. A lever which opens the jaws is manually operated for this action. • (3) attempt to cut, deburr or thread tubes which have been bent. Another variation of the machine is required for this.	• Capable of threading up to 2″ diameter tubes. • Has an attachment to enable the manufacture of nipples. • Cutting and deburring are carried out on the machine by rotation of the pipe. • In order to cut, deburr or thread tubes that have been bent, you will need a variation of this machine that has a rotating die mechanism rather than a rotating vice.	
Insulated screwdrivers	Always: • ensure that you carry out the safe isolation procedures before use • select the correct type of screwdriver • replace damaged or broken tools • use the right tool for the job. Never: • carry out electrical work unless you are competent to do so • misuse screwdrivers. They are for the removal or tightening of screws and should not be used as chisels.	• There are three main types of screwdriver tip: flared slotted (for general use), parallel slotted (where the head is the same size as the shaft) and cross-head (Phillips or Pozidriv®). • Other types of screwdriver tip are available for specialised jobs. • Stubby screwdrivers have their uses but provide only a limited amount of purchase. Always select the longest screwdriver length possible.	• Protective footwear • Protective overalls • Barrier cream

Table 4.01: Plumbing hand tools and power tools and their uses (*Possible, depending on method used) (continued)

Continued ▶

Tool	Care, operation and maintenance	Usage	Recommended minimum PPE
Power drills	Always: • take care not to drop or damage the equipment • inspect electrical lead and plug assembly for damage regularly • take instruction on how to use machinery you have not used before • use 110 V or battery-operated equipment where possible. Never: • suspend the drill by its lead (e.g. while working on a ladder) • expose the drill to wet conditions.	• Power varies depending on size of motor. Typical power ratings range from 620 W to 1400 W. • Some drills have a 'hammer action': as the drill rotates, it also moves fractionally backwards and forwards at high speed. This makes it easier to drill through masonry. • Available as 230 V or 110 V. • On sites, 110 V is required to meet health and safety requirements. • For domestic use, 230 V types are available, as 110 V types require a 110 V supply or suitable portable transformer. • Available with a traditional drill chuck (requires a chuck key to tighten the bit) or a special direct system (SDS) chuck. • SDS is a quick change chuck that does not need tightening or loosening to change the bit – simply push fit into position and release by pulling down a housing shroud. • SDS is very popular and gives the drill direct drive through to the drill bit. • Not all drill bits and accessories are available with an SDS end, so an SDS chuck adaptor is available to convert them to traditional drill chuck and key.	• Protective footwear • Protective overalls • Barrier cream • Eye protection • Ear protection* • Dust mask*
Typical 110 V extension lead and plugs	Always: • keep electrical equipment out of wet or damp conditions (e.g. never lay cables through puddles) • keep leads at high level to avoid trip hazards • check condition of leads and electrical equipment before use.	• Common on larger construction sites. • Identified by yellow cables, plugs and sockets. • Supply taken from mains 230 V and transformed down to 110 V. This offers greater protection to the user in the event of a fault/electric shock. • The transformer is portable and reduces the mains 230 V down to 110 V. On larger sites and premises, a dedicated 110 V supply is made available from a central transformer.	• Protective footwear • Protective overalls • Barrier cream
110 V transformer			
Selection of wood bits	Always: • ensure the bits are sharp and undamaged • keep hands away from the bits when in use • regularly oil the surface of the bits to prevent rust.	• Used for drilling holes in joists or other timber constructions for the passage of pipes. • You will also need other types of bit depending on the type of work you do (e.g. twist bits, augers, masonry bits and core drills).	• Protective footwear • Protective overalls • Barrier cream • Eye protection • Dust mask*

Table 4.01: Plumbing hand tools and power tools and their uses (*Possible, depending on method used) (continued)

Continued ▶

Continued ▶

Chapter 4

Tool	Care, operation and maintenance	Usage	Recommended minimum PPE
Hole saw set	Always: • make sure hole saws are sharp and undamaged • keep hands away from the bits when in use • regularly oil the surface of the bits to prevent rust. Never: • attempt to use without the pilot bit in place.	• Carrier for hole saw blade has a built-in pilot drill bit. This drills a hole in the material, which centres and holds the hole saw in position while the hole is cut.	• Protective footwear • Protective overalls • Barrier cream • Eye protection • Dust mask*
Battery-operated (cordless) hammer drill	• Powered by batteries. • Usually supplied with two batteries and a charger, so one battery is working and the other is charging.	• Much safer to use than drills with cords. • Useful if there is no electricity on site.	• Protective footwear • Protective overalls • Barrier cream • Eye protection • Ear protection* • Dust mask*
Powered circular saw	Always: • disconnect from the electrical supply before making any adjustments to the saw • make sure any electrical cable is out of the way of the cutting blade • let the blade stop running before putting the tool down • make sure the blade is kept sharp and the teeth are undamaged. You may need to seek professional help if circular saw blades need to be sharpened. • check condition of tool and any cable before use • ensure guards are in place during use.	• More commonly used by trades such as carpentry and joinery. • Useful for removing timber flooring materials. Depth may be set to the thickness of the material, which helps prevent damage to pipework and cabling beneath the floor.	• Protective footwear • Protective overalls • Barrier cream • Eye protection • Ear protection • Dust mask
Jigsaw		• More commonly used by trades such as carpentry and joinery. • Depth cannot be altered, so if this saw is to be used on flooring, you will need to inspect the underfloor area before you begin so that you can identify the positions of cabling, pipework or other potential obstructions.	
Electric pipe freezing kit	Always: • take care when handling the kit as the refrigeration unit is easily damaged • check electrical lead and plug before use and look for any obvious signs of damage to the equipment • wear gloves when handling leads and their connections because of the extreme low temperatures.	• Designed to save time when draining and refilling a system. • Will only work if there is no movement of water in the pipe. • More environmentally friendly than kits that use disposable canisters. Servicing and maintenance of the refrigeration unit should be carried out by a specialist refrigeration engineer.	• Protective footwear • Protective overalls • Barrier cream • Gloves

Table 4.01: Plumbing hand tools and power tools and their uses (*Possible, depending on method used) (continued)

Continued ▶

Tool	Care, operation and maintenance	Usage	Recommended minimum PPE
Canister and sleeve pipe freezing kit	Always: • wear gloves when handling leads and their connections, because of the extreme low temperatures • regularly check sleeves or clamps, adaptor and tube for signs of wear or fracture • replace canister when required.	• Designed to save time when draining down a system and refilling. • Will only work if there is no movement of water in the pipe. • The disposable canister is less environmentally friendly than an electric pipe freezing kit.	• Protective footwear • Protective overalls • Barrier cream • Gloves
Crowbar (wrecking bar)	Always: • ensure the claw is undamaged before using to extract nails • wear a hard hat if the tool is to be used as a wrecking bar.	• Used to prise up floorboards and nails. Once a board is partially lifted, push it down slightly to reveal the nail heads. Then you can remove the nails using a claw hammer or the claw on the end of the bar.	• Protective footwear • Protective overalls • Barrier cream • Eye protection* • Dust mask* • Gloves* • Hard hat*
Hydraulic pressure tester	Always: • check the pressure hose before use to make sure there is no sign of damage or deterioration • make sure there is enough water in the reservoir to carry out the test • keep the piston arm slightly greased at all times • empty and dry the reservoir when not in use. Never: • overpressurise: this could burst the hose burst or dislodge the fittings.	• Used to check the pressure of installations within buildings. • BS 6700 sets out separate procedures for testing rigid pipes and plastic pipes.	• Protective footwear • Protective overalls • Barrier cream • Eye protection

Table 4.01: Plumbing hand tools and power tools and their uses (*Possible, depending on method used) (continued)

Tool	Care, operation and maintenance	Usage	Recommended minimum PPE
Side cutting pliers (electrician's pliers)	Always: • keep lightly oiled, particularly the pivot joint. Never: • cut installed electric cable or flex without carrying out the correct safe isolation procedures first.	• The most common type of pliers used by plumbers. • As a plumber, you may also use wire cutters, circlip pliers (required for certain components such as tap assemblies) and long nose pliers. • Jaws are useful for holding small items or tightening small nuts and bolts. • Side cutter is used to cut small diameter electrical cable/flex or even thin diameters of steel wire or metal. • Various lengths of handle are available. Longer handles will be manufactured with more robust jaws so that larger diameters can be cut.	• Protective footwear • Protective overalls • Barrier cream
Crimping tool		• Many electrical cables are terminated using metal lugs, where part of the lug slides over the conductor and is squeezed tightly onto the conductor using a crimping tool. • Operated by hand or hydraulics.	
Side cutters	Always: • keep lightly oiled. The pivot joint in particular requires lubrication. Never: • cut installed electrical cable or flex without carrying out the correct safe isolation procedures first.	• Designed to cut hard materials such as wires and conductors	• Protective footwear • Protective overalls • Barrier cream
Insulation strippers	Always: • keep lightly oiled. The pivot joint in particular requires lubrication. Never: • strip installed electrical cable or flex without carrying the correct safe isolation procedures first.	• Place a cable or flex in the jaws of the stripper and use the calibrated markings to measure the amount of insulation to be removed. • Simply close the handles with one hand and hold the cable with the other and the insulation will be removed.	• Protective footwear • Protective overalls • Barrier cream

Table 4.01: Plumbing hand tools and power tools and their uses (*Possible, depending on method used) (continued)

Additional common plumbing tools

- **Files** – smooth and coarse, including a rasp type for chamfering plastic pipework for push-fit connections
- **Immersion heater spanner** – for tightening immersion heaters to cylinder bosses
- **Knife** – for trimming materials
- **Masonry drill set** – for drilling holes in masonry wall surfaces
- **Pointing trowel** – for making good to masonry wall surfaces

Checklist

Power tool use

- ☐ Make sure you have been properly instructed on the use of the tool before using it on site.
- ☐ Inspect tools visually for damage before use – do not use them if damaged.
- ☐ All electric tools should be double insulated or incorporate an earth cable.
- ☐ Battery-powered tools are preferable to mains-operated tools, as they are safer. If you use the mains or a temporary power supply, always use 110 V tools as opposed to 230 V.
- ☐ If 230 V power tools are used, the supply to the tool must be protected by a residual current device (RCD); see Chapter 2: *Electrical principles*.
- ☐ Check that electrical cables are not damaged or worn.
- ☐ Check that plugs are not damaged.
- ☐ Equipment should be **PAT** tested in accordance with your employer's procedures. PAT tests are maintenance records of all portable electrical equipment to ensure it is in safe working order. The maximum interval between tests for equipment used for construction is usually three months.
- ☐ Cartridge-operated tools are covered in Chapter 1 on health and safety. You must receive full instructions before using them.
- ☐ Make sure you use all the necessary PPE.

Key term

PAT – stands for portable appliance testing and is a legal requirement under health and safety legislation.

Working practice

You have a budget of £180 to purchase your own toolkit.

- Decide which tools you will buy, thinking carefully about quality, cost, safety and probable use (as discussed on pages 122–123).
- Create a simple table, identifying the tools you will buy, the manufacturers and the cost. Use the Internet to find these details.
- Make sure you have included VAT and, if you are considering shopping online, remember to include delivery charges as well.

Identify common faults found on power tools

Before using portable power tools:

- Always check the flex for any damage to the outer sheath. Is there any evidence of the lead fraying or are the inner cables showing? Has any part of the lead been covered over with insulation tape?
- Never suspend a drill by its cable at any time.
- Check the condition of the plug. Is it intact? Is there any visible damage? Has it been taped together?
- Is the plug the correct type for the supply you are using?
- Check the tool itself for signs of mechanical damage. Has it been mistreated or dropped? Have any repairs been attempted to the outer casing?
- Many power tools are fitted with safety guards. For example, circular saws have guards fitted to prevent the user's hands/fingers coming into contact with the blade.
- Under **no** circumstances should any guard be removed or a tool used without the safety guards in place.

Describe safe working practice when using hand and power tools

This and other areas concerned with workplace safety are covered in the table of tools and equipment on pages 124–135 and in Chapter 2 on health and safety, so refer back as necessary.

Progress check 4.01

1 What is the main disadvantage of buying cheap tools?
2 Why is it necessary to deburr a copper tube after tube cutters have been used?
3 Why is it important to ensure that the teeth on pipe grips and wrenches are free from the build-up of jointing compounds?
4 What type of surfaces would you use a hammer drill on?
5 What does the abbreviation PAT stand for? How often should this usually be carried out on electrical equipment used on building sites?
6 What essential piece of PPE must you wear when using electric drills and saws?
7 Make a list of essential hand tools (non-electrical) used by plumbers.
8 Make a list of essential power tools used by plumbers.
9 What checks should you make before using power tools?
10 Produce a general list of care and maintenance requirements for plumbing hand tools.

Figure 4.01: Some commonly used screws

KNOW FIXINGS AND COMPONENTS USED IN COMMON PLUMBING PROCESSES

This section covers the various types of fixing devices used to secure pipework so that it looks neat and is kept in its proper position. Fixings should provide sufficient support to withstand possible accidental damage, for example from people treading on pipework or pulling at it.

As a plumber, you will need to fix pipework, sanitary ware and appliances to various surfaces. You will also need to know how to refit boards and access traps in timber floors.

Pages 140–143 cover the measuring and marking out of fixings for pipework, and the clip and bracket types used for particular types of pipework. Refer to this section to see when some of these fixings are used.

List different screw heads used for fixing during plumbing activities

There are three main types of screw. These are referred to by the type of head into which the screwdriver is inserted. They are slotted, Phillips and Pozidriv, and a screwdriver with a different end is required to suit each type. There are also different shapes of screw head for different applications and different types of screw for different materials/components.

Screws also come in various shapes and sizes, and are specified by their length in inches and their gauge. The usual lengths and gauges used in plumbing range from 5/8 (16 mm) × 8 for fixing 15 mm saddle clips to 2½ (64 mm) × 12 for fixing radiator brackets. This is just a rule of thumb, and you will often have to make a decision about the right length and gauge for a particular situation. For example, if a wall is in poor condition, you might have to use a longer screw and a thicker gauge. Experience, trial and error will help you select the correct screw for the job.

Figure 4.01 shows some of the more common types of screw.

Describe the rationale for using screws made from different materials

Most screws are made of steel. However, it is also possible to buy sheradised (zinc-coated), stainless steel, enamel-coated or brass screws, as well as other types. In some cases, the material of the screw is chosen for looks or even hygiene reasons, but generally the material is chosen to prevent corrosion or to extend the life of the fixing.

- When securing bathroom equipment, because of the damp conditions, it is advisable not to use steel screws as they will quickly corrode. When securing a WC pan to a wooden floor, brass screws should be used. This is not just for aesthetic reasons, but again because of the high moisture content in the area of the pan. Also, brass is more resilient against urine, ensuring extended longevity and preventing the sanitary ware becoming loose or unsafe to use.
- Securing guttering and downpipe is another instance where safety and cost-effectiveness must be considered. Because of the exposed

location of such items, you should use sheradised screws to prolong the life of the installation. These are also cheaper than brass screws.

- Brass or alloy screws are used internally where they could be affected by moisture, including use in sanitary appliances. They are also used externally (mostly in alloy form because of the cost) for soil and rainwater fixings and wall-plate elbows for hose union bib taps.
- Self-tapping screws are used when fixing into metal sheet. This is not typical in domestic installations, but could apply if you need to clip to a metal stud partition. This involves drilling a pilot hole in a smaller gauge than the screw, using a steel drill bit, and then screwing into it.
- Steel countersunk screws are used for general purposes, such as fixing clips and radiator brackets.
- Chipboard screws, as the name suggests, are used for chipboard fastenings – for example, if you have to fix an access trap in a chipboard floor following some installation work.
- Mirror screws are used where appearance is important, and are often used for fixing timber or plastic bath panels.

Identify fixings used in plumbing activities

Table 4.02 describes some of the more common fixings you might use.

Fixing	Description and application	Used to fix into...
Wood screw	• Various uses and applications. • Screws directly into wood and, with a plastic wall plug, into masonry surfaces. • Available in a huge range of lengths and diameters and protective finishes/structures.	• wood • masonry surfaces (when combined with a wall plug)
Plastic wall plug	• Used with wood screws to allow fixing into the building fabric – by drilling the correct diameter hole (specified by the manufacturer). • Depth of hole drilled must be slightly longer than screw or plastic plug (whichever is longer) and plug must be taped below finished surface of wall or right through tiled surface. This prevents damage to wall surface (e.g. prevents tile surface from spelching/cracking) and stops plug being pulled out when screw is fastened. • Colour-coded for ease of selection (see Table 4.03 on page 140 for matching plastic plug and screw sizes). • The use of plastic wall plugs is shown in Figure 4.02 on page 139.	• block • ceramic tiles • brick (masonry)
Spring toggles	• A hole of the correct size (specified by manufacturer) is drilled through plasterboard. Toggle is inserted in its closed position, then a spring opens the toggles to grip the inner surface of the partition. • The disadvantage is that if the screw is withdrawn for any reason, the toggle is lost into the wall void.	• plasterboard • stud partitions/hollow walls
Masonry anchors (Fischer)	• Specialist fixing developed for securing washbasins into position. • Could be used on wood, but plastic plugs would not be required.	• wood • block • ceramic tiles • brick (masonry)

Table 4.02: Common plumbing fixings and their uses

Continued ▼

Fixing	Description and application	Used to fix into…
Wall bolt	• A hole of the correct size (specified by manufacturer) is drilled into masonry surface. • The bolt assembly is pushed into the prepared hole and tightened onto a washer by turning the bolt head with a suitable spanner. This drives a wedge on the end of the bolt, spreading the fixing walls hard into the masonry.	• block • ceramic tiles • brick (masonry)
Rawl bolt	• Works on the same principle as the wall bolt, with a very similar action. • Leaves a thread sticking out of the fixing surface, which is tightened using a nut and washer.	• block • ceramic tiles • brick (masonry)

Table 4.02: Common plumbing fixings and their uses *(continued)*

Screw size (gauge)	Drill size (mm)	Plug colour code
6–8	5	yellow
8–10	6	red
10–14	7–8	brown
14–18	10	blue

Table 4.03: Size matches for screws, drill bits and plastic plugs

Activity 4.1

There are many other types of fixing device available. Use the Internet to identify at least ten other forms of fixing device and find out where they would be used.

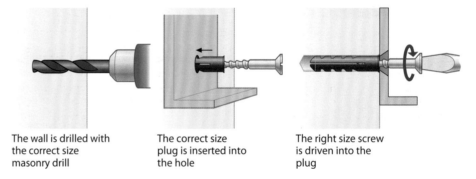

The wall is drilled with the correct size masonry drill

The correct size plug is inserted into the hole

The right size screw is driven into the plug

Figure 4.02: Fixing method using plastic plug and screw

Identify different clips used for plumbing installations

It is likely that the majority of your work will take place in domestic dwellings where copper or plastic clips are adequate for supporting copper and plastic pipework. There is a range of sizes of clips available. However, you might also work on other buildings: schools, hospitals or small industrial units, for example, where the clips or brackets need to be strong and robust.

On large jobs, clips or brackets will be specified, but often you will have to decide what type of fixing to use. Tables 4.04–4.06 on pages 141–142 describe the most common clips and brackets for different installations.

Copper saddle clip

- Rarely used today.
- Usually secured using roundhead screws.
- The main problem with this type of fixing is that the pipe is clamped to the fixing surface, creating a lack of air circulation around the pipe, which increases the risk of corrosion in damp conditions and harbours dust etc.
- Unsuitable for the application of pipe insulation.

Two-piece copper standoff clip

- Similar to the saddle clip in terms of construction and fixings.
- Does not have the same disadvantages as the copper saddle clip.

Plastic single pipe clip

- Single screw fixing.
- The pipe clicks into position, creating a rigid fixing that still allows for expansion and contraction.
- May be used to fix some types of insulation, depending on their wall thickness.

Wrapover plastic single pipe clip

- Similar to the click-in type but with the advantage of a plastic wrapover piece which clicks into place to surround the tube fully.
- The disadvantage is the difficulty of opening the wrapover for maintenance etc. – this may damage the clip.
- Clips may be colour-coded to show what the pipe is carrying: blue = cold water, red = hot water, yellow = gas supply.

Nail-in pipe clip

- Quick and easy to use.
- Same disadvantages as the copper saddle clip.
- Commonly used today.
- Predominantly used to fix small tubes (e.g. 6/8/10 mm R220 soft coils microbore (grade/table W).

Brass school board clip

- Decorative and fairly robust.
- Easy to use for pipework installation and decommissioning.
- The standoff allows air movement and the application of some pipework insulation materials, depending on thickness.

Brass two-piece standoff pipe clip

- Similar to school board clip.
- More widely used because of the cost.

Brass Munsen ring

- A very substantial pipe fixing bracket.
- The backplate is available as shown (external thread) or with an internal thread. This allows the centre of the bracket to be extended to the required centre by using threaded brass rod.

Table 4.04: Copper tube pipe clips and brackets

Saddle pipe clip

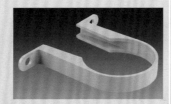

- Made from mild steel and protected with a bright zinc-plated finish.
- Suitable for steel and cast iron pipe and tube.
- As with copper saddle clips, there are disadvantages regarding air flow etc.

Malleable iron school board clip

- Easy to use for pipework installation and decommissioning.
- The standoff allows air movement and the application of some pipework insulation materials, depending on thickness.

Table 4.05: LCS tube pipe clips and brackets

Continued ▼

Malleable iron Munsen ring and backplate

- A very substantial pipe fixing bracket.
- The backplate is available as shown (internal thread), allowing the centre of the bracket to be extended to the required centre using threaded mild steel rod.
- The backplate is also available with an external thread, allowing the bracket to be secured into the backplate directly.
- Galvanised versions are also available.

Double-bossed malleable iron Munsen ring

- A double-bossed version of the malleable iron Munsen ring, allowing tubes to be fixed or hung vertically above or below each other.
- Galvanised versions are also available.

Table 4.05: LCS tube pipe clips and brackets (continued)

Nail-in clip

- This is the same as the nail-in clip which could be used on copper tubes.
- Mainly used on pressure pipework for hot and cold water and central heating systems.

Plastic single pipe clip

- Similar to the clip used for copper pipe; single screw fixing.
- The pipe clicks into position, creating a rigid fixing that still allows for expansion and contraction.
- Mainly used on pressure pipework for hot and cold water and central heating systems.

Plastic waste pipe clip

- Take care when positioning the screw centres and tightening the screws, otherwise the clip will distort because of the two screw fixing and the flexibility of the clip.

Soil pipe bracket

- A wraparound pipe bracket requiring two securing screws into the building fabric.
- The wrapover strap clicks into place underneath the screw housing cover to secure the soil pipe.

PVC-coated steel soil pipe bracket

- A sturdy bracket used in industrial and commercial installations.
- Also used in domestic situations where plastic clips could be damaged easily (for example, by vandalism or vehicle parking).

Plastic 'squareline' downpipe clip

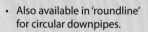

- Also available in 'roundline' for circular downpipes.
- The term 'fall pipe' is also used instead of downpipe.
- The downpipe connects a guttering system to the drainage system.

Table 4.06: Plastic pipework clips and brackets

Describe the reasons for using different clipping distances

Table 4.07 lists the recommended clipping distances (spacing) to use between internal pipework fixings. Clipping distances vary according to the tube material because of rigidity. For example, low carbon steel tubes are more rigid than copper tubes and so require fewer clips to secure them in position. Additionally, vertical clipping distances will be less than horizontal distances, as vertical tubes cannot distort as horizontal tubes can.

Pipe size		Copper		LCS		Plastic pipe	
mm	in	Horizontal (m)	Vertical (m)	Horizontal (m)	Vertical (m)	Horizontal (m)	Vertical (m)
15	½	1.2	1.8	1.8	2.4	0.6	1.2
22	¾	1.8	2.4	2.4	3.0	0.7	1.4
28	1	1.8	2.4	2.4	3.0	0.8	1.5
35	1¼	2.4	3.0	2.7	3.0	0.8	1.7
42	1½	2.4	3.0	3.0	3.6	0.9	1.8
54	2	2.7	3.0	3.0	3.6	1.0	–

Table 4.07: Recommended spacing between pipework fixings

Describe which fixing to use on different surfaces

This is covered in Table 4.02 on pages 139–140.

Working practice

Nadil has received a delivery from a plumber's merchant, which includes copper tube, fittings, bathroom appliances, etc., plus a box of copper saddle clips. Nadil really needed plastic single screw saddle clips. However, he did not notice that the clips were incorrect until after he had signed for the delivery and the driver had left.

He phoned the merchant but they refused to exchange the clips because they fit the original order and, the merchant said, 'A pipe clip is a pipe clip!'

- Who is to blame for this situation – Nadil, the merchant, or both?
- What did Nadil do wrong when the materials were delivered?
- Nadil has asked for your advice on why the clips are unsuitable. Produce a bullet point list indicating the disadvantages of the copper saddle clips so that he can call the merchant and ask them to exchange the clips.

Progress check 4.02

1 Name the three common types of countersunk screw head.

2 Which metals are recommended for internal use where they could be affected by moisture (e.g. for fixing sanitary appliances)?

3 Make a list of the different types of fixing device that are commonly used in the plumbing industry.

4 What are the disadvantages of using saddle or nail-in clips to secure metal pipework?

5 What advantage does a Munsen ring and backplate have over a one-piece clip?

6 Name the type of fixing that should be used to secure components to plasterboard walls.

UNDERSTAND THE PROCEDURES FOR MEASURING AND BENDING PLUMBING TUBES

Measuring, marking out and bending are essential basic skills for any plumber. We will concentrate here on bending copper pipe, which can be done by hand or by machine, and low carbon steel pipe, which is usually bent using hydraulic machines.

Identify equipment used for measuring and bending

Table 4.08 lists the tools and equipment required and their uses.

Tool	Care, operation and maintenance	Usage	Recommended minimum PPE
600 mm folding/split rule	Always: • keep lightly oiled to prevent rusting • lubricate the pivot pin but do not allow it to slacken • prevent dints and abrasions from appearing on the tool. Never: • use as a lever • misuse the calibrated tool.	• Folds to approximately 300 mm for ease of storing/carrying in overall ruler pocket. • Used for measuring pipe bends and lineal lengths. • Also useful to set the angle of a bend. The pivot pin must be tight and secure for this.	
600 mm steel rule	Always: • keep lightly oiled to prevent rusting • prevent dents and abrasions from appearing on the tool. Never: • use as a lever • misuse the calibrated tool.	• Used for measuring pipe bends and lineal lengths. • Very accurate.	
Steel measuring square (roofing square)		• Very useful for measuring around the centre of a square bend or between the centres of an offset bend, etc. • Very accurate way of measuring pipework, rather than estimation.	
Steel tape measure	Always: • keep lightly oiled to prevent rusting • prevent dints and abrasions from appearing on the tool • let the rule run back into the case steadily. This will extend the life of the tool and prevent you being cut by the metal. Never: • bend the tape, as this may damage it so that it cannot retract back into the case.	Advantages: • flexible and pocket size • can measure short or long lengths • internal and external measurement easily achieved. Disadvantages: • steel band fractures easily if misused • accuracy is not good over long lengths because of the natural bowing of the tape due to its thickness.	

Table 4.08: Tools and equipment for measuring and bending tubes

Continued ▶

Continued ▶

Tool	Care, operation and maintenance	Usage	Recommended minimum PPE
Pencils, French chalk and fine point marker pens	Always: • keep pencils and French chalk sharp (a wedge shape is most accurate for the chalk) • keep tops on marker pens when not in use.	• Essential for accurate measurement and marking out. • Marker pens are excellent for marking out copper tube but poor when marking on black LCS tube. • French chalk is harder than normal board chalk and is easily seen on the tube surface.	
Slide bevel and protractor	Always: • protect the bevel blade from damage • keep the blade slightly lubricated and the tightening bolt lightly greased • ensure the protractor edges are free from damage. Wooden types are more robust than smaller plastic types. Never: • misuse the blade of the bevel • store the slide bevel without refitting the blade into the tool body.	• Carpenters and joiners traditionally use a slide bevel. • By using a protractor, the bevel may be set to the desired angle. This angle is then easily transferred to the pipe, either while in the machine or as a check on a completed bend.	• Protective footwear • Protective overalls • Barrier cream
Hand bender/scissor bending machine	Always: • keep the pivot bolt lightly oiled • oil the inside of the tube formers and back guides periodically. Never: • handle the bender roughly as this could damage the tube formers • bend anything other than copper tube.	• The small hand-held bender is used for bending copper tube sizes of 15 mm and 22 mm. • Light and very portable. • The back guide is required to support the tube while it is being bent. • The machine does not require any manual adjustment before bending different diameters.	• Protective footwear • Protective overalls • Barrier cream
Small mini-bore bender (6/8/10 mm)	Always: • keep the pivot bolt lightly oiled • oil the inside of the tube formers periodically. Never: • handle the bender roughly as this could damage the tube formers • bend anything other than copper tube.	• Small mini-bore benders are designed to bend small radius bends on tube diameters 6/8/10 mm. • Bend angles are easily achieved as they are indicated along the wheel. • The top handle is used as a guide to enclose the tube, ensuring that bending is free from ripple and throating.	• Protective footwear • Protective overalls • Barrier cream

Table 4.08: Tools and equipment for measuring and bending tubes

145

Tool	Care, operation and maintenance	Usage	Recommended minimum PPE
Freestanding (stand) bender Pipe vice Tube Stop Rubber protection to feet Adjustable roller Back guide Tube former	Always: • keep the pivot bolt lightly oiled • oil the inside of the tube formers periodically • when moving, plan your route and apply the kinetic lifting procedure: these machines are heavy and awkward to carry because they are unbalanced • store the tube formers and back guides safely and securely. Never: • drop the tube formers or back guides • bend anything other than copper tube.	• Can handle pipes up to 42 mm by changing the back guides and formers to suit the respective pipe sizes. • It is important to set the machine up properly. If the roller is too loose, this will cause rippling on the inside of the pipe radius. • If the roller is too tight, it will cause throating (reduced diameter at the bend).	• Protective footwear • Protective overalls • Barrier cream
Internal/external bending springs	Always: • keep lightly oiled • slightly overbend the tube and then pull it back to the required angle. This relaxes the coils in the spring, making it easier to remove. Never: • stretch the spring if it is stuck inside the tube, as this will render it unusable for future work • try to pull a tight radius bend without any bending spring, as this will usually put a kink in the spring, rendering it unusable.	• Rarely used nowadays because it is quicker and easier to use a portable scissor bend. • When using a bending spring, the operative's knee is used to form the shape of the bend. Over time, this may result in damage to the knee joint. • Remember, only grade R250 copper is suitable for bending. • External bending springs are normally used on microbore because of their much smaller diameter – 6, 8 or 10 mm. **Safety tip** Be careful when using this type of spring: if the spring is suddenly released from the tube, the coils may get caught in your hair.	• Protective footwear • Protective overalls • Barrier cream
Electric tube bender	Always: • keep moving parts well oiled/greased • check electrical lead and plug for visible signs of damage or wear. Never: • use near water or in damp conditions • drop or treat the tool roughly.	• Very portable. • Suitable for most tubes. • Instantly ready – no need to set tensions manually etc. • Has both fast and creep speed operations for exact bending. • Requires bending and back formers. • Up to 180° bends are possible.	• Protective footwear • Protective overalls • Barrier cream
Hydraulic bending machine	Always: • keep moving parts well oiled/greased • ensure the hydraulic fluid level is correct before use • take care when lifting or moving this equipment as it is very heavy • take care when removing the bent LCS tube from the tube former.	• Needed to bend low carbon steel tubes, owing to the strength of the material and the thickness of the pipe. • Pipe does not need to be fully supported with a back guard. • Used to form most bends, including 90° and offsets. • The hydraulic mechanism is usually oil-based. Liquids are incompressible, so they can exert a considerable force on the pipework once under pressure.	• Protective footwear • Protective overalls • Barrier cream

Safeguarding tools and equipment

It is important to look after your toolkit in order to:

- keep the tools in good condition – clean, free from rust/corrosion, sharp (where applicable)
- prevent loss or theft.

You should keep your kit in a good-quality toolbox or bag. This will keep the tools safe when not in use safe and also means not having to search around for items as and when you require them.

Never leave tools unattended in busy areas. At night use the site facilities to secure tools and equipment or remove anything portable from the site. If using a vehicle, it is a good idea to take your tools out overnight: it is now common to see signs on vans stating, 'No tools kept in this vehicle overnight'. This also helps to protect the vehicle from damage.

> **Progress check 4.03**
>
> 1 What four factors are you recommended to consider before purchasing tools?
>
> 2 What is the minimum recommended PPE required when carrying out any plumbing operation?
>
> 3 List seven points to follow in order to care for and maintain tools and equipment.
>
> 4 What is the benefit of having a slide bevel and protractor in a plumbing toolkit?

Working practice

The company you work for has won a large contract that will involve extensive use of copper tube. Your employer wants you to help him make a decision about the spending budget for this contract. He has asked you to advise him of the alternatives available for bending copper tubes and their purchase cost.

- Prepare a list of the methods available to bend copper tube and use the Internet to find out the cost of each of them.
- Write a short summary for your employer indicating your recommendation, giving a reason for your decision.

Identify common materials used for plumbing tubes

The main materials used are copper, steel and plastic. These and other materials you may come across are covered in detail in Chapter 3 on pages 86–90.

Plastic pipe can be bent, but its use in plumbing is usually restricted to small-bore polythene pipes. These can be positioned into large-radius 90° bends, or offset by hand and then clipped into position. The pipe can also be clipped into preformed 90° steel brackets for tighter bends.

Identify different angles

There are several types of angles (bends):

- **90°** or **square** – this can be varied to form other angles such as 30° or 45°.
- **Offset** – this is used to negotiate a pipe around an obstacle or object. The set bend is usually 30° or 45° but this can be varied. A good example is when pipework is clipped to a wall surface and requires connecting to a washbasin tap.
- **Kickover** (or **passover offset**) – this is normally used when tubes are fixed parallel to one another and a branch or change in direction is required to enable the pipe which is the obstacle to be negotiated.
- **Crank passover** – this has a similar use to the kickover (or passover offset), but is designed to be used where the pipe run is not connected into a plumbing fitting.

Bending pipe, rather than using ready-made fittings, has the following advantages:

- It produces larger radius bends than elbow fittings (larger radius bends have less frictional resistance).
- Using bends costs less than using fittings.
- Long sections of pipework can be prefabricated before installation, saving time.
- There is less chance of leaks.

Describe the procedure for bending different angles

Although you may not actually be carrying out bending at this stage, this section will cover the procedure for different materials and using the different equipment described in Table 4.08 on pages 144–146.

Bending methods – copper pipe

The plumbing industry uses four main types of copper tube, which are described in more detail on page 87.

- R250 half-hard is suitable for bending by machine and (although little used nowadays) internal or external bending spring.
- R290 hard should not be bent, as the wall thickness is too thin.
- R220 soft coils may be bent with care to large radius bends but is not suitable for bending machines or spring bending.
- R220 soft coils micro-bore may be bent using mini-bore bending machines, or internal or external bending springs.

Bending by hand

This is a method of bending pipe when carrying out maintenance and repair work. You may be working in a loft space, for example, where it would be much quicker to fabricate the pipe in situ. This is best done by using a bending spring.

Bending springs can be used externally or internally. In either case you pull the bend against your knee to get the desired angle (Figure 4.03). When using an internal spring, there are a few things to remember:

- Don't try to bend a piece of pipe that is too short. You would have to apply a lot of physical pressure, which could lead to injury, or kink the pipe, or both.
- Don't make the radius of the bend too tight, or you might find that you can't remove the spring.
- Always pull the bend slightly further than the required angle, and then pull it back to the required angle. This will help to release the spring.
- The spring has a 'hook' at the end. This can be twisted clockwise and will tighten the spring coil inwardly, making it easier to remove the spring.

Explain the method of measuring tube

Bending tubes using 'x' and 'y' dimensions involves calculating the exact measurement required to place a bend accurately into position. This reduces waste dramatically.

Allowance is made for each fitting (the 'x' dimension) and also, if applicable, an allowance for a pipe clip (the 'y' dimension). Figure 4.04 shows a simple

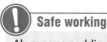

Safe working

Always use padding to protect your knee from the pressure of the bending action.

Figure 4.03: Use the same technique for internal and external bending springs

example using 15 mm copper tube, one tee piece and one plastic standoff clip. Both 'x' and 'y' dimensions will vary depending on the tube diameter, type of fitting and clip. You will need to take measurements or you may be able to look them up in the manufacturer's published 'x' and 'y' dimensions. The same method is adopted when using low carbon steel tubes, but the fittings allowances are referred to as the 'z' gauge.

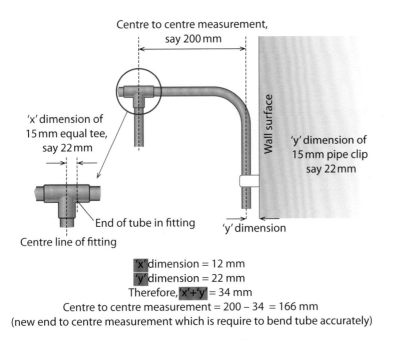

'x' dimension = 12 mm
'y' dimension = 22 mm
Therefore, 'x'+'y' = 34 mm
Centre to centre measurement = 200 – 34 = 166 mm
(new end to centre measurement which is require to bend tube accurately)

Figure 4.04: Calculations for bending tube using 'x' and 'y' dimensions

Activity 4.2

Choose a variety of plumbing fittings which are used on copper tubes, making sure you select a range of types and sizes. Calculate the 'x' dimension of each fitting. What do you notice?

The procedure is not used just to calculate bends on tubes; it is also good practice to carry this out when cutting straight pipe runs. Applying these simple rules prevents wasting these very valuable resources.

Machine bending

This is the most common method used for bending copper tube. Bending machines can be either hand-held (hand bender) or freestanding (stand bender), and they work on the principle of leverage. It is important to set up the stand bender properly. If the roller is too loose, this will cause rippling on the inside of of the pipe. If it is too tight, it will reduce the pipe diameter at the bend. This is called throating.

You will need to your prove your competence in bending and joining tube for your plumbing qualification.

As described on page 147, the three main types of bend are 90° or square, offset and passover.

90° or square bend

1 **Taking the measurement:** You may be given a measurement from a site drawing, but it is more likely that you will take a site measurement yourself. You should take measurements from a fixed point to the back of the bend (see Figure 4.05).

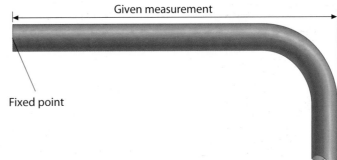

Figure 4.05: Taking the measurement

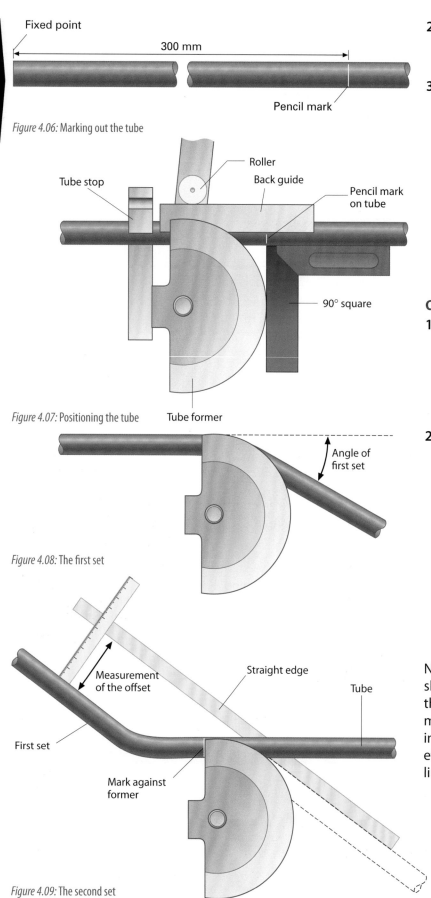

Figure 4.06: Marking out the tube

Figure 4.07: Positioning the tube

Figure 4.08: The first set

Figure 4.09: The second set

2 Marking out: Make a pencil mark from the fixed point (end of tube) to the required length (see Figure 4.06).

3 Positioning the tube in the bender. The tube is set up in the machine with the mark squared off from the outside of the former (see Figure 4.07). Then position the back guide, correctly adjust the roller and pull the bend. (It is not necessary to adjust the roller on a hand/scissor bender.) You can make return bends using the same technique, only now the back of the first bend becomes the fixed point.

Offset

1 The first set: This is made to the desired angle (see Figure 4.08). The first angle is not critical, but will usually depend on the profile of the obstacle you want to offset the pipe around.

2 The second set: After making the first set, reverse the tube and return it to the machine. Place a straight edge against the former, parallel to the tube, and take the measurement for the offset from the inside of the tube to the inside edge of the straight edge. Once the tube has been adjusted against the stop of the machine, it is a good idea to mark the edge against the former in case of any movement (see Figure 4.09).

Note the dotted line on Figure 4.09 showing the finished position of the second set. This shows that the measurements are taken from the inside to the back of the tube, so in effect it is a centre-to-centre line measurement.

> **Remember**
>
> You can place a piece of pipe alongside the former to help you visualise and measure where the pipe will be when it has been bent to its finished form.

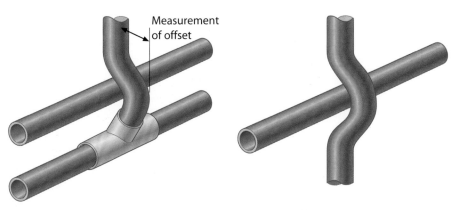

Figure 4.10: Passover offset (kickover)   Figure 4.11: Crank passover

Passover

This is used to clear obstacles, such as other pipes, and could be a passover offset (kickover) (Figure 4.10) or a crank passover bend (Figure 4.11). You take the measurements for a passover in the same way as for an ordinary offset. The angle of the first pull will be governed by the size of the obstacle it has to pass over. Make sure the first bend is not too sharp or it will be difficult to pull the offset bends.

Place a straight edge across the bend at the distance of the obstacle and mark the pipe (see Figure 4.12). This will be the back of the finished offset.

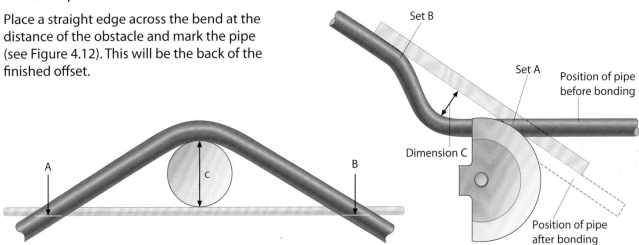

Figure 4.12: Measurement of crank passover   Figure 4.13: Final set in a crank passover bend

Return the pipe to the machine and, when the first mark lines up with the former, you can pull it to the position shown. Then turn the pipe around in the machine, line up the second mark in the former and pull the pipe to complete the passover (see Figure 4.13).

Passover offset (kickover)

Figure 4.14 shows a completed passover offset or kickover bend passing over a 15 mm copper tube with an allowance for a 15 mm clearance. For this you will need a 600 mm folding/split rule and a 600 mm rule or other straight edge (see Table 4.08 on page 144).

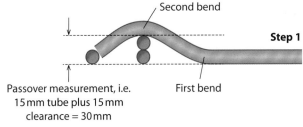

Figure 4.14: Completed passover offset or kickover bend

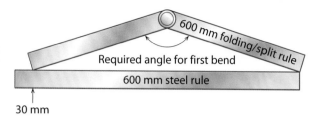

Step 1: Use a folding/split rule along a 600 mm rule to gain the required angle of the first bend, say 30 mm required.

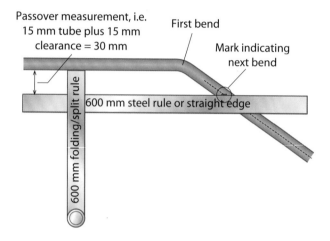

Step 3: Hold a steel rule or any straight edge parallel to the pipe and measure off the required passover dimension.

Then draw a line on the remaining tube where the straight edge touches the centre line.

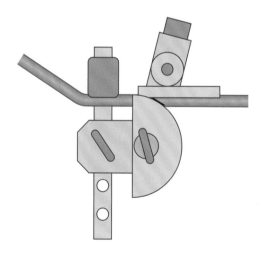

Step 5: Reposition the tube in the bending former, ensuring that the line drawn in Step 4 glances off the edge of the former. Remember to look at the alignment of the first bend before proceeding.

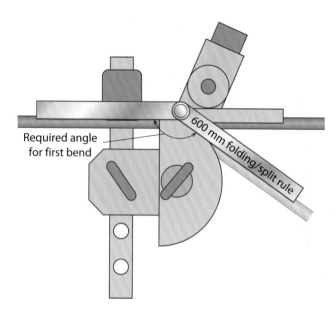

Step 2: Make the first bend to the angle of the folding/split rule.

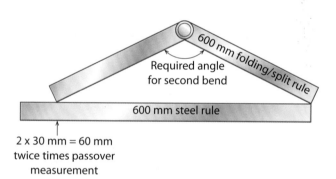

Step 4: Close the folding rule again, this time to twice the required gap (60 mm).

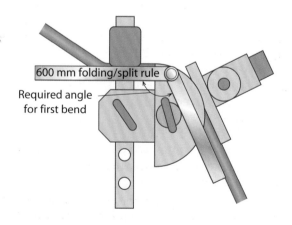

Step 6: Bend to the required angle set in Step 4.

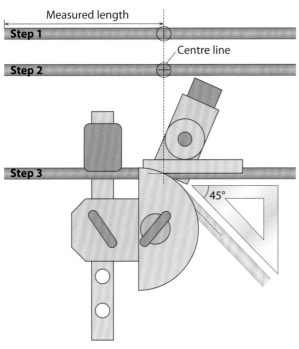

Figure 4.15: Machine bent 45° bend

45° bend

This is illustrated on Figure 4.15.

1 Mark the desired measurement on the tube.

2 Mark a line representing the centre line that crosses through the first mark. Where the two lines cross, position this at the edge of the bending former.

3 Bend the tube to an angle of 45° and check the angle using a set square, protractor or preset slide bevel. Note: the same rule used to indicate a 45° bend applies to bend any other angle required.

Working practice

Mark has six 15 mm copper pipes. They have to rise vertically in the corner of a room and then turn through 90° and travel horizontally at ceiling level across the top of the wall.

He has bent three so far, but the customer then comments that they 'look awful'. The customer accepts that the pipes will have to be on show but he is unhappy about the distances between the bends because they vary so much.

Mark has been careful in his setting out and has allowed 50 mm centre to centre to each pipe, but the customer is right, they do look a mess where all the bends are. They certainly are not 50 mm apart at that point.

Note: the bends must be 'pulled bends'. It is an unacceptable response to use elbows to resolve the issue. Also, the pipework may not be boxed in.

- What is the problem here? Why are the pipes at different centres where the bends occur? Suggest a solution that would both satisfy the customer and enable Mark to be proud of his work.

- Indicate the start and finish of the centre line and end markings that would be required to make your solution work on six pieces of scrap tube or paper.

- The note points out that it would not be acceptable to use elbows. Why would elbows be an unacceptable solution to the problem? Do not consider cost in this case.

Figure 4.16: Using a hydraulic bending machine

Low carbon steel pipe

LCS pipes are usually bent using a hydraulic pipe bender. Only medium and heavy grade tubes are used in plumbing systems; they are manufactured to BS 1387, and are supplied in black-painted or galvanised coatings. You are more likely to work on medium-grade tube. Galvanised pipe should not be bent, as this will cause it to lose its coating. In any case, you are unlikely to come across this material outside maintenance work.

Bending with a hydraulic machine

You need to use a hydraulic machine (see Figure 4.16) to bend low carbon steel tubes, owing to the strength of the material and the thickness of the pipe. For these reasons, the pipe does not need to be fully supported with a back guard, unlike copper pipe. Hydraulic bending machines are used to form most bends, including 90° and offsets. The hydraulic mechanism is usually oil-based. Liquids are incompressible, so they can exert a considerable force on the pipework once under pressure.

90° bend to be produced

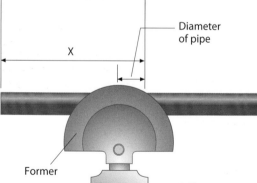

Diameter of pipe

Former

Figure 4.17: Forming a 90° bend

90° bend

This is illustrated in Figure 4.17.

1 Mark a line on the pipe at a distance from the fixed point, where the centre line of the finished bend is required.

2 From this measurement you deduct the nominal bore of the pipe. This is because there is a gain in length of one pipe diameter when bends are made.

3 Make sure the correct size of former is in the machine.

4 Put the pipe in the bending machine and line up the mark with the centre of the former (see Figure 4.17).

5 You can then work the machine to apply pressure and bend the pipe to 90°.

6 Because of the elasticity of the metal, you need to take it to approximately another 5° over 90° to allow it to 'spring back'.

Offset

It is a good idea to make a template from steel wire to help you achieve the required offset profile. The method of marking out the offset is similar to that of copper, as follows:

1 Mark off the required measurement for the first set onto the pipe.

2 Place the pipe in the machine but do not make any deduction. The measurement *X* mm is from the fixed end of the pipe to the centre of the set (see Figure 4.18 on page 155).

3 Pull the first set to the required angle.

4 Take the pipe from the machine and place a straight edge against the back of the tube. The measurement of the offset is marked at Point A (see Figure 4.19).

5 Replace the tube in the machine and line up the mark with the centre of the former. Pull the second set and check against the wire template. Again, allow a 5° overpull for the spring back. You will need to level the pipe in the machine to prevent any twist or distortion in the set.

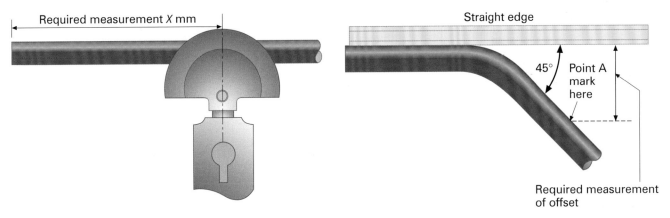

Figure 4.18: Locating the pipe for the first bend

Figure 4.19: Setting out for the second set

Faults and defects

The main problems encountered when bending low carbon steel pipe are usually caused by incorrect marking out, such as not making the deduction for the nominal bore, so the bend does not fit when it is pulled. In the case of offsets, if the pipework is not set up level in the machine, then the resulting offset will be screwed or twisted.

> **Remember**
>
> During the bending operation, the pipe can become wedged into the former. Do not hammer the former or the pipe to remove it. You can place a timber block on the pipe and give it a sharp tap with a hammer. Alternatively, remove the pipe and former. Place a wooden block on the floor and strike the end of the pipe on the block. Try to hold the end of the former so that it does not fall to the ground.

> **Working practice**
>
> An insurance company has asked your company to survey the workmanship of plumbing and heating systems in an existing property. You notice straight away that nearly all the bends are malformed, having been produced in a machine that was incorrectly adjusted or set up. However, before you can produce the report, you need to do a little homework to make sure you describe the problems using the correct terminology.
>
> - Make a list of the types of bends that are used and produced on copper tube.
> - The poor quality of the bends is the result of a machine being wrongly adjusted or set. Which type of machine would have been used? Which two terms are used to define poor adjustment and what would be the cause of each one?
> - In some instances, elbows have been included in the system where machine bends could have been used. What is the effect of this practice on the movement of water through the system?
> - How much lineal length does the use of a 15 mm elbow add to a system?

1 List the tools and equipment you would use to create an offset bend on a copper tube.

2 List the tools and equipment you would use to create a 90° bend on LCS tube.

3 When using a freestanding bender, what would cause 'throating' of the pipework?

4 What method could you use to bend LCS tubing, apart from hydraulic bending?

5 Produce a step-by-step guide describing how to bend a 45° angle on copper tube.

6 Why is it so important to be able to measure pipework and the position of bends accurately?

UNDERSTAND HOW TO JOINT COMMON PLUMBING MATERIALS

Pipes, fittings and jointing materials acceptable for Water Regulation purposes are listed in the *Water Fittings and Materials Directory*. BS 6700 also states the minimum requirements for pipe joints and fittings. At Level 2, you must consider the following factors when selecting materials for use in plumbing systems:

- effect on water quality
- compatibility of different materials
- vibration, stress or settlement
- ageing, fatigue and durability
- internal water pressure
- mechanical factors
- internal and external corrosion
- **permeation**.

Key term

Permeation – air entering the water supply through the external wall of a pipe.

Identify common plumbing fittings

There is a huge range of plumbing fittings available, covering just about every eventuality. Most will look very similar to those used on different tube materials. Even when selecting fittings for one type of tube, there are many different types to choose from. For example, when choosing a tee piece (sometimes called a branch) for copper tubes, you need to decide whether you need an equal tee, a one end reduced tee or a branch reduced tee. You must also decide whether it should be fitted using an end feed solder capillary, an integral solder ring capillary or compression.

Table 4.09 on page 157 indicates just a sample of the plumbing fittings you will come across. There are many more to be aware of. Most of the common fittings are available to suit the type of tube or material they are to be applied to. When you consider that each one has a full range of diameters to meet the different needs, the list is vast.

Activity 4.3

Look at Table 4.09 on page 157 and Table 4.10 on page 158, covering tubes, fittings and jointing methods. Then use the Internet to identify as many others as you can, to familiarise yourself with the many fitting types, shapes and materials and to see if you are able to identify their possible uses.

Tubes and groups of fittings	Copper tube	Low carbon steel (LCS) tube	Plastic pressure pipework	Plastic soil and waste pipework	Plastic rainwater gutter and downpipe
Adaptors – internal/external thread	✓	✓	✓		
Bends (elbows) – socket to socket	✓	✓	✓		
Bends (elbows) – spigot to socket	✓	✓	✓	✓	
Bends (elbows) – socket to socket union		✓			
Bends (elbows) – socket to socket reducer	✓	✓	✓		
Bent/straight tap connectors	✓		✓		
Black or galvanised		✓			
Bushes (reducing)		✓			
Compression		✓	✓		
Crosses – equal	✓	✓	✓		
Crosses – branches reduced	✓	✓	✓		
Crosses – ends reduced	✓	✓	✓		
Cylinder unions	✓				
Flanges – screwed/welded		✓			
Metric–imperial converter	✓				
Nipples – barrel/hexagon		✓			
Offset/passover/kickover	✓				
Push-fit	✓		✓	✓	✓
Reducing fitting or bush	✓	✓	✓		
Return bends	✓	✓			
Slip coupling	✓				
Sockets (straight couplers/joints)	✓	✓	✓	✓	✓
Solvent weld			✓	✓	
Stop ends, caps or plugs	✓	✓	✓	✓	✓
Strap-on boss				✓	
Tank connectors	✓	✓	✓		
Tees – branch reduced	✓	✓	✓	✓	✓
Tees – end reduced	✓	✓	✓		
Tees – both ends reduced	✓	✓	✓		
Threaded – parallel or tapered		✓			
Unions – socket to socket (equal/unequal)		✓			
Unions – spigot to socket (equal/unequal)		✓			
Welded		✓			

Table 4.09: Some common fittings for different types of pipework

Describe methods for joining plumbing materials using common jointing techniques

Refer to Table 4.10 and to Table 4.09 on page 157 throughout this section.

Tubes and types of jointing method	Copper tube	Low carbon steel (LCS) tube	Plastic pressure pipework	Plastic soil and waste pipework	Plastic rainwater gutter and downpipe
Compression – malleable cast iron		✓			
Compression type A (non-manipulative)	✓				
Compression type B (manipulative)	✓				
End feed capillary soldered joint	✓				
Flange – welded or threaded (screwed)		✓			
Fusion weld			✓		
Integral solder ring	✓				
Mechanical joint	✓	✓	✓	✓	
Press-fit fitting	✓				
Push-fit (O-ring)				✓	
Push-fit – plastic/copper/brass	✓		✓		
Silver solder	✓				
Snap-lock gutter joint system					✓
Solvent weld			✓	✓	
Threaded – (**BSPT**) (screwed)		✓			
Welded (bronze)	✓				
Welded (steel)		✓			

Table 4.10: Common jointing methods and applications

Copper tube

There are three main methods of jointing copper pipework:

- compression joints
- soldered (capillary) joints
- push-fit joints.

Compression joints

Compression joints are used to connect sections of copper tube, and can be:

- manipulative (or type B) – where you form the end of the tube
- non-manipulative (type A) – the end is not formed.

Manipulative joint (Type B)

Figure 4.20 shows a typical manipulative fitting detail.

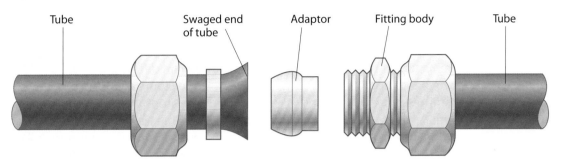

Figure 4.20: Manipulative compression joint

The jointing process is as follows (see Figure 4.21):

1 Cut the tube to the required length, using either tube cutters or a hacksaw.

2 Deburr the inside of the pipe (if cut with tube cutters) or the inside and outside of the pipe (if cut with a hacksaw).

3 Place the nut and compensating ring onto the pipe. This acts as a washer when the nut is tightened against the flared tube.

4 Use a swaging tool or drift to flare out the end of the tube so that it fits the angle of the adaptor.

5 Apply this process to each pipe end being used in the fitting (two for a straight coupling or elbow, three for a tee).

6 Assemble the fitting. The angled side of the adaptor fits into the flared end of the tube and the other side fits into the body of the fitting (see Figure 4.22). The joint should be finger tight, with all the fitting components fully engaged. The fitting can then be completely tightened using an adjustable spanner or other suitable smooth-jawed tool.

> **! Safe working**
>
> Make sure the hacksaw end is square by using a file.

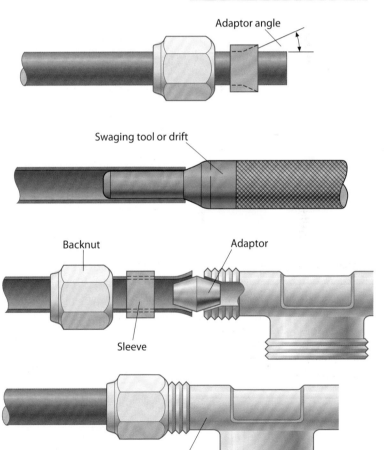

Figure 4.21: Assembling a manipulative compression joint

Figure 4.22: Coupling a manipulative compression joint

This type of copper compression fitting is rarely used because of the increased amount of time required to form the joint. However, compression fittings used on underground copper cold water supply pipework must always be manipulative (Type B) joints.

Non-manipulative joint (Type A)

This is similar to the manipulative fitting. The main difference is that the pipe end is not flared to receive an adaptor (see Figure 4.23). Instead, a compression ring or olive is used to provide a watertight seal on the pipe end. The cutting and deburring process is the same as for a manipulative fitting. Again, the joint is hand-tightened first, making sure the pipe is fully pushed into the body of the fitting, before tightening is completed with an adjustable spanner or other suitable smooth-jawed tool (see Figure 4.24).

> **Remember**
>
> This type of fitting is suitable for use on R250 half-hard lengths, R290 hard lengths and R220 soft coils micro bore. Its use is not allowed on underground services.

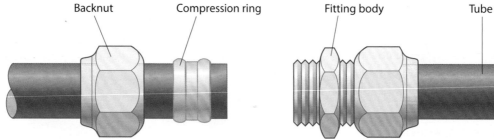

Backnut   Compression ring   Fitting body   Tube

Figure 4.23: Detail of non-manipulative joint

Figure 4.24: Coupling a non-manipulative joint

Brass fittings

There is an enormous range of brass fitting designs available for different pipe sizes, each with several variations. These include adaptors, couplers, tap connectors, tees, elbows, etc.

Soldered capillary joints

Capillary jointing simply means soldering. Capillary joints are the most common plumbing joints. Soldered joints can be classified as soft-soldered (see Figure 4.25) or hard-soldered. Hard-soldered joints, for example using silver and silver alloys including copper, require heating to a much higher temperature than soft-soldered joints. Hard-soldered joints are rarely used in domestic plumbing installations so will not be discussed here.

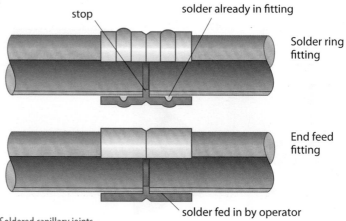

stop   solder already in fitting

Solder ring fitting

End feed fitting

solder fed in by operator

Figure 4.25: Soldered capillary joints

Soft-soldered joints are made using two types of fittings:

- integral solder ring
- end feed.

The difference between the two jointing methods is that the solder ring has solder contained in a raised ring within the body of the fitting (integral). When the fitting is heated, the solder melts (between 180°C and 230°C) and the solder is drawn into the fitting by capillary action (capillarity). The same principle is used for the end feed, except that the solder is end-fed separately from a spool of wire solder. See Chapter 3 pages 96–97 for more on capillary action.

The jointing process is fairly straightforward.

> **Key term**
>
> *Flux* – a paste (or liquid) used in soldering techniques (see page 162).

The jointing process

① Cut the tube square and deburr it. Clean the fitting inside and out to a bright, shiny, light orange colour. This is because the solder will not adhere correctly if these surfaces are oxidised.

② Apply **flux** to the tube and fitting using a brush – not your fingers.

③ Assemble the joint and apply heat to the body of the fitting.

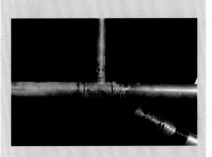

- If you are using an integral solder ring, wait until you see a complete film of solder around the end of each fitting. If using an end feed, apply solder from a wire spool and feed it in until the joint is complete.
- When the fitting is cool, clean off any excess flux residue with a damp cloth. You should also flush the system to remove flux residue from inside the pipework, as this could corrode the inside of the pipes.

> **Remember**
>
> Do not use too much solder as it will look unsightly when it cools. Only use lead-free solder (a mixture of tin and copper) on hot and cold water supplies. The spool should be clearly marked 'lead-free'.

Table 4.11 summarises what types of fitting can be used on the various grades of copper tube.

Jointing method	R250 half-hard lengths	R290 hard lengths	R220 soft coils	R220 soft coils microbore
Capillary	✓	✓	✓	✓
Type A compression	✓	✓	✓	✓
Type B compression	✓*		✓	

*Not underground

Table 4.11: Correct fittings for different grades of copper tube

Flux

Soldering flux used in the plumbing industry is usually in paste form. It can be active (it cleans the pipe when you apply heat) or non-active (it does not clean the pipe when heated). Flux is usually chemical-based, so you must read the COSHH advice on the label (see Chapter 1 page 7). When using flux, you should make sure it is non-acidic, non-toxic and WRAS-approved (approved by the Water Regulations Advisory Scheme). Flux is applied to:

- clean the pipe
- prevent oxidisation while heating
- assist with the **whetting** process
- float away impurities
- assist heat transfer through the joint.

Capillary fittings

As with compression fittings, there is a vast range of soldered capillary fittings which come in the same two types as compression fittings: solder ring and end feed. See Table 4.11 for examples.

Push-fit joints – pressure pipework

This method should not be confused with the jointing processes used on low pressure pipework, such as uPVC and ABS, which are associated with soil and waste pipework installations. Push-fit joints are made from plastic or metal. A grab ring locks the pipe in place, and a neoprene O ring makes it watertight (see Figure 4.26). You must ensure all pipe burrs are removed as these can cut the seal, causing leaks. These fittings are bulky, so they do not look attractive where they are exposed. Take care when using these fittings, as the earthing integrity of the electrical system may be compromised.

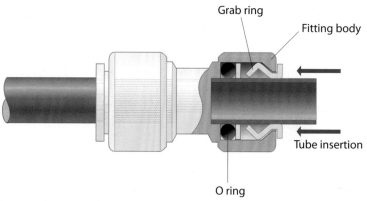

Figure 4.26: Typical push-fit joint

Press-fit copper tube fittings

This is a relatively new method of jointing copper and other materials. It is a flame-free application, which makes it particularly useful where a blowtorch cannot be used. The fittings consist of a socket with a rubber seal and stainless steel grab ring inside it. When the prepared tube is inserted, the grab ring temporarily secures it in place until the press tool is used to crimp the fitting mechanically onto the tube to make a watertight seal.

Certain steps are needed to ensure a leak-free joint, as shown on page 163.

Certain steps are needed to ensure a leak-free joint, as shown on page 163.

Key term

Whetting – sharpening using friction or grinding.

Did you know?

Materials used for pipe fittings include copper and brass. The design will depend on the specific purpose for which the pipe is intended and the ease of manufacture.

How to press-fit fittings to copper tube

1 Measure, mark and cut the copper tube to the required length.

2 Deburr the tube internally (to prevent turbulence and build-up of substances in the water), and externally (to prevent damage to the O ring inside the fitting).

3 If an exiting tube, it is best to clean the pipe with emery cloth, sandpaper or a soft scouring pad.

4 Offer a fitting up to the edge of the tube and mark the insertion depth required. When inserting the tube into the fitting, make sure the shoulder of the fitting lines up with the mark on the tube.

5 Locate the correct sized pressing jaw around the bead of the fitting and start the pressing process. This is usually automatic at the touch of a switch, although some earlier tools were manually operated.

6 Visually inspect the fitting. Some press tools have an indicator to inform you of a successful joint. If the joint is unsuccessful or the fitting needs to be removed for any reason, discard it and replace it with a brand new one.

Methods of jointing LCS pipe

For smaller installations, the two main jointing methods are threaded or compression.

Threaded joints

Jointing LCS pipe can be done by cutting threads into the end of the LCS pipe to give a BSPT, then jointing them with threaded steel or malleable iron fittings. The jointing process is as follows:

- Hold the LCS pipe securely in a pipe vice and cut to length using heavy-duty pipe cutters or a large frame hacksaw.
- Remove any burrs from inside and outside the pipe.
- **Chamfer** the end of the pipe that is being threaded to provide a leading edge for the dies to catch the pipe wall.

Key term

Chamfer – smooth a cut edge by slightly bevelling or angling it.

Figure 4.27: Pipe assembly using a vice

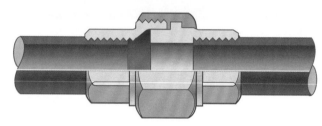

Figure 4.28: LCS union connector

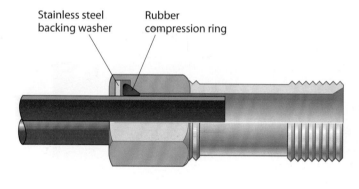

Stainless steel backing washer   Rubber compression ring

Figure 4.29: LCS compression joint. The length of tube located below the compression joint is removed, as it serves no purpose

- Apply cutting compound to the end of the pipe before starting the thread-cutting operation. The thread should be cut approximately 1½–2 threads longer than the length of the inside of the fitting.
- Wipe off the excess cutting compound.
- Then apply threaded pipe sealant (hemp and paste, PTFE sealing tape or gas thread sealing tape), to the thread. Then screw the fitting into place.
- Join the pipe and fitting using adjustable pipe grips or by using a short length of pipe as a lever (see Figure 4.27).

Because of the way screwed joints are made and installed (rotated on the pipe), it is sometimes difficult to remove or assemble lengths of pipework. Where this is the case, a union connector, which allows the pipework joint to be 'broken', should be used (see Figure 4.28).

Threaded pipe fittings for LCS

These fittings can be made of steel or malleable iron. Steel fittings can withstand higher pressure but are more expensive than malleable iron. They are manufactured to BS EN 10241 for steel and BS 1256 for malleable iron.

Compression joints

There are a number of designs for LCS compression joints, and Figure 4.29 shows a typical one. The fitting is designed to allow steel pipes to be installed without threading. Made of malleable iron, compression couplings use locking rings and seals which are tightened onto the pipe. They can be used on gas supplies and water and, although they are more expensive than threaded joints, they do save time on installation.

Methods of jointing plastic pipe

- **Fusion welding** (polythene and polypropylene): This requires specialist equipment and is used mainly for gas and water mains, so will not be discussed here.
- **Mechanical jointing**: This applies to polythene pipework, uPVC and ABS.

Polythene pipe: compression joint

Used for underground services, polythene pipe is identified by blue colour coding. It is also available, coded black, for internal use on cold water services, although for domestic situations it is mainly used underground. The joints are made using gunmetal or brass or plastic fittings.

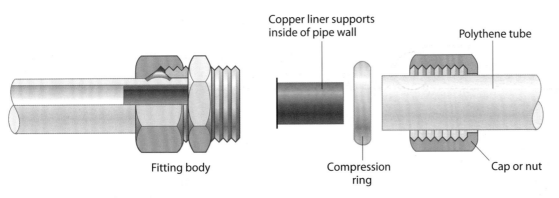

Copper liner supports
inside of pipe wall

Polythene tube

Fitting body

Compression
ring

Cap or nut

Figure 4.30: Typical polythene pipe fitting

The pipe is cut to length and deburred. The nut and compression ring are slid onto the pipe, and the copper liner inserted into it. The compression ring applies pressure, while the liner stops the plastic pipe from being squashed when the nut is tightened onto the body of the fitting. The pipe is fully inserted in the fitting, and hand-tightened, before the tightening process is completed using an adjustable spanner or other suitable smooth-jawed tool (see Figure 4.30).

uPVC and ABS: compression joints

These joints are restricted to waste pipes. Traps are a typical example, using a rubber O ring or plastic washer to make the seal.

uPVC and ABS: solvent-welded jointing

This method is used to joint both uPVC and ABS, using solvent cement. The cement temporarily dissolves the surface of the pipe and fitting, causing the two surfaces to fuse together. The joint sets initially within 5–10 minutes but will take 12–24 hours before it is fully set. This method is used primarily for joints on soil or waste pipes.

uPVC and ABS: push-fit joints

These are used mostly on uPVC or ABS soil and waste applications. The pipe is cut to length, making sure it is square and deburred. The outside edge of the pipe is chamfered to make it easier to push into the fitting. This also prevents the O ring being dislodged. A suitable lubricant (usually silicone spray or paste) is applied to the pipe and fitting, and the pipe is pushed home. Never use oil-based substances or washing up liquids, etc. as these will deteriorate the O ring and possibly the fitting. The pipe must be withdrawn from the fitting to a length of about 10–15 mm to allow for expansion (see Figure 4.31).

Push-fit connectors and pipework

These are used for low pressure plastic pipework, as shown in Figure 4.31. See also Table 4.12 on page 167 for typical jointing methods for plastic pipes.

The cutting and assembly sequence of the Hep$_2$O flexible plumbing system is given on page 166.

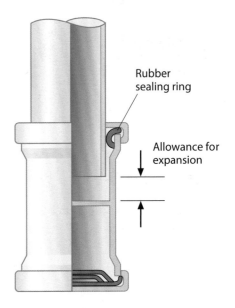

Rubber
sealing ring

Allowance for
expansion

Figure 4.31: Push-fit joint for low pressure installations

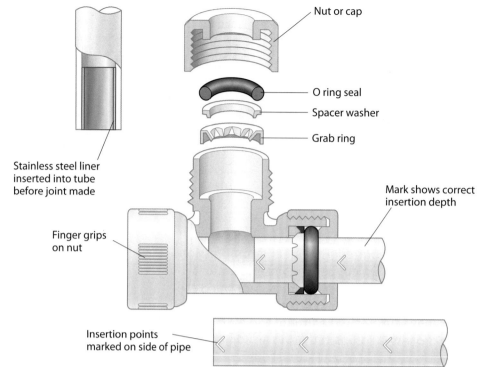

Nut or cap

O ring seal

Spacer washer

Grab ring

Stainless steel liner
inserted into tube
before joint made

Mark shows correct
insertion depth

Finger grips
on nut

Insertion points
marked on side of pipe

Figure 4.32: Typical demountable fitting for polybutylene pipe

The cutting and assembly sequence

1 Use only Hep$_2$O specialist pipe cutter to cut the pipe – never a hacksaw. Wherever possible, cut the pipe at the 'V' marks provided. To ensure a clean, square cut, rotate the pipe while maintaining pressure on the cutter until the pipe is severed.

2 Push the pipe firmly into the pre-lubricated fitting. A secure joint has been made when the end of the retaining cap has reached the next 'V' mark on the pipe. Never knock the fitting onto the pipe, or slacken the retaining cap prior to pipe insertion: these actions will not ease jointing.

3 Ensure the pipe end is clean and free from burrs and surface damage. Insert an Hep$_2$O support sleeve into the pipe end.

4 Tug back on the pipe to ensure the grab wedge engages correctly. Do not undo the retaining cap after pipe insertion.

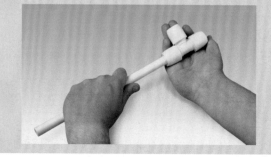

Types of plastic	Mechanical joints	Solvent welding	Fusion welding	Push-fit O ring*
Polythene	✓		✓	
Polyethylene	✓			
Polypropylene	✓		✓	✓
uPVC	✓	✓		✓
ABS	✓	✓		✓

*Discharge pipes and overflows only

Table 4.12: Typical jointing methods for plastic pipes

Identify different sealing material used for tubes

Various paste jointing compounds are available. Any jointing compound used on hot and cold water services must be for use on potable water and should be non-setting.

Before selecting a jointing compound, always make sure that it is approved by the Water Regulations Advisory Scheme (WRAS). Oil-based jointing compounds are not suitable for use with plastic cisterns or pipework – the paste will break down the plastic as both materials are oil-based. This may happen in a very short period of time.

PTFE tape is now probably the most common form of jointing to plumbing threads. Note that it should only be used on plumbing pipework systems and not gas. Fibre or rubber washers are another alternative.

Identify hazards associated with soldering copper tubes

There are several important safety points that you must observe when carrying out soldering operations. Also refer to Chapter 1, pages 30–31.

- Remove flammable objects where possible.
- Check to see whether the site you are working on is a designated place of work requiring a 'hot work permit'.
- Replace lids on any fluxes, solvents, etc. in the vicinity.
- Use a heat-resistant mat and protect any flammable materials not removed.
- Be aware of and protect any electrical equipment or cabling.
- Have a suitable fire extinguisher available.
- Check soldering equipment regularly for damage, wear and tear.
- **Always** ensure adequate ventilation while carrying out soldering work, and make sure gas equipment is turned off when not in use.

There are risks associated with inhaling soldering flux fume. Never carry out soldering activities in confined spaces unless ventilation is provided. Even in well-ventilated areas, never breathe in the fumes being produced.

Most plumbers use liquefied petroleum gas (LPG) to carry out soldering, usually propane. Propane is heavier than air and so will sink to a low level if released or leaked to the atmosphere, creating an explosive risk. Good ventilation will help prevent this risk if the unignited gas escapes for any reason.

Activity 4.4

Visit the Health and Safety Executive website and type in 'Plumbing solder fume' to find out more. Create a set of guidelines explaining how to use flux safely.

Progress check 4.05

1 List 10 fittings in general use for copper tube.

2 List 10 fittings in general use for LCS tube.

3 List 10 fittings in general use for plastic pressure pipework.

4 Identify four methods of jointing copper tube using fittings.

5 Which type of plumbing jointing compounds (not brand names) should not be used on plastic materials? Give a reason for your answer.

6 Make a list of potential hazards when carrying out soldering activities in an occupied property.

7 Why is good ventilation important during soldering?

Working practice

While carrying out a cold water storage cistern (CWSC) installation in a roof space, Gary was solvent welding the warning (overflow) pipework. However, he did not put the lid back on the solvent cement.

Later, when soldering some capillary fittings, Gary knocked over the solvent cement, which ignited. As he had no fire extinguisher, he panicked and beat out the fire with his hands. The roof space was undamaged, but Gary suffered painful burns and needed two weeks off work.

1 Make a bullet list of the safety precautions Gary failed to put in place while working.

2 Were any regulations contravened? If so, which ones?

3 The accident resulted in Gary being off work for two weeks. What would his employer need to do, if anything, in this case?

4 What should Gary's employer do to prevent a similar incident occurring?

KNOW COMMON PLUMBING PREPARATION TECHNIQUES

This section discusses work on flooring, so make sure you have the correct tools and personal protective equipment before you start any work. Refer back to Table 4.01 on pages 124–135 as necessary.

Describe methods of preparing flooring materials

Tools and equipment needed

- circular saw
- pad saw
- jigsaw
- wood chisel
- claw hammer
- screwdrivers
- nail punch
- bolster chisel
- floorboard saw
- crowbar (wrecking bar)
- hand drill and suitable wood bits

All power tools should be 110 V or battery-operated.

Personal protective equipment needed

- eye protection
- ear defenders
- dust mask
- suitable footwear
- overalls
- gloves

Access

- Check access is safe.
- Check lighting levels are adequate.
- Warn others of the work.
- Erect warning signs.

Occupied/unoccupied property

- If the property is occupied, check whether any furnishings or personal belongings need to be moved or removed from the work area. If so, inform the occupier.
- If the property is unoccupied, there may be equipment belonging to other tradespeople in the area. Ask before moving or removing anything that does not belong to you.

Measuring and marking out

- Check if it is possible to cut along the centre of a joist. This will make reinstallation easier.
- If not, allow for cutting at the sides of the joists, meaning fixing cleats will be fitted to the joists for reinstatement.
- Mark out the area to be removed.

Preparing to remove flooring

- Ensure you wear all necessary PPE at all times.
- Safely set any tools to the depth of the flooring material to be cut.
- Visually reconsider potential existing pipework and cable runs before beginning to cut.
- If possible, turn off power supplies.
- Where possible, use manual means to remove flooring rather than using electrical equipment. Safety first!

Removing flooring material

- Turn off and remove any power tools.
- Remove the flooring using the appropriate tools.
- Remove any nails or screws from joists and flooring.
- Check for damage in the floor void.

Finally

- Prepare the pipe runs, ensuring any notching meets the Building Regulations requirements and allowance is made for expansion or contraction.
- Use pipe guards to protect the pipes.
- Commission and test pipework.
- Prepare the area to replace the flooring you have removed, e.g. remove sawdust, fix cleats to joists if necessary.
- Replace the flooring using screws (for ease of removal later).
- Use a pencil to write on the flooring, indicating that pipework lies beneath the reinstated flooring.
- Remove your tools and equipment and clean and tidy the area.
- Inform the occupier and replace removed items to the area if required.

Identify risk factors to consider when removing flooring materials

Visually check the area.

- Look closely for evidence of nails or screws in the existing flooring.
- Look for electrical wall sockets, which may suggest electrical cables are below the floor.
- Check the layout of existing pipework which may be below the flooring level (e.g. heating pipework).

Describe different types of flooring materials

The two basic flooring materials are:

- tongue and groove – slotted timber (wooden) boards
- chipboard – commonly used in modern buildings.

Tongue and groove

Older buildings may have suspended wooden floors on some or all of the ground floor. Other floors will be similarly constructed on timber joists. In both cases, the flooring material is normally tongue and grooved timber flooring.

One edge of each board has a protruding tongue, while the opposite edge has a groove – a recess cut into the timber. The tongue on one board fits into the groove on the next board. In very old buildings you may find that the boards are wider and do not have a tongue or groove. The tongue and groove board was conceived to stop dust falling between the boards, and to cut down on draughts and noise between floors.

Tongue and groove flooring is squeezed together using a floorboard cramp. Sometimes this may still be evident, making removal of such boards difficult without damaging them. Always take great care when removing boarding not to damage the flooring. It's a good idea to use a nail punch to knock the floor nails (sometimes known as brads) into the joist to make removal easier.

Chipboard

Chipboard sheets have largely replaced tongue and groove in modern buildings. The sheets are normally 2400 x 600 x 18 mm, resulting in the flooring being laid much more quickly and cheaply.

The sheet edges have tongue and groove edges, for the same reasons as traditional flooring. Gaining access under the flooring presents similar problems to those described for tongue and groove. In addition, the sheet material is more fragile to remove and physically much wider than traditional floorboarding.

Describe the process of replacing flooring materials

By the end of this section you should be able to:

- describe the procedures for lifting and replacing floor surfaces
- state the requirements for cutting holes and notching timber joists
- describe the procedures for cutting holes and chases through a range of building materials
- describe how to make good masonry surfaces.

In existing properties, gaining access to work under the floor will mean lifting floor surfaces, so you need to know how to do that. Often, both on new work and maintenance work, you will be required to cut or drill holes into brickwork, blockwork, concrete and timber, so that will also be covered.

Lifting floorboards to run pipework through joists

This usually involves lifting a single length of floorboard. Lifting a full length is easier because you won't have to cut across the board.

Using a hammer and sharp bolster, carefully cut the tongue and groove joint down either side of the floorboard. Alternatively, you could use a pad saw.

Good practice requires punching down the nails to enable the board to be removed. Alternatively, you can use a wrecking bar, or draw bar, to prise up the floorboard and nails. Once the board is partially lifted, pushing it down slightly will reveal the nail heads, allowing you to remove them with a claw hammer.

 Safe working

When sawing across a joist, hold the blade at a slight angle to help you avoid the nails.

Pay particular attention to what is underneath the boards you are cutting; especially with a circular saw. And remember – never use a circular saw unless you have been trained to do so.

If you need to lift only part of a board, you will have to make one or possibly two cuts across a joist, so that when the board goes back it has a firm fixing point. If you cannot locate a joist, you will have to insert timber cleats. You can locate a joist by finding the floorboard nails.

You can make cross-cuts on a floorboard using an extremely sharp wood chisel or a purpose-made floorboard saw.

Lifting floorboards using power tools

This is done using a circular saw. Remember the following when using a circular saw:

- Use a 110 V supply only.
- The depth of the cut must not exceed the depth of the floorboard.
- If possible, remove a trial board using hand tools to check for electrical cables or hidden pipework.
- A guard must always be fitted to the saw blade.

The saw can be used to cut down the full length of the tongue and groove on each side of the board. Make a cross-cut over a joist, making sure that the blade does not hit the nails.

Cutting traps in floorboards

Cutting a trap uses the same method as removing a single floorboard, using either hand or power tools. For a trap you will need to cut more boards, but in shorter lengths (see Figure 4.33).

Replacing floorboards and traps

Screw the floorboard length or trap back into position to make future inspection or maintenance easier. Use countersunk wood screws. When refixed over pipework, the board surface should be marked accordingly, e.g. 'hot and cold pipework'.

If it is not possible to find a joist to refit the board or trap, you must use cleats to support the board end. Figure 4.34 shows a trap or board replacement over joists and using cleats.

Removing and replacing chipboard

This is more difficult than removing wooden floorboards, as chipboard is laid in wider sheets. The best way to remove a chipboard section is with a circular saw. If a power supply is not available, you can cut a section of board with a floorboard saw. Mark out the section to be removed across the board, creating guidelines to follow for the cut (see Figure 4.35). If you are using a pad saw, it is helpful to drill holes in each corner of the area to be lifted in order to start the cutting process. If you are using this method, you will need a new piece of chipboard to replace the removed section.

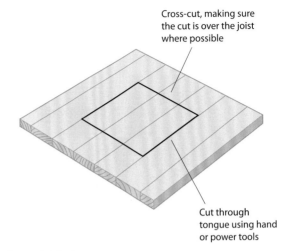

Cross-cut, making sure the cut is over the joist where possible

Cut through tongue using hand or power tools

Figure 4.33: Cutting a trap

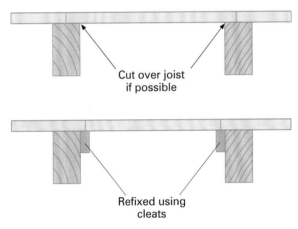

Cut over joist if possible

Refixed using cleats

Figure 4.34: Replacing floorboards using cleats

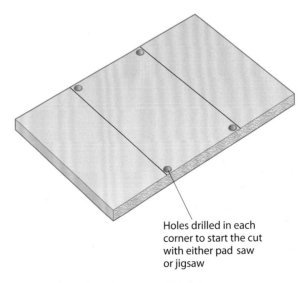

Holes drilled in each corner to start the cut with either pad saw or jigsaw

Figure 4.35: Preparation for removing chipboard

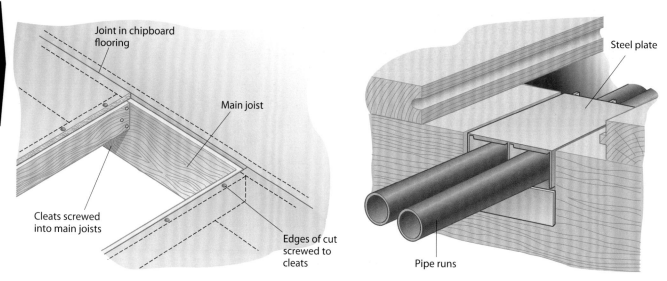

Joint in chipboard
flooring

Main joist

Cleats screwed
into main joists

Edges of cut
screwed to
cleats

Steel plate

Pipe runs

Figure 4.36: Refixing chipboard

Figure 4.37: Pipe guards

As with floorboards, you should screw chipboard back into position
(see Figure 4.36).

Pipe guards

Pipe runs under floors should be marked. A more effective way of
protecting pipes that pass through joists is to use pipe guards. Figure 4.37
shows a typical example of their use.

Identify the requirements to consider when carrying out notching

Notching means cutting grooves in joists to allow pipework runs under
timber floors. In many newly constructed dwellings, you will not be allowed
to drill or notch the joists as this would have a serious impact on the

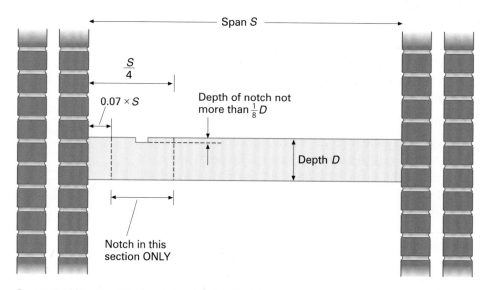

Span S

$\frac{S}{4}$

$0.07 \times S$

Depth of notch not
more than $\frac{1}{8}D$

Depth D

Notch in this
section ONLY

Figure 4.38: Building Regulations requirements for notching joists

structure of the building. In this type of dwelling, the joists will have knock-outs in them which have been designed for use by the manufacturer.

The preferred method of notching is to drill the joist in the centre of its depth, as this is where there is least stress. However, in practice, apart from when using plastic hot and cold water supply pipe, this tends to be impractical. The main thing to remember, for either notches or holes, is not to weaken the joists. This also applies to the distance from the wall where the joist is notched or drilled.

The Building Regulations set out requirements for notching or drilling joists, and you must follow these at all times (see the example in Figure 4.38).

Worked example for notching a joist

Joists are normally cut by hand, using a hand or floorboard saw. They are cut to the required depth and width, and the timber notch removed using a wood chisel and hammer. The width should be enough to give freedom of movement to allow for expansion and contraction, particularly in the case of hot water supply or central heating pipework.

Consider a joist 200 mm deep (D) and 2.5 m long (the span, S).

Any notch must have a maximum depth of $\frac{D}{8}$ mm. For our joist the depth of the notch is $\frac{200}{8} = 25$ mm.

The minimum distance from the wall is $7 \times \frac{S}{100}$ mm.

For our joist, this is $\frac{(7 \times 2500)}{100} = 175$ mm.

The maximum distance from the wall is $\frac{S}{4}$ mm from its bearing, giving a maximum span for our joist of $\frac{2500}{4} = 625$ mm.

Drilling holes in timber floor joists

There are similar requirements laid down for drilling holes in joists, although the calculation figures are slightly different (see Figure 4.39):

- The maximum hole diameter is one-quarter of the depth of the joist.
- Holes should be positioned no closer together than 3 times their diameter.
- The minimum distance from the wall is 0.25 × the span.
- The maximum distance from the wall is 0.4 × the span.

> **Remember**
>
> When drilling a joist, make sure that what you are doing will not weaken it.

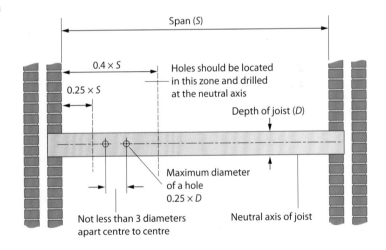

Figure 4.39: Acceptable positions for drilling holes in floor joists

Describe installation techniques for fitting pipework in concealed locations

Ducting

The term duct can mean several things in construction. In this context it means a void that conveys air in warm air and air conditioning systems or that is used to describe a space (or void) to locate building services – water, gas, telecommunications, heating pipework and electrical services.

Chapter 4

Key term

Fabric – what the building is made of: brickwork, concrete, blockwork or timber.

Access to ducting is gained via screwed-on panels or, in large installations, via access plates in the floor. These may be large enough to crawl or even walk through, but you will need specialist training to work on these installations as they are classified as a confined space.

Clipping

All pipework must be clipped in order to secure it to the **fabric** of the building. The type of clip, bracket or fixing will depend on the:

- type of building
- weight of pipework
- pipework material
- building fabric where the fixing is to be made.

Different pipework materials need to be clipped at different centres, horizontal centres being different from vertical ones, depending on the rigidity of the pipe when installed.

Sleeving

Where a pipe passes through the building structure, it must be sleeved. You can do this by drilling the hole one diameter larger than the pipe to be installed and inserting a piece of tube cut to the thickness of the wall concerned. You can then slide the smaller tube into the larger one. Sleeves:

- protect pipes from damage caused by movement within the building structure
- help prevent surface damage to walls caused by expansion or contraction
- protect pipes from corrosion. For example, if copper pipes come into contact with mortar, over time the mortar (which contains cement) will cause the copper to corrode and then leak.

After testing has been carried out, the sleeve must be sealed to prevent draughts, smells or insects gaining access to the adjoining room. The sealant should also be fire retardant, as fire could spread between rooms via an unprotected sleeve.

Cutting holes in the building fabric

Power tools

The selection of drill will depend on the job you are doing (see Table 4.13). Also refer to Tables 4.01 (pages 124–135) and 4.08 (pages 144–146) for other tools and equipment needed.

Rotary hammer drill	500 watt	850 watt	1400 watt diamond core drill
Drill bit required for brick/block	30 mm	42 mm	152 mm
Drill bit required for concrete	24 mm	30 mm	120 mm

Table 4.13: Choosing appropriate drill bits

- **Drill bits** are designed for specific tasks and are purpose-made for brick, block, concrete, steel or wood. A diamond core drill uses either diamond- or tungsten-tipped bits. Core drills are excellent for drilling through brickwork, blockwork or concrete where you need to pass large-diameter waste or soil pipes or flue pipework. They provide a much neater finish to the job, and less making good is required than with hand tools.
- **Wood bits/wood-boring bits** are useful for drilling holes in joists or other timber constructions for the passage of pipes.
- **Hole saws** are handy when drilling through kitchen units in order to pass waste pipes through the side or back of cabinets. They are also used to drill holes in plastic cold water storage cisterns.

Hand tools

Hand tools for brickwork, blockwork and concrete include cold chisels, brick bolsters (preferably with guard), plugging chisels and club hammers. Tools for timber include claw hammers, wood chisels and brace and bits. However, power tools have generally replaced the brace and bit. A brace and bit, or cordless drill, is used to bore holes through timber in the same way that power tools are used. However, power drills can use much larger-diameter bits and do not require as much hard work.

Hand tools such as club hammers, cold chisels and bolsters are used to cut holes and chases in brickwork, blockwork or concrete. You must take care to cut out only the minimum of material required for the passage of the pipe, in order to keep making good to a minimum. Plugging chisels are used to chase out mortar between brickwork; plumbers do this to let in sheet lead flashings or timber-fixing plugs.

Making good

For most jobs, a mortar mixture of four parts sand to one part cement will be adequate for pointing any brickwork joints disturbed while doing the job, and for making good the gap around the pipe penetration. Mastic sealant, either clear or close to the colour of the pipework, can be used as an alternative.

Working practice

The occupant of a new-build house has complained about creaking floors and noises from beneath the bedroom floor when the central heating system is running.

You walk around the bedroom and notice that the flooring in the centre of the room gives slightly under your weight. You lift the carpet and see that the chipboard has been cut poorly to form a trap in the floor. The original nails are still showing and it is obvious that the boarding has not been cut over a joist.

You try your best but the nails which have been used to secure the trap cause even more damage to it. When the chipboard trap is removed, the following issues are revealed.

- Cleats have been used to trim the joist. These cleats have been nailed to the sides of the joist using nails just slightly longer than the thickness of the cleat.
- The notches cut into the joist are only just wide enough to accept the two 15 mm central heating tubes and they are touching each other. This is the cause of the noises when the central heating is running.

1 Produce a comprehensive list identifying the mistakes the original installer has made. State what should have been done during the original installation to correct these errors.

2 Have any regulations been contravened? If so, which ones?

3 Who is to blame for the shoddy work you have discovered?

Progress check 4.06

1 What should you do when preparing to remove wooden flooring materials?

2 Make a list of all the safety requirements you must consider when removing wooden flooring materials.

3 Name the two most common types of wooden flooring found in domestic premises.

4 The preferred method of cutting flooring is along the centre of the joist. If this is not possible, where should you make the cut and what must you do to the joist before refitting the flooring material?

5 When refitting flooring materials, how should you secure them? Give reasons for your answer.

6 Where would you find information about the position and depth of notches in joists to locate pipework?

7 Briefly explain what the term 'ducting' means in the context of plumbing work.

KNOW SYMBOLS USED FOR IDENTIFYING PLUMBING PIPEWORK AND FITTINGS

Identify different plumbing symbols

Drawing symbol	Reference and explanation	Drawing symbol	Reference and explanation
————	**Visible pipe** *Surface mounted pipework*	– – – – – –	**Pipe behind duct** *Pipework hidden from view behind or in a duct*
—··—·—··—	**Pipe at high level** *Pipework running at high level around a room*	··—·—·—·—	**Pipe above suspended ceiling** *Pipework hidden from view in a suspended ceiling*
— — — —	**Soil waste (below ground)** *Underground soil and waste pipework system*	🔵 WATER	**Meter** *Water meter*
—·—··—·—	**Cold water** *Cold water system pipework*	—··—··—	**Hot water** *Hot water system pipework*
——···——	**Hot water (return)** *Hot water secondary return pipework*	WH	**Water heater** *Water heater or storage cylinder*
CW	Clothes washer	DW	Dishwasher
—G——G—	Gas system pipework	GAS	Gas meter
➤	**Direction of flow** *Direction of water flow through a pipework system*	●	**Pipe perpendicular to plan detail** *Pipework from high level to low level or vice versa*
—‖—	**Union** *Type of pipework joint, usually associated with LCS tubes*	—‖—	**Flange** *Type of pipework joint, usually associated with LCS tubes*
]	**Capped end** *Pipework fitted with cap or stop end (this will create a dead leg)*	⊣	**Hose connection** *Hose union connection for the purpose of draining the system (plans or elevations)*
⋈	**Drain-off valve** *Drain-off valve with hose union connection for the purpose of draining the system*	⋈	**Valve** *Isolation or service valve*
⋈	**Three-port valve** *Valve associated with central heating systems. Flow water enters the bottom connection; one of the top ports feeds the hot water system, the other feeds the heating system.*	⋈	**Straight two-port valve** *Valve associated with central heating systems to control the flow of water electronically to hot water storage, central heating or a heating zone*
⋈	**Angled two-port valve** *Valve associated with central heating systems to control the flow of water electronically to hot water storage, central heating or a heating zone*	⋈	**Wheel-headed valve** *Usually a gate valve, used on the low pressure feeds from a CWSC or to act as an isolation or service valve on low pressure feeds to appliances*
⋈	**Pressure reducing valve** *Used to reduce mains water pressure to suit equipment, or for other similar purposes*	⋗	Strainer
⋈	**Lockshield headed valve** *A valve intended for use by the installing or maintenance plumber and not intended for use by the consumer*	⌐	90 degree elbow

Table 4.14: Pipework symbols

Continued ▼

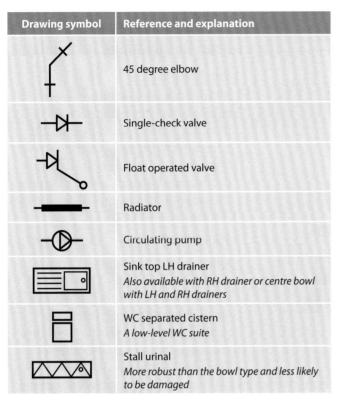

Drawing symbol	Reference and explanation	Drawing symbol	Reference and explanation
	45 degree elbow		Tee
	Single-check valve		Draw-off point (tap)
	Float operated valve		Pressure relief valve
	Radiator		Heated towel rail
	Circulating pump		Sink
	Sink top LH drainer *Also available with RH drainer or centre bowl with LH and RH drainers*		Wash basin
	WC separated cistern *A low-level WC suite*		Bowl urinal *Sometimes called a 'bean' urinal*
	Stall urinal *More robust than the bowl type and less likely to be damaged*		Bath

Table 4.14: Pipework symbols (continued)

Identify colour coding of plumbing pipes and tubes

The colours in Table 4.15 are indicative only. For true reference colours, please refer to BS 1192:2007.

Pipe contents	Basic ID colour	Colour code ID			Basic ID colour
Water					
Drinking	Green	Auxiliary blue			Green
Cooking (primary)	Green	White			Green
Boiler feed	Green	Crimson	White	Crimson	Green
Chilled	Green	White	Emerald	White	Green
Central heating <100°C	Green	Blue	Crimson	Blue	Green
Central heating >100°C	Green	Crimson	Blue	Crimson	Green
Cold down service	Green	White	Blue	White	Green
Hot water supply	Green	White	Crimson	White	Green
Fire extinguishing	Green	Red			Green
Others					
Natural gas	Yellow ochre	Primrose			Yellow ochre
Compressed air	Light blue	Light blue			Light blue
Steam	Silver grey	Silver grey			Silver grey
Drainage	Black	Black			Black

Table 4.15: BS 1192:2007 pipework colour codes

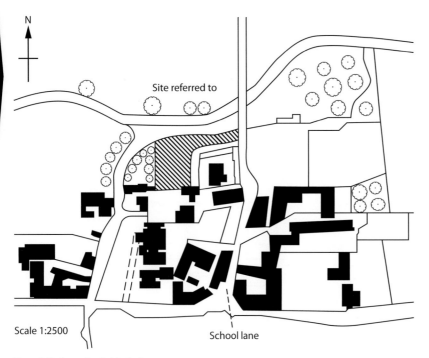

Scale 1:2500

Figure 4.40: Example of a block plan

Describe drawing methods used for plumbing installations

Block plans

A block plan is sometimes referred to as a location plan or even a site plan. The scale is small to enable a large area to be reduced onto a piece of paper. A scale rule is used when producing drawings of buildings and the position of the appliances etc. (see Figure 4.40).

Floor plans

A floor plan – sometimes referred to as just a plan – does exactly as it states. It is a plan view (a bird's eye view) of the layout of a building, showing the location of everything you need to know (see Figure 4.41). Plans are also available showing the exact location of fitments: baths, sinks, WCs, washbasins, radiators, boilers, etc. We need all this exact information to estimate the costs of work and, once given the go-ahead, to know where things have to be placed.

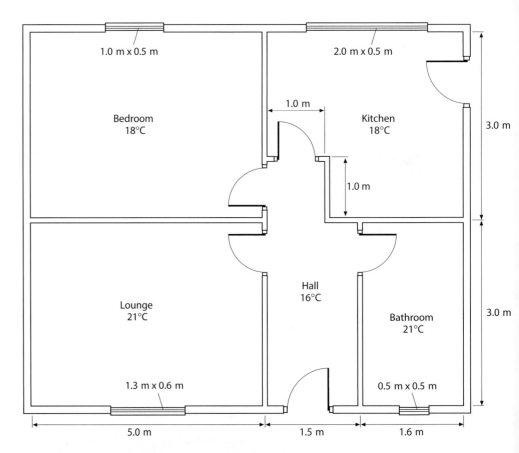

Figure 4.41: Example of a floor plan

Isometric views

Isometric projection is a means of visually displaying three-dimensional (3D) objects in two dimensions (see Figure 4.42). The object is turned (tilted) through 30 degrees. An isometric view of something – for example a house – provides us with a much more in-depth look at the outside of the house as well as the inside layout.

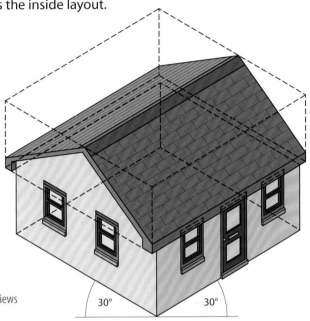

Figure 4.42: Isometric views are tilted through 30°

30°   30°

Working practice

You are installing a bathroom in a recently sold new-build property. The new owners arrive and ask you to stop work because they want to change the layout of the appliances.

1 Who decides where items such as bathroom fittings are to be positioned in a room?
2 How is this information usually conveyed to the installer?
3 Should you alter the bathroom layout as the owners ask? If not, why not?
4 If you decide not to meet their request, who should you advise them to contact?
5 Should you tell anyone about the incident? If so, who?
6 If you were to alter the bathroom layout, what could be the potential repercussions:
   •   to you?
   •   to your employer?
   •   to the owners?

Progress check 4.07

1 Which symbol represents a pipe located above a suspended ceiling on a drawing?
2 Which symbol indicates a drain-off valve on a drawing?
3 What would be the contents of a pipe that was colour-coded green/auxiliary blue/green, according to BS 1192:2007?
4 According to BS 1192:2007, how many times should the colour code ID be used on each label, once or twice?
5 What information is found on a block plan?
6 What is the term for a drawing produced by viewing through a 30 degree angle?

Knowledge check

1 Which of the following is **not** a consideration when selecting tools?

a quality
b safety
c appearance
d cost

2 Which one of the following bending machines is used to bend low carbon steel tubes?

a hydraulic bender
b mini-bore bender
c freestanding bender
d scissor bender

3 Which of the following statements about tool and equipment care is incorrect?

a Keep tools in good condition.
b Always use the right tool for the job.
c Keep tools which should be sharp blunt for safety.
d Secure tools and equipment at all times.

4 Which of the following references identifies general purpose, half-hard copper tubes?

a R220
b R250
c R290
d Table W

5 What is the correct colour coding for heavy grade LCS tube?

a white
b brown
c blue
d red

6 Which of the following statements describes the threads on LCS pipework and fittings?

a diminishing thread
b British Standard Pipe Thread (BSPT)
c parallel thread
d tampered thread

7 Which of the following statements about safe soldering is incorrect?

a Leave the torch running to save re-ignition.
b Remove flammable objects where possible.
c Use a heat-resistant mat and protect any flammable materials not removed.
d Have a suitable fire extinguisher available.

8 What should you use to remove the burr from the inside wall of copper tube?

a pipe slice
b wood chisel
c spiral ratchet reaming tool
d pipe reaming tool

9 On which type of material would you use a brass Munsen ring, a brass two-piece standoff or a single plastic pipe clip?

a galvanised LCS tube
b black LCS tube
c R250 copper tube
d plastic barrier pipe

10 Which type of drawing will show the position of appliances and pipework?

a block plan
b floor plan
c location plan
d sketch

Cold water systems

This chapter will cover the following learning outcomes:

- Know the requirements for water distribution to domestic dwellings

- Understand the requirements of the cold water supplies into domestic dwellings

- Know the components used in domestic cold water systems

- Understand the requirements for pipework installations in domestic cold water systems

- Understand the key requirements of testing and decommissioning of domestic cold water systems

- Understand the basic maintenance requirements of domestic cold water systems

Key term

Wholesome water – good-quality water for human consumption.

Introduction

Without water humans would not exist. Our bodies are well over 50 per cent water, and our brains are about 85 per cent. We need to drink an average of around two to three litres per day to remain healthy and to prevent dehydration.

Despite constant complaints about rain, it is vital to the Earth's water cycle, which is a gigantic natural purification system. Many industrial processes and technological advances could not take place without water.

Water used for cooking, washing and for leisure activities needs to be of high quality. The UK is lucky in having a clean water supply to virtually every building, and stringent regulations help ensure that this remains the case. Central heating systems also use water – and keep plumbers busy. This chapter will help you gain an understanding of installing and maintaining cold water systems.

KNOW THE REQUIREMENTS FOR WATER DISTRIBUTION TO DOMESTIC DWELLINGS

The water we use must be clean and fit for purpose in order to prevent unpleasant and sometimes lethal diseases. This section introduces the rules and regulations governing water treatment and distribution. It also covers sources of water and its journey from source to consumer.

Identify the key sources of information related to the installation of cold water systems

Water Supply (Water Fittings) Regulations 1999

These are national regulations that apply to England and Wales, made by the government's Department for Environment, Food and Rural Affairs (DEFRA). They cover water installations and require water companies to supply **wholesome water**. They replaced the Water Byelaws, which were made in and applied to local areas.

The purpose of the Water Regulations is to prevent:

- contamination of a water supply
- the waste of water
- the misuse of a water supply
- undue consumption of water
- erroneous measurement (fiddling the meter).

The regulations are also designed to permit the introduction of new products and ideas, as well as supporting environmental awareness.

Enforcing the regulations

The water supplier is responsible for enforcing the regulations. Under the Water Regulations, water companies are encouraged to set up approved contractor schemes. Approved contractors must certify that the water fittings installed comply with the regulations.

Water Regulations Advisory Scheme (WRAS)

Formerly known as the Water Byelaws Scheme, WRAS carries out fitting testing and advises on water regulations. The scheme produces a *Water Fittings and Materials Directory* of all approved fittings. This is an important guide for anyone aiming to comply with or enforce the Water Regulations. Products approved by WRAS carry the symbol shown in Figure 5.01.

Building Regulations 2000

The Building Regulations 2000 Part G covers cold water systems, including the requirements of the supply, and water efficiency of fittings and appliances. The Regulations' *Water Efficiency Calculator for New Dwellings* also states that the consumption of wholesome water should not be more than 125 litres per person per day. The Building Regulations are enforced by local authorities.

Manufacturer's instructions

You should always follow the manufacturer's instructions as they detail approved methods of installation, repair or maintenance.

- The manufacturer's instructions will override all other documents you may use concerning a particular component.
- The instructions will comply with all current regulations.
- If manufacturer's instructions are not available, then you should follow British Standards, Water Regulations and Building Regulations information.
- Manufacturers are not liable if their goods have not been installed, serviced or maintained correctly – in this case, any insurance or warranty will be invalid.

Describe the rainwater cycle

Water is clear, tasteless and colourless. It is defined in chemical terms as a compound of the two gases, hydrogen and oxygen, in the ratio of two parts hydrogen to one part oxygen (H_2O).

Where does water come from?

Water evaporates from the sea, rivers, lakes and the soil. It forms clouds containing water vapour, which condenses and falls as rain. Some of the rainwater runs into streams, rivers and lakes. Some soaks into the ground, where it may collect temporarily and evaporate, or form natural springs or pockets of water that can be accessed by wells. This sequence of evaporation and rainfall is called the water cycle (see Figure 5.02 on page 184).

The classifications of water

The most important property of water is probably its ability to dissolve gases and solids to form solutions; this is referred to as its solvent power and has a bearing on how soft or hard the water eventually becomes. Solvent power can be a problem as it corrodes and blocks up systems and components.

Did you know?

Although there are no legal requirements for people working on water services to be qualified, anyone who carries out such work can be prosecuted for an offence against the Water Supply (Water Fittings) Regulations. If convicted of an offence, they can be fined.

Figure 5.01: WRAS-approved product symbol

Remember

BS 6700 is the main British Standard detailing the installation requirements for hot and cold water systems – so you will come across it a lot.

Chapter
5

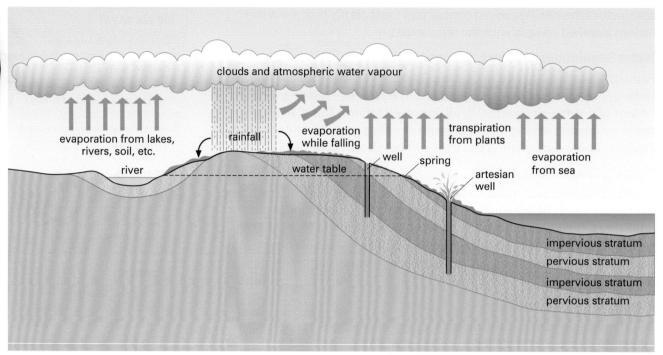

Figure 5.02: The rainwater cycle

Figure 5.03: Limescale build-up in pipework

Water type	Dissolved gases (mg/litre)
Soft	0–50
Moderately soft	50–100
Slightly hard	100–150
Moderately hard	150–200
Hard	200–300
Very hard	Over 300

Table 5.01: Water hardness scale

Water is classified according to its hardness. Hard water can result in limescale build-up, as seen in Figure 5.03. This can be tackled using water treatment methods, which are covered on pages 275–277 of Chapter 6: *Domestic hot water systems*.

The hardness of water is measured in parts per million (ppm) or milligrams (mg) per litre of dissolved gases. Table 5.01 summarises how hardness is classified. This depends on the factors shown in Figure 5.04.

Describe the different sources of water supply

Water for public consumption comes from two main sources, which depend on rainfall:

- **surface sources:** upland surface water, reservoirs, rivers and streams
- **underground sources:** wells and springs.

Surface sources

Upland surface water

This category covers lakes, streams and natural and impounding (artificial) reservoirs. Such sources are mostly found in northern parts of the UK where the landscape is hilly or mountainous. Here lakes form naturally and the damming of streams allows the formation of impounding reservoirs. The quality of upland surface water is good because it is generally free from human or animal contamination. It is usually classified as soft as it runs directly off the ground surface and into the water source, so it is not affected by passing through a particular soil type. Where water comes into contact with peat, it can become acidic.

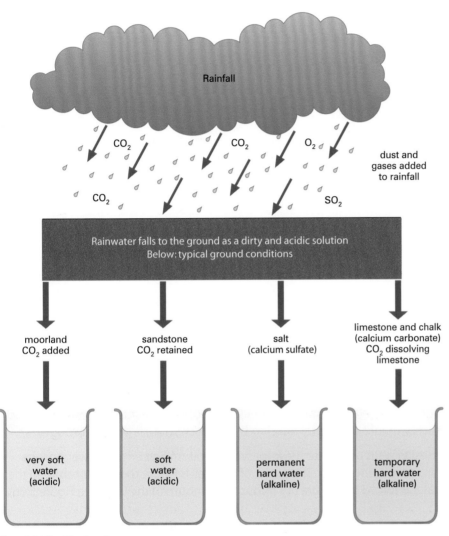

Figure 5.04: Classification of water

Artificial reservoirs

These are built to meet ever-increasing demands for water for domestic and industrial use. They are also used in areas lacking sufficient natural resources. Reservoirs are created by flooding low-lying areas of land, normally by damming a water course. The water is classified as soft.

Rivers and streams

The quality of water from rivers depends on the location. Water from moorland rivers and streams tends to be relatively wholesome compared with water from further downstream, where it could be polluted by natural drainage from farmyards, road surfaces and industrial waste. Water from upland river sources is generally soft compared with that of the lower reaches, which is usually hard.

Underground sources

Wells and boreholes

Before the existence of water authorities (water companies), wells supplied water to dwellings or small communities. Wells can be shallow or deep.

Shallow well

Deep well

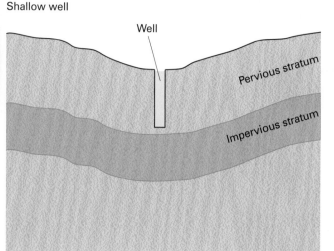

Figure 5.05: Shallow and deep wells

However, this classification may not refer to the actual depth of the well, but to whether or not it penetrates the first **impervious** stratum (layer) of the Earth (see Figure 5.05).

Some water companies still retain deep wells to back up supplies from other sources, or as a standby in case of drought. A shallow well has a greater risk of contamination, whereas water from a deep well should be pure and wholesome. Water from deep wells and bore holes is usually of quite good quality thanks to natural filtering by the rock strata it passes through.

Artesian wells penetrate the impervious stratum and enter a lower **porous** zone containing water. The outlet of the well is below the **water table**, so the water is forced out by gravity through the mouth of the well (see Figure 5.06).

> **Key terms**
>
> *Impervious* – does not allow fluids to pass through.
>
> *Porous* – allows fluids or air to pass through.
>
> *Water table* – the natural level of water in the ground.

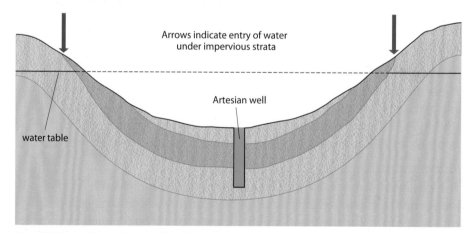

Figure 5.06: Head of water forces supply out through artesian well

Springs

The quality of spring water depends on the water's route to the surface. If it has travelled for a long distance through rock formations, it will probably be free from contamination but is likely to be hard.

Table 5.02 summarises the water sources and their levels of contamination before treatment.

Classification	Source	Contamination level
Wholesome	• spring • deep well • upland surface	Very palatable
Suspicious	• stored rainwater • surface water from cultivated land	Moderately palatable
Dangerous	• river • shallow well	Non-palatable

Table 5.02: Summary of water source classification

Water storage and supply

Water is becoming more costly and water shortages during the summer are increasing. This places more importance on finding ways to save and reuse water. There are two main methods:

- **rainwater harvesting** – collecting rainwater
- **grey water harvesting** – using waste washing water.

Rainwater and grey water are not considered wholesome. However, they can be used for flushing toilets, washing cars and watering gardens. This is covered in more detail later in the chapter.

Water companies store water either in its untreated state in impounding reservoirs or lakes or as wholesome water in service reservoirs. They usually aim to store enough drinking water in service reservoirs to maintain supplies for about 24 hours in an emergency. This safeguards against failure of pumps or **water mains** and allows time to repair any faults before supplies run out.

Identify treatment methods of water supply prior to its distribution to properties

The water companies are responsible for ensuring that the water they supply is fit to drink, so all water must be treated before it enters the supply system. Its treatment depends on its source and the impurities it contains. Some impurities are actually essential to health and will be retained; others are harmful and must be removed.

Deep wells and bore holes

The quality of water from deep wells and bore holes is already quite good, thanks to the natural filtering process as the water passes through the rock strata. The only treatment needed is **sterilisation** to keep both the water and the supply pipework to our homes free from bacteria.

Rivers, lakes and impounding reservoirs

Rivers, lakes and impounding reservoirs are our main sources of supply. However, this water is usually dirty and polluted, so treatment needs to be extremely thorough. The treatment process usually involves the water being strained, filtered, chemically treated and sterilised.

Key terms

Palatable – pleasant to taste.

Water mains – the network of pipes that supply wholesome water to domestic and commercial properties.

Sterilisation – purification by boiling, or, within the industry, dosing the supply with chlorine or chlorine ammonia mix.

Treatment methods

Water treatment involves several distinct processes, each designed to remove different types of impurity or contaminant. These steps are described in Table 5.03 and illustrated in Figure 5.07.

Method	Purpose/process
Screening	Removes floating debris from raw water sources. **Primary:** removes leaves, wood, dead creatures, waste paper, etc. **Secondary:** uses fine filters or micro-strainers to remove suspended algae and plankton.
Coagulation	Removes dirt or particles missed by screening. Uses chemicals, which may include alum. Matter is collected as 'floc', which attracts other impurities. The floc becomes heavy and is then removed during the next process.
Sedimentation	Water is allowed to settle in huge tanks. Floc and remaining heavier suspended matter (sludge) sink to the bottom, leaving clear water on the surface. Sludge is collected and removed from the tank for safe disposal.
Filtration	The water is passed through filters to ensure removal of particles that have escaped the previous processes. Filters are usually layers of sand, gravel and charcoal, but other systems are also available.
Aeration	Removes or reduces unwanted compounds, including hydrogen sulfide, carbon dioxide and traces of iron.
Disinfection	Probably the most important of the water treatments. The water is finally treated to kill off or deactivate any pathogenic bacteria. Various methods used include chlorination, chloramines, ultraviolet (UV) light and nanofiltration.
Ionisation	The acid and alkaline content found in water are segregated by subjecting the water to electrolysis using the natural electrical charge found in magnesium and calcium ions. This process is normally carried out by specialist equipment in the consumer's property.

Table 5.03: Summary of water treatment methods

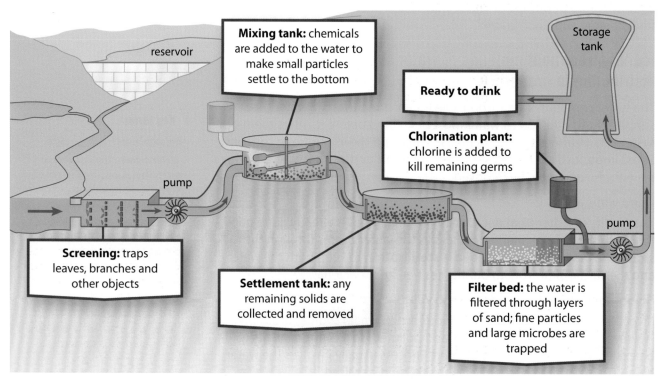

Figure 5.07: Making water safe to drink

Describe typical distribution pipework systems from treatment works to properties

Water is supplied to our homes via a network of pipes known as mains. Mains pipes vary in diameter depending on their purpose and demands on the supply. Figure 5.08 shows an example of a typical water supply system layout.

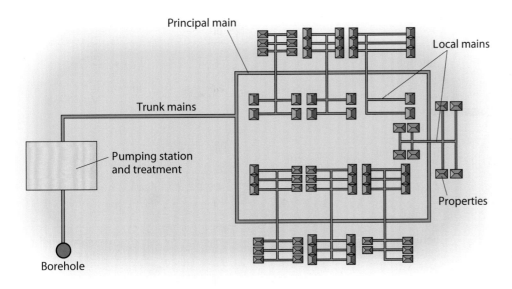

Figure 5.08: Mains water distribution from source to consumer

Figure 5.09: HPPE trunk mains pipe is coloured blue

Treated water leaves the pumping station to be stored in a **service reservoir**. The service reservoir is protected from external contamination and will either be found elevated (low-lying areas) or below ground (in high ground areas). Both locations are to give the water a **head** of pressure.

The water travels from the storage reservoir through trunk mains. These are large-diameter, high-performance polyethylene (HPPE) pipes. This is important because water in dead legs may become stagnant, potentially causing a health risk.

When the trunk main reaches the area to be supplied with water, a grid formation is laid using principal mains and local mains supply pipework. The grid formation ensures that no **dead legs** are formed. The grid also provides an even distribution of pressure and availability.

Finally the treated wholesome water from mains supply is conveyed to the property through an underground service pipe (see the following section). The local mains provide the final leg of the supply of water to homes.

Mains connection

Water mains may be constructed of asbestos cement, steel, PVC or cast iron. PVC is now used extensively on new installations and mains replacements. The connection to the mains is the responsibility of a water company. New connections to existing mains, operating under live pressure, require the use of specialist equipment.

> **Key terms**
>
> *Service reservoir* – where cleaned and treated water is stored ready for use by the consumer. The water is protected from contamination, including sunlight.
>
> *Head* – the vertical distance between the reservoir and the consumer outlet. This determines the water pressure at the consumer outlet. Water is stored at a higher level than the consumer outlet and is delivered by gravity in most areas, rather than being pumped.
>
> *Dead leg* – a water pipe with no draw-off point. In the case of hot water supply, a pipe over the recommended length.

Key term

Ferrule – a metal fitment used to connect to a main to allow isolation.

Figure 5.10 shows an example of a connection to a cast iron main using a mains tapping machine, where the new connection can be made without the need to isolate the live main.

The connection to the main is made using a **ferrule**, which is in fact an isolating valve. Once the threaded section has been inserted into the mains, the valve can be isolated, allowing the mains tapping machine to be removed and the service pipework to be connected to the ferrule outlet (see Figure 5.11).

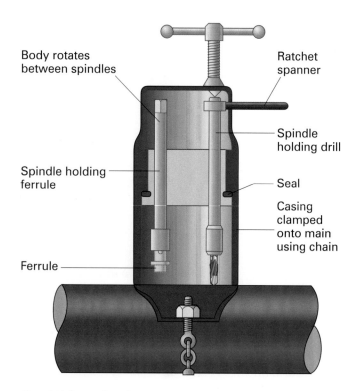

Body rotates between spindles

Ratchet spanner

Spindle holding drill

Spindle holding ferrule

Seal

Casing clamped onto main using chain

Ferrule

Figure 5.10: Section through a mains tapping machine

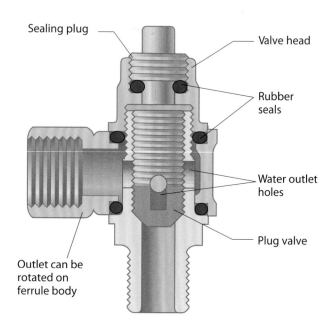

Sealing plug

Valve head

Rubber seals

Water outlet holes

Plug valve

Outlet can be rotated on ferrule body

Figure 5.11: Section through a brass ferrule

Activity 5.1

A customer has asked you to explain what precautions are taken to ensure a clean and safe supply of water from a reservoir to their property.

Produce a simple flow chart, with a brief explanation of each step, describing how water is treated and conveyed to the consumer's property.

Progress check 5.01

1   Which regulations replaced the Water Byelaws?
2   What does the abbreviation WRAS stand for?
3   Identify the five criteria that the Water Regulations are designed to prevent.
4   Briefly describe what is meant by the 'rainwater cycle'.
5   List and give a brief explanation of the different sources of water supply.
6   Produce a simple layout drawing to indicate the distribution pipework system from treatment works to the consumer's properties.
7   What does the term 'head' mean in the context of plumbing systems?
8   Define the term 'dead leg'.

UNDERSTAND THE REQUIREMENTS OF THE COLD WATER SUPPLIES INTO DOMESTIC DWELLINGS

Describe medium-density polyethylene (MDPE) pipework

MDPE pipe is used for underground water services (see Figure 5.12). It has replaced materials such as alkathene (black) or copper tube that you may come across when working on older installations. Originally the copper was unprotected and was covered by BS 2871. The tube is now protected by a blue outer sleeve (indicating water) and is covered by European Standard EN 1057. MDPE is supplied in coils of 25, 50, 100 or 150 metres to reduce the risk of unnecessary jointing underground and potential leakage.

Figure 5.12: MDPE pipe is coloured blue

Explain the difference between a communication pipe and a service pipe

Communication pipe

Once the ferrule has been fitted to a water main to enable a supply to be connected to a property, a pipe is taken from the ferrule and terminated with an external stop valve located in a purpose-made chamber (see Figure 5.13). The pipework connecting the two valves is known as the communication pipe.

The communication pipe and the external stop valve are the property of the water authority, which will therefore carry out any service or repair work required. The external stop valve is usually at the boundary of the property. The communication pipe should be installed in a **gooseneck** shape (see Figure 5.14 on page 192). This allows for lineal movement if the ground conditions change (settlement) without rupturing the pipe.

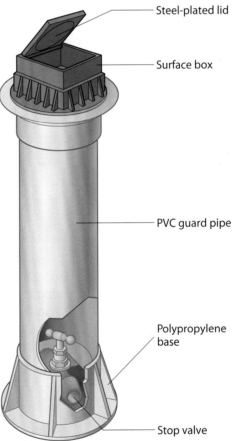

- Steel-plated lid
- Surface box
- PVC guard pipe
- Polypropylene base
- Stop valve

Figure 5.13: PVC stop valve chamber without meter

> **Key term**
>
> *Gooseneck* – looking down into the trench, the pipework should be in the shape of a letter P.

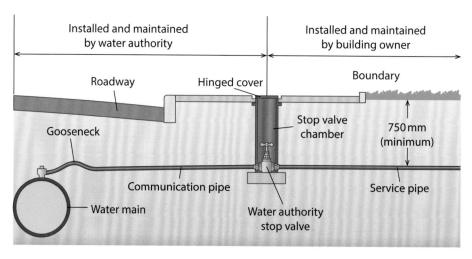

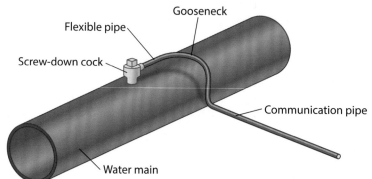

Figure 5.14: Gooseneck in communication pipe

Service pipe

The service pipe is the length of underground pipe from the external stop valve to the internal stop valve of a property. The property's owner is responsible for the care and upkeep of the service pipe. The pipe should be **snaked** in its trench, for the same reasons as the gooseneck, to allow for lineal movement in case of settlement or if the pipe will be subjected to point loads (cars, lorries, etc.).

Safe working

Lead would never be used for any form of pipework today. However, you may come across it in older installations. If you do, you should inform the customer of the hazard to their health and the pipe should be replaced.

Case study

A local authority was resurfacing a road in a village. The road roller, which weighed around 9 tonnes, was stored overnight in a local farmer's yard.

When the roadworkers arrived the next morning, the farmer told them that the farm had been without water since the previous evening. The road roller was moved away, revealing a large dent in the yard where it had been parked. A plumber was called and excavations made.

It turned out that the service pipe was made of lead and had only been buried around 300 mm below the surface. The weight of the roller had stretched the lead so much that very little water could pass through it; but the pipe had not fractured. When the service pipe was fully excavated, it was discovered that it had not been snaked, so there was no allowance for movement.

Explain the key requirements related to the supply pipework into a domestic dwelling

- Mains and service pipes should be at least 750 mm beneath the ground in order to insulate and protect them from frost damage.
- The maximum depth of cover should not exceed 1.35 m as this would make access difficult.
- The minimum size for a cold water service pipe to a dwelling is 15 mm. Service pipes of 25 mm internal diameter now usually supply new dwellings in order to meet the flow rate demands of higher-pressure mains systems.
- If the service pipe is uninsulated, it must enter the building at least 750 mm away from the outside wall surface.
- Metal pipes should be protected against corrosion, particularly from acidic soils. This can be done by using plastic-sheathed pipe, wrapping the pipe with anti-corrosive tape or installing the pipe inside a duct.

You should also consider the type of cold water system being installed and the **coincidence of demand** locally to the installation. For example, demand for water is much greater at peak times of the day, such as mealtimes. Large numbers of consumers drawing off the water supply at the same time will affect the pressure and flow rate of water available.

As the service pipe enters the building, it is placed through a duct or sleeve, which extends through the external wall of the building and is left flush with the finished floor level (FFL) inside the building. The duct or sleeve is normally 75 or 100 mm in diameter and has a slow radius bend. The purpose of the sleeve or duct is to enable ease of installation and/or withdrawal of the service pipe should this be required. The sleeve or duct must be suitably sealed at both ends to stop the ingress of vermin or insects into the property.

As the service pipe enters the building and emerges from the duct or sleeve, it must be terminated 150 mm above the FFL with a stop valve. Immediately after the stop valve, a drain valve must be fitted. Some stop valves incorporate a drain tap. Stop-and-drain valves are covered in more detail later in this chapter, on page 204.

Identify pipework isolation points

As Figure 5.15 on page 194 shows, the service pipe can be isolated from the mains by using the external stop valve at the boundary of a property. Ease of access to the stop valve is very important, so it should be located in a stop valve chamber, made from 150 mm PVC or earthenware pipe, sited on a firm base and finished off at the top with a stop valve cover. The cover is usually made from steel plate or a combination of steel plate and plastic.

> **Key term**
>
> *Coincidence of demand* – when many consumers try to use the same services (e.g. water, gas or electricity) at the same time. For example, there may be surges in electricity demand during the commercial breaks in popular television shows, as large numbers of people try to make tea at the same time. This may cause problems to the providers of these services.

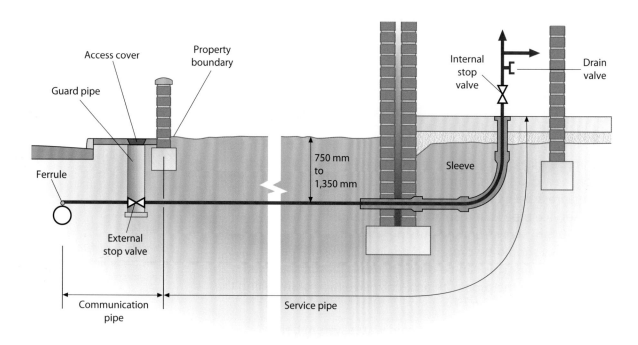

Figure 5.15: Supply pipework to a building, showing isolation points

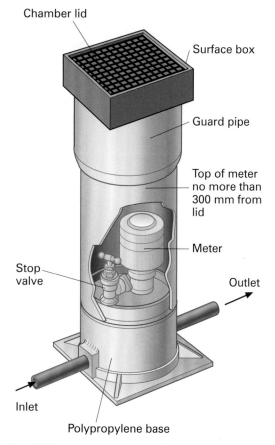

Figure 5.16: External water meter installation

Figure 5.13 on page 191 shows a PVC stop valve chamber commonly found on older installations. Figure 5.16 shows its current replacement, which is found on new installations. The guard pipe is cut to the desired length to suit the installation as part of the fitting process. The pipework from the mains up to and including the external stop valve is the responsibility of the **water undertaker**.

Identify the different types of water meter installations

Water meters in domestic premises are becoming more widespread, particularly in new housing developments. They measure how much water is used. The occupants then pay only for the water used. Water meters can be installed internally or externally and are the responsibility of the water undertaker.

- **External** water meters usually incorporate a service stop valve and usage meter, and are found on new domestic installations (see Figure 5.16). The parts come ready assembled, and the guard pipe is cut to length during the fitting process carried out by the water undertaker.
- **Internal** water meters are normally seen in commercial or larger domestic properties, but may be found elsewhere. They can be exposed or concealed (see Figures 5.17 and 5.18).

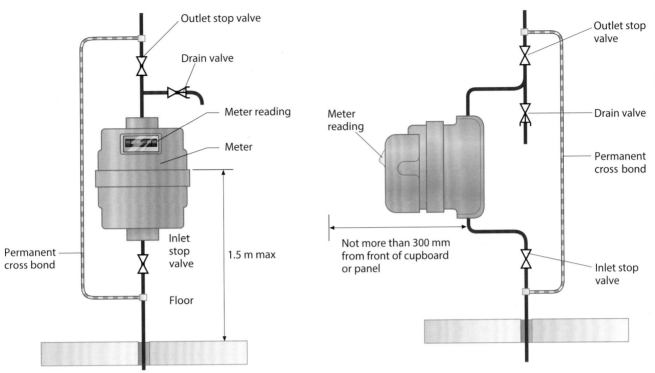

Figure 5.17: Exposed internal water meter

Figure 5.18: Concealed internal water meter

Describe key aspects of incoming mains into domestic dwellings

Key aspects such as pressure and flow rate are covered later in this chapter, on page 224. Figure 5.15 on page 194 shows a full installation from the mains to the stop-and-drain valve wherever it enters the building. A plumber's work usually starts from the external stop valve.

Identify factors that may impact on measurements

Fluctuations in the mains pipework may or may not affect cold water supply availability and pressure. If a property is experiencing problems, you should consider the following, among other factors.

- **Time of day:** Mealtimes will dramatically reduce water pressure and flow due to heavy demand, although over a fairly short period.
- **Location:** A property's distance from the main provision or even an excessive length of service pipework to the property could affect the flow, pressure and availability of water.
- **Incoming size of main:** The service pipework to the building may be inadequate. This could be because of the size used for the type of building, or it may just be undersized by modern standards.
- **Local demand:** The property may be in an area of recent housing development or industrial premises may have been built nearby without upgrades to the local water main. Change of use developments may also have an unexplained impact on the availability of water flow and pressure to existing properties.

> **Remember**
>
> Copper and lead may also be found on older buildings, where they were used to connect underground service pipework to a property.

> **Key term**
>
> *Water undertaker* – the legal term for a company that supplies domestic water.

Identify basic domestic cold water systems

Cold water systems can be direct or indirect. With the increasing demand for conservation, other water systems are being introduced to supplement our water usage. These systems may include:

- **grey water** – waste water generated by domestic activities such as bathing, laundry and dishwashing
- **rainwater harvesting** – the gathering and storage of rainwater.

Direct cold water systems

In a direct system, all the pipes to the draw-off points (sink, bath, hand basin, WC, etc.) are taken directly from the rising main and operate under mains pressure (see Figure 5.19). Direct systems are permitted in domestic properties in medium- to high-pressure areas where the supply can provide adequate quantities of water at sufficient pressure to meet the building's needs.

The direct system is the most commonly installed cold water system in domestic properties because it is cost-effective and the pressure of supply available is usually relatively high. It is a form of direct system that normally provides the cold water supply to an instantaneous water heater or a combination boiler.

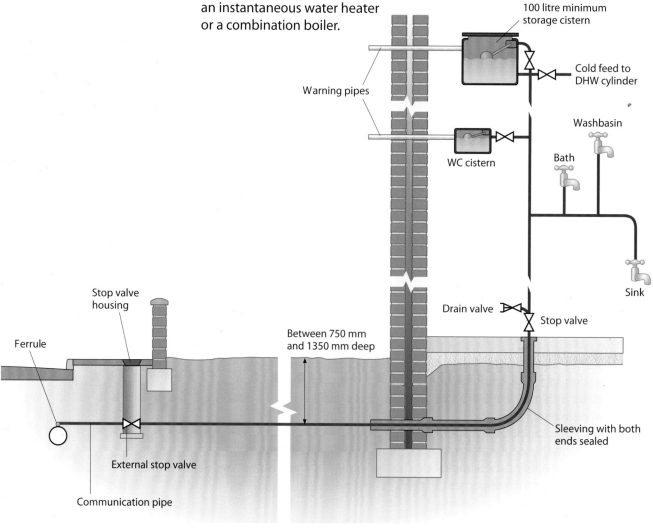

Figure 5.19: Pipework layout for a direct cold water system

Advantages of the direct system

- It is cheaper to install because:
  - less pipework is required
  - the storage cistern is smaller (100 litre minimum); BS 6700 recommends that the storage cistern to a direct system should be at least the same size as the hot water storage cylinder.
- Drinking water is available from all draw-off points.
- There is less risk of frost damage because of the smaller amount of pipework.
- Less structural support is required for a smaller cistern.

Disadvantages of the direct system

- Higher pressure may make the system noisy.
- There is no reserve of cold water if the mains or service supply is shut off or interrupted.
- There is more wear and tear on taps and valves because of the high pressure.
- There is higher demand on the mains at peak periods.

Indirect cold water systems

In an indirect system, one point – usually the kitchen sink – is fed directly from the rising main, which then supplies the cold water storage cistern (see Figure 5.20). The remaining

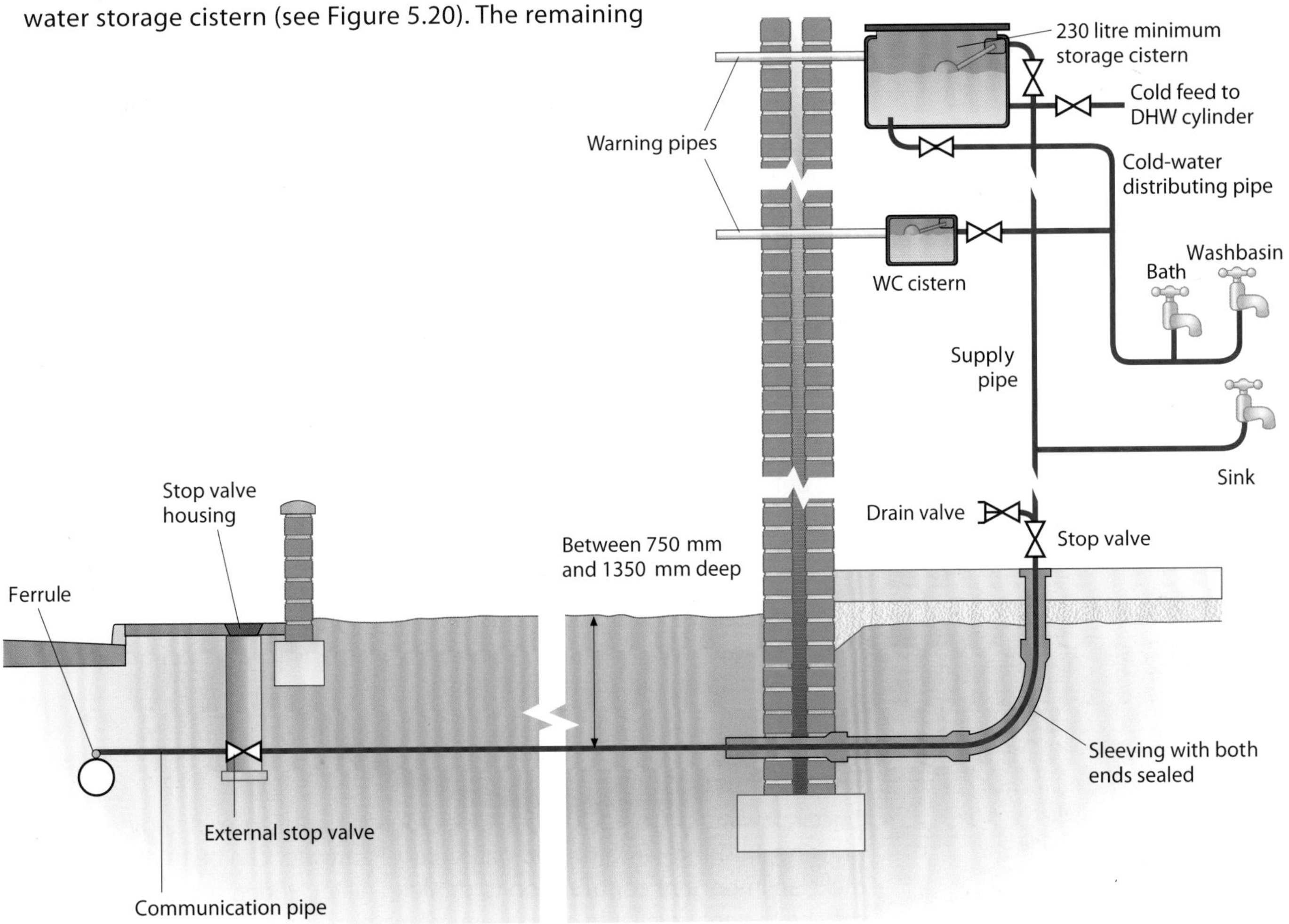

Figure 5.20: Pipework layout for an indirect cold water system

draw-off points are fed from the cold water storage cistern – hence the term 'indirect'. The draw-off points in an indirect system are fed indirectly from the cold water storage cistern. The system is designed to be used in areas of low water pressure where the mains supply pipework cannot supply the full requirement of the system. This type of system also has a reserve of stored water in case of mains failure.

Advantages of the indirect system

There is:

- a reserve of water should the mains supply be turned off
- reduced risk of system noise because of lower pressures
- reduced risk of wear and tear on taps and valves, again because of lower pressure
- lower demand on the main at peak periods.

Disadvantages of the indirect system

- There is a higher risk of frost damage due to more pipework.
- More space is occupied by the larger storage cistern (230 litre minimum – the minimum size for supplying both hot and cold outlets in dwellings as laid down by BS 6700).
- The storage cistern and pipework cost more.
- Water might not have been potable before the 1999 Water Regulations.

Grey water recycling

With activities such as bathing, laundry and dishwashing, up to half of the water used in a house can end up as grey water. Grey water does not include waste water from toilets but, with water reserves dwindling, it seems sensible to put it to better use wherever possible.

As grey water is less contaminated than waste water, common recycling involves treating it and then using it for flushing toilets or for watering gardens. Some toilets now have inbuilt washbasins to feed grey water straight to them without the need for additional pipework.

Advantages of grey water recycling include the following.

- It can reduce water consumption without changing consumer behaviour.
- It is easy to install and maintenance free.
- It removes the need for complex water treatment.
- If properly designed, a grey water system can lower sewage costs.
- It can reduce ground water use for irrigation.
- Less water will enter a city's sewage systems. This saves building new, or extending old, treatment plants.

The main disadvantage is that the water is not suitable for drinking.

Rainwater harvesting

Once rainwater is stored, it can be used for any normal use, including drinking, flushing, washing and gardening. Rainwater can be treated to make an important contribution to drinking water, and in some cases, it may be the only available or economical water source.

Safe working

Water gathered from a roof can pick up anything that is on the roof. This could include bird droppings, dead animals or substances in the roof material.

Most rainwater harvesting systems are simple and inexpensive, as they use the existing guttering and downpipes on a house. Water then flows via a filter to an underwater storage tank (see Figure 5.21).

Advantages include:

- A rainwater harvesting system is usually free, simple and low maintenance.
- Large volumes of water are kept out of the stormwater management system, helping to reduce flooding risks.

Chapter 5

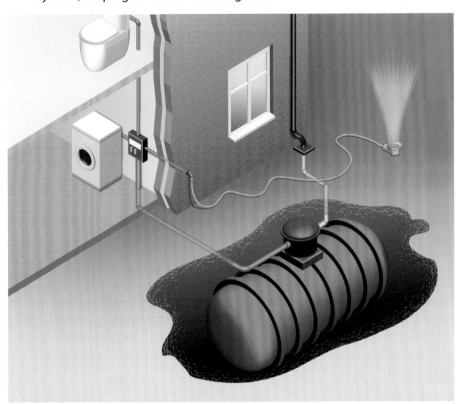

Figure 5.21: Rainwater harvesting systems can be simple and efficient.

Working practice

The owner of a self-build site is constructing a two-storey detached house and is supplying a lot of the materials himself.

Your job is to install the underground service pipe. You have been supplied with several lengths of 15 mm copper tube (R250 half-hard lengths), some Type A compression fittings and a stop valve.

The shallow trench the customer has prepared is approximately 300 mm below ground level, is very uneven and contains brick rubble. The service pipe is to enter the building through a hole in the brickwork and will terminate through the internal concrete floor slab, directly under the kitchen window.

Produce a list of recommendations for the customer, outlining what steps you will need to take and why. Include:

- how the area needs to be prepared
- what materials you will need
- why you cannot carry out the work until he has provided what you need.

Progress check 5.02

1 What does MDPE stand for? Where would it be used?

2 What is a communication pipe?

3 What is the purpose of a gooseneck?

4 Where is the external stop valve usually located on a property?

5 Identify two places where a water meter could be located on a cold water supply.

6 What should the service pipe be passed through as it enters a building? Why is this?

7 What are the maximum and minimum depths below ground for installing a service pipe?

8 As a service pipe rises into a building, what should be fitted 150 mm above the finished floor level?

9 What is the measurement from the outside wall surface of the building to the rising main if the service pipe is uninsulated?

10 State the main differences between direct and indirect cold water systems.

KNOW THE COMPONENTS USED IN DOMESTIC COLD WATER SYSTEMS

State reasons for using approved water fittings in cold water plumbing systems

You will be using and working on a wide range of components, so it is important that you know what to use, where and when. By using approved fittings, you can be confident that you are complying with the Water Regulations and any other relevant legislation. The Water Regulations are designed to prevent waste, undue consumption, misuse, contamination and inaccurate measurement of water, so it is important for you to use approved fittings in all your work.

Legal requirements

The Water Regulations state that all fittings must be:

- suitable for their purpose
- made of corrosion-resistant materials
- sufficiently strong to withstand normal and surge pressures
- capable of working at appropriate temperatures
- easily accessible to allow the renewal of seals and washers.

Backflow and contamination

You must ensure that the system remains hygienic at all times and is not subject to contamination that could endanger health. The two major ways in which water in a system may become contaminated are through:

- the use of a non-approved material such as lead
- backflow in the system, which can cause contaminated water to be drawn off inside the system.

These topics are covered in more detail later in this chapter (see pages 217–218).

Taps and valves

Taps and valves are used to:

- isolate supplies
- reduce the flow rate through pipework
- permit drainage from systems
- provide an outlet to an appliance.

They are used in appliances such as washbasins, baths and washing machines.

Taps and fittings for both hot and cold water systems should conform to BS 1010 Parts 1 and 2, BS 1552 and BS 5433. They are usually made of brass pressings or castings, and are chrome-plated when appearance counts, which also makes them easier to clean. Plastic taps and valves are also available.

Corrosion

The increased use of plastic instead of metal pressure pipework in modern plumbing systems means that corrosion is becoming less of a problem. However, plastic materials are also at risk from **degradation** and **permeation**.

As metals are still used in plumbing systems, corrosion is a very serious threat to these components and indeed any metallic pipework they are connected to.

Brass, an alloy of copper and zinc, is still commonly used for valves. However, brass can be subject to **dezincification** in certain types of water. You should therefore choose corrosion-resistant components which do not contain zinc, such as gunmetal or bronze. They are marked 'CR' in the UK or 'DRA' in the rest of Europe.

Describe methods for operation of key isolation valves

Stop valve

Figure 5.22 on page 202 shows a typical screw-down valve. The washer and jumper fit into the threaded part of the spindle, which is raised or lowered by turning the crutch head. When lowered to its maximum extent onto the seating of the valve, the incoming supply is isolated. The stop valve is usually located on high-pressure pipelines such as between the incoming mains and rising main. It is therefore used as a main supply isolation valve.

The Surestop® valve is a stop valve that allows the consumer to control the mains water supply simply by flicking a switch. The switch is fitted remotely, so it can be installed in a location where it is easily accessible, even above a worktop or inside a kitchen cabinet. This is a significant advantage over traditional stop valves.

The valve is becoming more popular and is now specified by many local authorities and housing associations, as well as by British Gas and HomeServe. It is also approved by the Water Regulations Advisory Scheme (WRAS).

> **Key terms**
>
> **Degradation** – breakdown of a material caused by exposure to ultraviolet (UV) light, usually in sunlight.
>
> **Permeation** – the penetration of a substance, such as a gas or liquid, through a solid material such as a pipe wall.
>
> **Dezincification** – the breakdown of zinc in brass fittings by the electrolytic corrosion between zinc and copper, making the components porous. This occurs in some soil and water conditions, especially in soft water areas. See Chapter 3, pages 99–100 for more on this.

 Safe working

Always look out for the arrow indicating the direction of the flow of water to ensure that the stop valve is installed in the correct way.

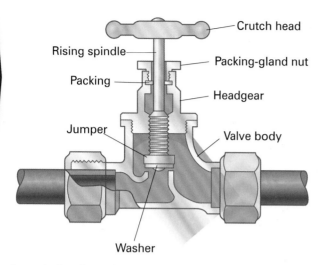

Figure 5.22: Typical stop valve

It is available in 15 and 22 mm diameters, is easy to install and operates by water pressure (no batteries or electrical connections). It is unaffected by limescale.

Gate valve

Gate valves are sometimes referred to as full-way gate valves because, when they are open, there is no restriction through the valve. The gate valve is fitted with a wheelhead attached to the spindle. When the head is turned anticlockwise, the threaded part of the spindle is screwed into the wedge-shaped gate, raising it towards the head (see Figure 5.24).

- A gate valve is classed as a service valve – it allows equipment to be serviced.
- When closing gate valves for servicing, beware of closing the valve too tightly. A common fault is that, on reopening, the spindle breaks and the gate stays wedged in the closed position.

Figure 5.23: The Surestop stop valve is limescale resistant

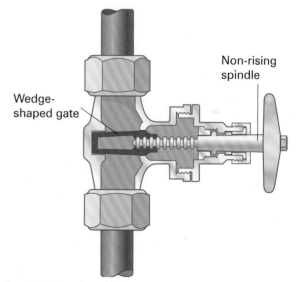

Figure 5.24: Gate valve

Gate valves are usually located in low-pressure pipelines, such as the cold feed from the cold water storage cistern (CWSC) to the hot water storage cylinder or the cold water distribution from the CWSC. You may also find them used on supplies to shower valves where the shower is fed from low pressure via a CWSC.

Full-bore valve

A full-bore valve provides the same function as a full-way gate valve but is easier to operate. It works in one quick movement by turning the handle a quarter of a turn, unlike a gate valve which takes a number of turns to operate. A sphere with a hole through it inside the valve, much like a service valve, is turned as the valve opens and closes. Unlike a service valve, in which the bore is reduced considerably by the sphere, the bore of the pipe in a full-bore valve is maintained through the valve just like in a full-bore gate valve, and so provides minimal restriction to the water.

Quarter-turn valves (spherical plug valves)

The 'ball valve'-type quarter-turn valve (spherical plug valve) can be used as a service valve to a cistern. Here, it will have a compression nut and olive on each side of the valve. It can also be used on the hot and cold supplies to a washing machine or dishwasher, when it will have a flexible hose connection. These valves are used to isolate the supply for the service, repair or maintenance of components and appliances within the hot and cold water systems.

The valves work by the quarter-turn operation of a square head (or slotted head) at the top of the valve, which aligns the hole in the valve with the hole in the pipe. When the head is in line with the direction of the pipe, the valve is open (see Figure 5.25). The quarter-turn head could be made from plastic and operated by hand – as used on washing machine supplies – or it could have a slotted head, operated by a screwdriver, as fitted on supplies to cisterns.

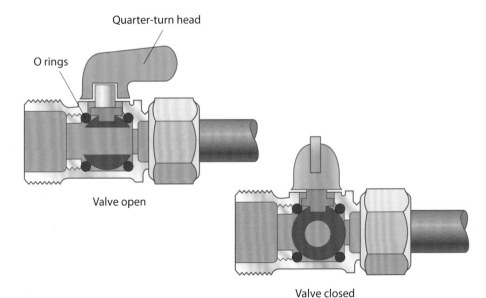

Figure 5.25: Lever-operated spherical plug valve

Figure 5.26: A reduced bore type valve

Remember

The supply pipe must contain an isolation valve as close as possible to the point where it enters the property, together with a drain-off facility at the lowest point. These arrangements can be made up using separate fittings as an alternative to the combined valve shown in Figure 5.27.

Quarter-turn valves can be used on low- or relatively high-pressure installations for servicing purposes. Care should be taken with the use of spherical plug valves on low-pressure installations; for this type of installation, the valve should be of the full-bore type rather than the reduced bore type shown in Figure 5.26. A reduced bore in the valve control mechanism will greatly increase the water pressure loss through the valve body, which will have an impact on the water flow rate available.

Stop-and-drain valve

Figure 5.27 shows a combined stop-and-drain valve as found inside a building. The purpose of the stop-and-drain valve is to:

- turn on/off the water supply to pipework fittings and components in order to enable system maintenance
- drain down all the system pipework fittings and components to enable system pipework repair or replacement.

Servicing valves

Servicing valves (either screw-down or spherical) should be located as close as possible to the inlets to all float-operated valves, cisterns, washing machines, dishwashers, water heaters, water softeners and any other similar appliances. This is to enable maintenance to be carried out and to provide emergency access.

Explain the method of operation of float-operated valves used in cisterns

Inlet controls

Pipes supplying water to a storage cistern must be fitted with an effective adjustable shut-off device, which will close when the water reaches its required level. For most domestic applications, a float-operated valve is used. This must comply with BS 1212 (Parts 2–4), and the following types are available:

- Portsmouth type (Part 1 valve)
- diaphragm valve made of brass or plastic.

Portsmouth equilibrium float-operated valve

The Portsmouth valve (see Figure 5.28), which complies with BS 1212 Part 1, is no longer used on new installations because it does not provide an effective air gap between the water level and the point at which the valve discharges. This means that there is a risk of backflow contamination, so this valve cannot be installed on potable water supplies. However, you may see it on existing installations.

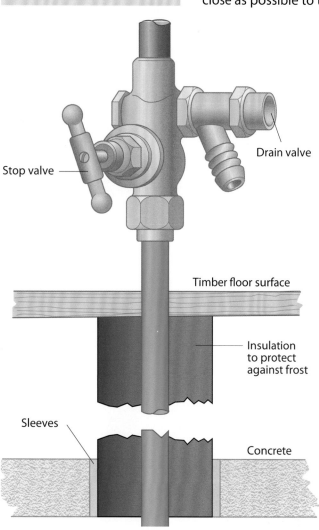

Stop valve

Drain valve

Timber floor surface

Insulation to protect against frost

Sleeves

Concrete

Figure 5.27: Combined stop-and-drain valve

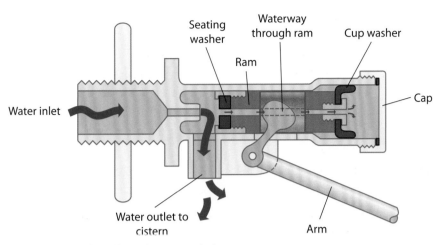

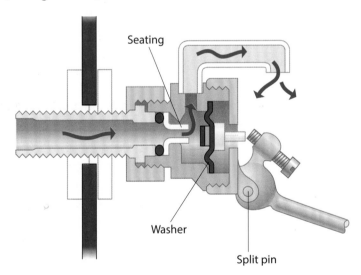

Figure 5.28: Portsmouth equilibrium float-operated valve

The concept of an equilibrium valve is that at peak non-use times, pressure within a cold water service will increase. At these times the water pressure created could be greater than the force being applied to the lever arm of the valve via the float. By using the equilibrium valve, pressure is forced back onto the piston by the use of a cup washer attached to it. This has the effect of equalising the pressure created by the water main and assists the lever arm and float to hold the valve closed.

Diaphragm valve

Brass or plastic diaphragm float valves, which comply with BS 1212 Part 2 or 3, are common in new storage cisterns. They provide an effective air gap between the water outlet from the valve and the water level in the storage cistern (see Figure 5.29).

Figure 5.29: Diaphragm float valves complying with BS 1212 Part 2 or 3

BS 1212 Part 4 refers specifically to diaphragm equilibrium float valves, which are designed primarily for use in WC flushing cisterns (see Figure 5.30 on page 206). Portsmouth and diaphragm valves work by leverage, while the equilibrium valve uses water pressure. However, equilibrium valves are available with a similar external appearance to Portsmouth and diaphragm valves.

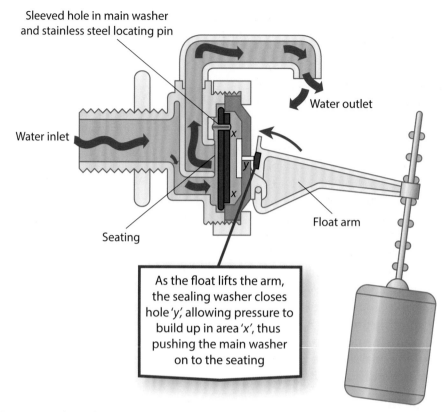

Sleeved hole in main washer and stainless steel locating pin

Water outlet

Water inlet

Float arm

Seating

As the float lifts the arm, the sealing washer closes hole 'y', allowing pressure to build up in area 'x', thus pushing the main washer on to the seating

Figure 5.30: Diaphragm float valve complying with BS 1212 Part 4 for WCs

Service valve

Inlet pipes to cisterns must have a servicing valve fitted immediately before the connection to the cistern, to enable maintenance to be carried out on the cistern without turning off the whole water supply. This also applies to WC cisterns. The valve usually installed is a spherical-type plug valve, suitable for use with higher-pressure cold water supplies.

Torbeck float-operated valve

The Torbeck float-operated valve is also of the equilibrium type, complying with BS1212 Part 4. Its most noticeable feature is the extremely small float, which enables the valve to fill the cistern rapidly and ensures a fast shut-off time.

When water is required, a very small hole in the diaphragm, which is located on a stainless steel pin, allows water to enter the void at the back of the diaphragm. A small hole in the valve body ensures that water pressure in the void is less than the water entering the valve on the opposite side of the diaphragm. As the water level rises, the arm closes off the water escaping from the void and seals the gap by means of a rubber washer. The effect is that the pressure within the void increases and pushes the diaphragm back onto the outlet seating of the valve and so closes off the supply.

Outlet pipes

Outlet pipes should be connected as low in the cistern as possible. The new Water Regulations recommend locating the connections at the bottom of the cistern rather than on the side. This prevents the build-up of sludge at the bottom of the cistern. Outlet pipes, such as cold feed and distribution pipes, should be fitted with servicing valves, and these should be located as close to the point of connection to the cistern as possible, while still being accessible. The valve type used in this position is usually a wheelhead gate valve, suitable for isolating lower-pressure cold water supplies.

Overflow and warning pipes

When water in a cistern rises above a preset level, usually caused by a faulty float-operated valve, the water is allowed to flow through a pipe away from the cistern. An overflow pipe discharges excess water safely to where it will not damage the building. A warning pipe is used to warn the occupiers of a building that a cistern is overflowing and needs attention.

Small cisterns of up to 1000 litres (i.e. domestic cisterns) must be fitted with a warning pipe and no other overflow pipe. Larger cisterns (1000–5000 litres) are fitted with both.

Describe the methods of operation of different taps

Draw-off taps, such as bib taps and pillar taps, usually work in a similar way to screw-down valves – such as a stop valve (see pages 201–202). However, modern types instead allow two polished ceramic discs to turn and align with two portholes through which the water can pass.

Hose union bib tap

The tap shown in Figure 5.31 includes a hose union attachment for garden hosepipes. This type of tap is a fluid category 3 risk and requires a double-check valve to prevent backflow contamination. See page 218 for fluid categories.

Bib tap (supatap)

You may come across this type of tap (see Figure 5.32) in older installations. A bib tap is likely to be installed over a sink that is a high risk (category 5) appliance, so an AUK3 air gap is required (see page 218). As this type of tap needs to be installed in a backplate, the height will be determined by the installer. However, there is little risk of contamination because, from a practical point of view, the tap will need to be above the minimum height for an AUK3 air gap (at least 275 mm from the tap outlet to the bottom of the bowl) in order for buckets etc. to be placed and filled in the sink.

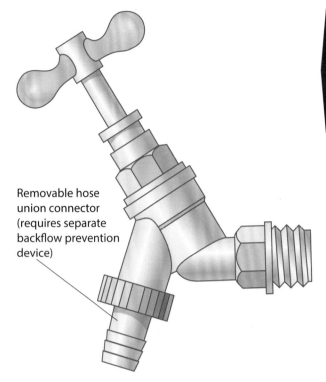

Removable hose union connector (requires separate backflow prevention device)

Figure 5.31: Hose union bib tap

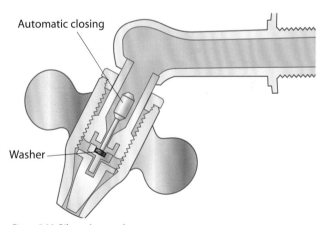

Automatic closing

Washer

Figure 5.32: Bib tap (supatap)

Activity 5.3

Obtain a varied selection of taps and valves and undo all the components so that they are loosely assembled. Dismantle the taps and valves and, using the information in this section, cross-reference the various parts to the diagrams in order to see what they look like and how they work.

Do the same for float-operated valves.

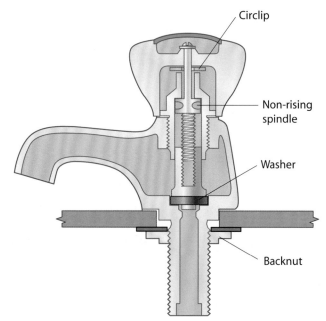

Figure 5.33: Pillar tap

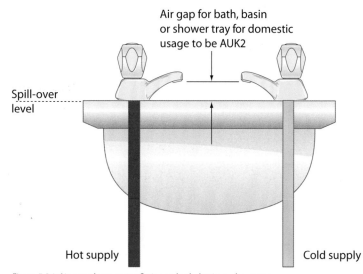

Figure 5.34: Air gap above water fitting to bath, basin or shower tray

Pillar taps

The pillar tap has a long vertical thread, which passes through pre-made holes in the appliance. It is held in position with a backnut and the supply pipe is attached to the thread via a tap connector. It works in a very similar way to a stop valve. The tap spindle is fixed in position by means of a circlip and does not rise when the tap is opened (see Figure 5.33).

Pillar taps are common on washbasins and baths, which are both category 3 risk appliances. The backflow prevention requirement is already built into the design of the tap and is referred to as an AUK2 air gap (see Figure 5.34).

A sink requires a high-neck pillar tap, as it is a category 5 risk appliance. An AUK3 air gap is required but, as with pillar taps for washbasins and baths, the manufacturer will have built the required termination height into the tap design.

Rising/non-rising spindles

Most modern taps (other than ceramic disc types) operate using a mechanism called a 'non-rising spindle'. When the tap is operated, the spindle carries a hexagonal barrel, which has a rubber washer fitted to it, inside the head of the valve. The spindle is fixed to the tap head using a brass circlip, which may need replacing during servicing of the tap.

Older types (though still available) use a rising spindle with a jumper and washer attached. When the tap is operated, the spindle rises out of the tap headgear.

Both types need a packing gland, which stops water inside the tap seeping up the spindle and causing unsightly damage to the outside of the tap covering or appliance. This gland may also need attention during servicing or repairs.

Taps and shower roses for baths, basins and shower trays

The contents of the bowl (a bath, basin or shower tray) are assessed as a fluid category 3 risk. To prevent the contents of the bowl being drawn back into the hot or cold water system, a type AUK2 backflow prevention device is usually installed. This uses a simple air gap between the outlet of the water fitting and the spill-over level of the bowl of the appliance. This fixed air gap prevents the contaminated water from being drawn back into the system. This is known as a non-mechanical backflow device.

The air gap must be at least:

- 20 mm for an outlet with a diameter of 15 mm or less
- 25 mm for an outlet with a diameter of 15–25 mm
- 70 mm for an outlet with a diameter over 25 mm.

If the air gap cannot be provided as above, then a double-check valve (category 3 protection) will need to be installed to both the hot and cold pipes feeding the individual appliance. This is known as a mechanical backflow device.

Globe taps

You may come across globe taps on old bath installations. Also, 'antique' or 'designer' ranges of reproduction bathroom suites are available, which will include globe taps (see Figure 5.35). Instead of a spigot thread connection to the supply, as in a bib tap, they use a socket thread. The spigot threaded outlet from the water supply is connected through holes in the bath, and the globe tap is then screwed to the fitting. Globe taps can pose a real threat of water supply contamination as they discharge below the spill-over level of the bath. They operate in a similar way to the stop valve.

A bath is a fluid category 3 appliance but, because globe taps are installed below the natural overspill of the bath tap, an AUK2 air gap is not available. Instead, a double-check valve must be fitted.

Mixer taps

There are numerous design variations for mixer taps, but they all work by allowing water from the hot and cold supply to flow from one outlet (see Figure 5.36). Mixer taps are similar to high-neck pillar taps and are usually installed in sinks. A sink is a fluid category 5 risk appliance, so an AUK3 air gap is required. However, as mentioned previously, the required termination to the tap design is built in.

There are two types of mixer tap:

- Single-flow outlet – mixing can take place and cross-contamination of water can occur between the hot and cold supplies.
- Twin-flow outlet – the water is not allowed to mix in the tap swivel outlet.

Drain-off tap

Conforming to BS 2879, the drain-off tap is located at the lowest point of any system and has a ribbed outlet to enable a good grip for a hosepipe connection. Drain-off taps can be supplied so that they can be soldered or screwed into a fitting or have an elongated tail enabling them to be used with compression or plastic push-fit fittings. You can also get a type fitted with a 'lockshield head', which means it can only be operated by a purpose-made key (see Figure 5.37). Combined stop valves with drain-off taps are also available (see Figure 5.38).

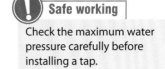

Safe working

Check the maximum water pressure carefully before installing a tap.

Figure 5.35: Globe taps

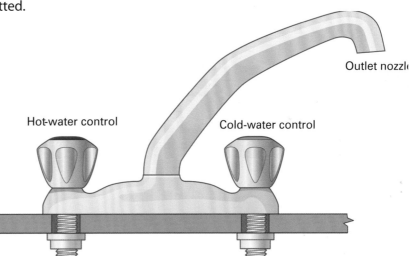

Outlet nozzle

Hot-water control

Cold-water control

Figure 5.36: Mixer tap

Did you know?

The twin-flow separation of supply is achieved using a specially designed valve body and swivel outlet, which includes a complete division in the casting so that the hot and cold water do not mix until they leave the swivel outlet.

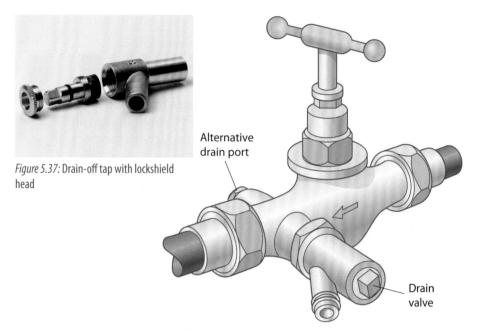

Figure 5.37: Drain-off tap with lockshield head

Alternative drain port

Drain valve

Figure 5.38: Combined stop valve with drain-off tap

Ceramic disc tap

The tap shown in Figure 5.39 uses quarter-turn ceramic discs. These taps are in common use, but they often have maximum water pressures within which they operate. You must follow the manufacturer's requirements during installation, or the discs could shatter under operating conditions.

Ceramic disc taps have a close-fitting piece of ceramic with a hole through it (see Figure 5.40). When turning the tap on, the disc is rotated until a hole in its side matches a hole in the tap body and the water flows out. When turned the opposite way, the hole in the body is blocked, stopping the water flow.

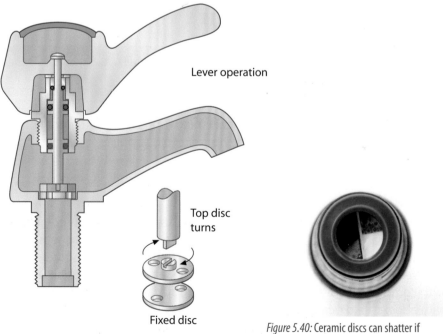

Lever operation

Top disc turns

Fixed disc

Figure 5.39: Ceramic disc tap

Figure 5.40: Ceramic discs can shatter if poorly installed

Ceramic disc taps work on a quarter-turn only, so they are much easier to operate than traditional taps and require much less maintenance. However, quarter-turn taps can often give rise to **water hammer** (see pages 234–235).

Explain the requirements for positioning drain valves in cold water plumbing systems

A drain tap is required at any low point of the system to drain all supply and distributing pipes in a building. This may be of the types described previously (pages 209–210) or of the spherical type. Drain taps must be accessible and should not be buried or covered with soil. They must not be installed so that they may be submerged or are likely to become submerged. There are two primary reasons for the positioning of drain taps in pipework systems:

- to drain the system as a frost precaution in uninhabited properties
- to ease maintenance/servicing or extension of the system procedures.

Explain the requirements for a cold water storage cistern (CWSC) in a domestic cold water plumbing system

Under the Water Regulations, a storage cistern supplying cold water outlets (or feeding a hot water storage system) must be capable of supplying wholesome water. Therefore, various protection measures are included in the cistern's design to ensure that the water supply does not become contaminated.

Cisterns must be:

- fitted with an effective inlet control device to maintain the correct water level
- fitted with service valves on the inlet and outlet pipes
- fitted with screened warning/overflow pipes to signal overflow
- covered to exclude light and insects or vermin
- insulated to prevent heat loss and undue warming
- installed so that the risk of contamination is minimised
- arranged so that water can circulate, preventing stagnation
- supported to avoid distortion or damage leading to leaks
- readily accessible for maintenance and cleaning.

Figure 5.41 shows a screened protected CWSC.

Materials for cisterns

Most cold water storage cisterns used to be made of galvanised low carbon steel and you may still come across these on maintenance jobs. However, most new installations use plastic cisterns – polyethylene, polypropylene or polyvinyl chloride (PVC). Most cisterns are polypropylene because it allows them to be:

- light
- strong
- hygienic
- resistant to corrosion
- flexible enough to be manoeuvred through small openings.

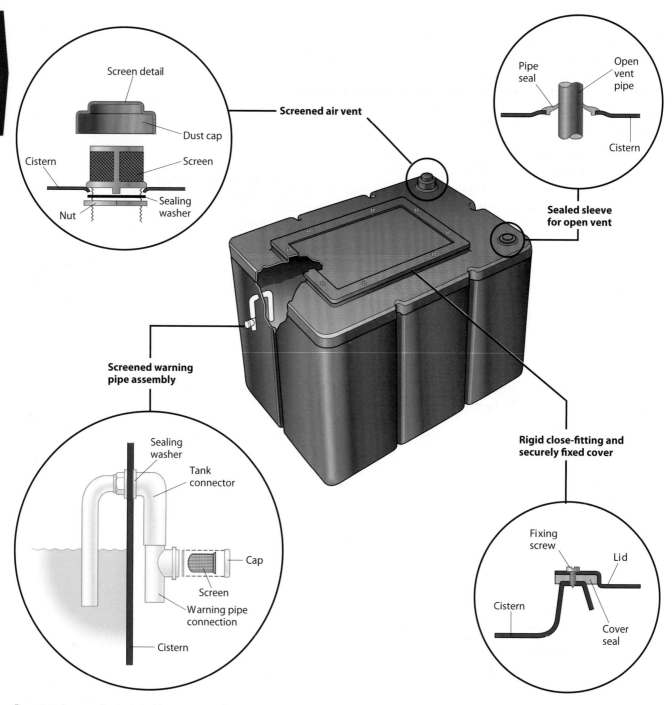

Figure 5.41: A screened protected cold water storage cistern

 **Safe working**

Never use oil-based pastes on plastic cisterns because they will degrade the plastic. In addition, you should not use a heated section of pipe to make the hole in a cistern, as this also degrades the plastic and will result in cistern failure.

Cisterns can be square, rectangular or circular and are black to prevent the growth of algae. Because they are flexible, the base must be fully supported throughout its entire length and width, as covered in the Building Regulations Part G. Holes for pipe connections should be cut using a hole saw. The joint between the cistern wall and fitting should be made using plastic or rubber washers.

Location of a warning pipe in a small cistern

The warning pipe must be located so that a minimum air gap is maintained between its point of discharge and the normal water level in the cistern. The position of the float-operated valve is also crucial to ensure that a minimum air gap is maintained between its outlet and the spill-over level at the warning pipe (see Figure 5.42). You should remember the following points.

- If the float-operated valve becomes defective, the warning pipe should be able to remove excess water without becoming submerged.
- The minimum pipe size is 19 mm internal diameter, although it may need to be larger for higher inlet flow rates to the cistern.
- The warning pipe should fall continuously from the cistern to the point of discharge.
- Warning pipes should discharge where the water will be noticed, usually outside the building.
- Warning pipes should be fitted with a screen or filter to exclude insects.

Pipes and cisterns in roof spaces

Figure 5.43 shows the correct installation of a cold water storage cistern and associated pipework in a roof space.

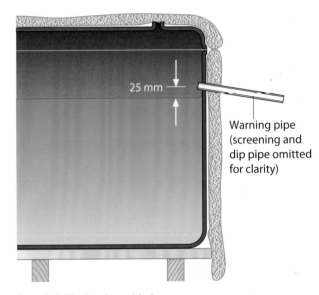

Figure 5.42: Warning pipe positioning

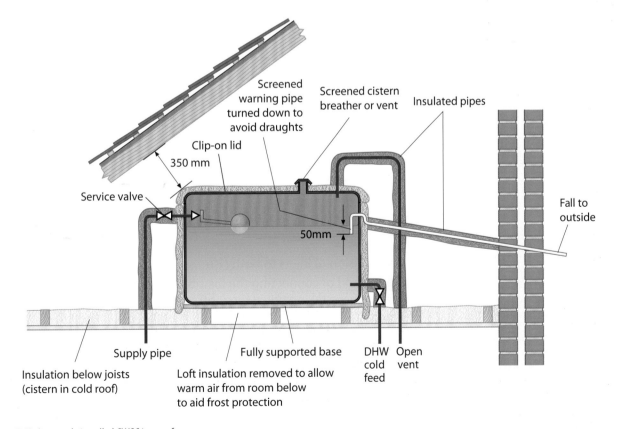

Figure 5.43: A correctly installed CWSC in a roof space

You should remember the following points.

- A 350 mm gap to the roof surface is needed to provide frost protection and to allow access for maintenance.
- Both the pipework and cistern must be fully insulated.
- The space under the cistern is left uninsulated to allow heat from the property to warm the cistern.
- The Water Regulations also require that cold water systems should not exceed 25°C. Using insulation on pipework and components acts to prevent stored water from becoming overheated in hot weather.

Explain the procedures for linking two small cisterns in a plumbing system

It is crucial to avoid contamination and stagnation in cisterns, which may occur if water cannot fully circulate throughout the cistern. This is more likely when two or more cisterns are joined together – perhaps because the access hatch was not big enough. When joining cisterns, you must ensure proper water movement through them to avoid stagnation, as illustrated in Figure 5.44.

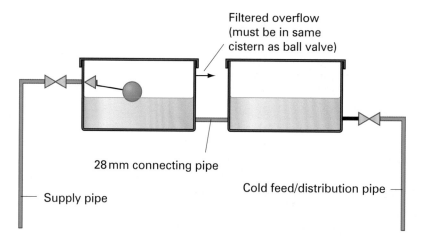

Figure 5.44: An example of how to link two or more cisterns

When linking larger cisterns in domestic properties, you should pay attention to:

- the provision of inlets to both cisterns fed via a float-operated valve
- the method of connecting the distribution pipes via a manifold arrangement taking equal draw-off from both cisterns simultaneously
- siting outlet connections at the opposite end to inlet connections to allow effective water distribution across the cisterns
- the method of linking the cisterns using lateral connections between them.

Describe the key contamination issues in plumbing systems

Water supplies can be contaminated by:

- the use of non-approved materials
- backflow
- back-siphonage
- cross-connection.

Non-approved materials

When installing a hot or cold plumbing system, you must ensure that it remains hygienic and not subject to contamination. Water can be contaminated by the use of non-approved materials, such as lead, in the system. You can deal with this by always installing approved pipework, components and fittings.

Backflow

Backflow occurs when water under positive pressure flows in the opposite direction to the direction intended. This could result in liquids or other substances being drawn back into the public water supply. There are two conditions that can make this situation occur: back-siphonage and back pressure. They should not be confused with each other as they are totally different conditions.

Back-siphonage

Back-siphonage occurs when the flow in a system is reversed by negative pressure, creating a vacuum or partial vacuum in the supply pipework. This could occur when there is a break or interruption in the supply of water from the water main. This could be created by:

- a burst water main
- maintenance/repairs by the water authority
- firefighting appliances drawing from the water
- a reduction in the supply pressure caused by peak time usage by other consumers.

Back pressure

Back pressure is the backflow of the normal flow in a system. This could be created by an increase in the downstream pressure, which is above the mains water supply pressure. This could be possible in some heating system installations, elevated CWSCs and pressure producing systems. Water flows in the direction of least resistance.

Cross-connection

In plumbing terms, a cross-connection is an actual or a potential connection between a potable water supply and a non-potable water supply. An unprotected cross-connection could lead to degradation of the water quality in a property or in neighbouring properties. This poses a serious public health risk. There are many, well-documented cases of cross-connections being responsible for the contamination of drinking water that has resulted in the spread of disease.

The issue is of grave concern, as even though a system may not have cross-connection issues when originally installed, alterations or additions may change the situation at any time.

Describe the fluid categories

Before selecting a suitable backflow prevention device, you must assess the fluid risk category that the water fitting may come into contact with. The UK uses a five-level risk system for assessment, summarised in Table 5.05.

Fluid risk category	Description	Examples in domestic properties
1	Wholesome water provided by the water undertaker	Water used for drinking taps
2	Water that has had its aesthetic quality changed (e.g. temperature, taste, odour)	Mixed hot and cold water discharged from mixer taps Water discharged from a base exchange softener
3	Fluid that represents a slight health hazard	Water in primary central heating circuits Water in the bowl of a bath, basin or shower tray Hand-held garden hose used with hose union tap Contents of a washing or dishwashing machine
4	Fluid that represents a significant health hazard	Water used in a mini-irrigation system in a garden (e.g. a porous hose or a sprinkler system with sprinkler heads less than 150 mm above ground level)
5	Fluid that represents a serious health hazard	Contents of a sink bowl, WC pan or bidet

Table 5.05: Fluid risk categories

Describe the need for point of use backflow protection

As mentioned on page 217, backflow is a major cause of contamination, but can be dealt with by installing point of use backflow devices at all water outlets (such as taps) and fitting connections. An extensive range of these devices is available. The selection of the correct type of device is based on an assessment of the risk presented by the fluid that the plumbing fitting may come into contact with. See Table 5.05 above for the fluid risk categories.

Non-mechanical backflow protection is provided by air gaps.

- Type AUK2 air gap to domestic basins and baths: see page 209.
- Type AUK3 air gap to domestic sinks: this must be 20 mm or twice the diameter of the inlet pipe supplying water to the fitting (whichever is greater).

Backflow protection may also be provided by a double-check valve (for example, to an outside tap supply). However, a double-check valve only provides protection up to fluid risk category 3.

Working practice

While carrying out some maintenance at a customer's house, you are asked to replace a faulty tap in the garden. The tap is connected through an outside wall directly into the rising main inside the property. The installation does not conform to current Water Regulations at all.

Produce a list of bullet points of the requirements for this part of the installation to meet the Water Regulations. Briefly indicate the reasons for each requirement.

Progress check 5.04

1   Other than providing frost protection and keeping heat in pipework and components, what other purpose does insulation serve?

2   Why is it recommended that loft insulation is removed from under the base of a cold water storage cistern?

3   List the five fluid risk categories.

4   Define what is meant by an AUK2 air gap.

5   Up to which fluid risk category does a double-check valve give protection?

6   Define the term 'cross-connection'.

7   Where should backflow protection devices be installed in relation to the appliance they are intended to protect?

8   Define the terms 'back pressure' and 'back-siphonage'.

9   If a sink mixer tap of the single-flow outlet type is fitted, which device should be installed to prevent a cross-flow condition potentially occurring?

10 Define the term 'dezincification'.

UNDERSTAND THE KEY REQUIREMENTS OF TESTING AND DECOMMISSIONING OF DOMESTIC COLD WATER SYSTEMS

Soundness testing on completed hot and cold water systems is essential to ensure that there are no leaks. Even the most competent plumber can make the occasional mistake, so testing is always required. BS 6700 provides the standard for soundness testing on hot and cold water systems.

Explain methods of testing cold water pipework systems

BS 6700 has separate procedures for testing rigid plastic and metal pipes (such as copper). The testing methods for plastic pipes (procedures A and B) and for metal pipes are covered on pages 221–222. (See also the Water Regulations for more details.)

Describe how to use hydraulic test equipment

This is covered later in the chapter, on pages 221–222.

Describe the requirements for flushing a system

Following any work on a system, it should be flushed to get rid of any particles or foreign matter that might cause damage or harm.

Filling cold water pipework with water at normal operating pressure and checking for leakage

When repairs or extensions to existing systems have been carried out, it may only be possible to test the system by turning the cold supply back on. This could be because the remainder of the system is old and you cannot isolate it from the new works. You will need to turn the existing supply back on and allow the system to refill. Testing should only be done at normal working pressure. Once the system is full and subject to normal pressure, you should make a visual check of all joints and connections.

> **Remember**
>
> Any particles left in the CWSC will get into the pipework and could cause a blockage.

Actions to take when inspection and testing reveal defects

If you find a leak, you should isolate the system to allow repair or replacement. Once isolated, you can carry out further investigations to see what action you need to take. If the water supply is not isolated before you start moving items around the leak, you may find that the water starts coming out much faster and causes more damage. Once the repair has been made, you should follow the flushing procedure below.

Flushing procedure for cold water systems and components

Figure 5.45 shows the procedure for flushing both hot and cold water installations.

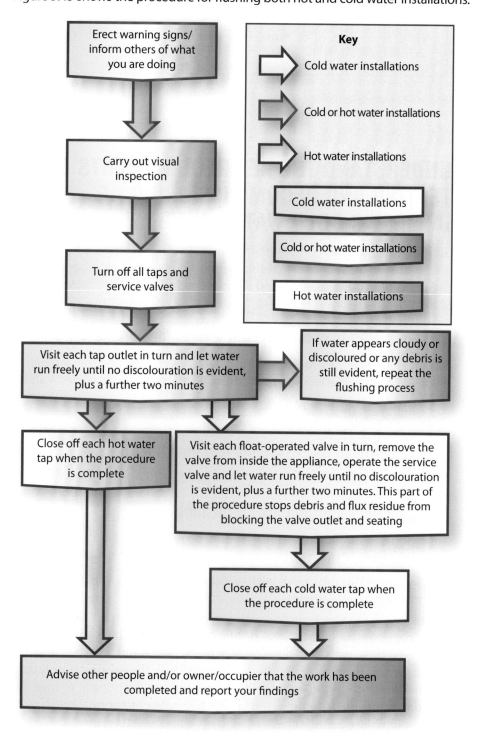

Figure 5.45: Flushing procedure for hot and/ or cold water systems

Identify the key phases in soundness testing a cold water system

Soundness testing of hot and cold water systems includes:

- visual inspection
- setting up hydraulic test equipment
- test period – testing for leaks; pressure testing
- final checks
- completion of documentation.

Visual inspection

You should carry out a thorough visual inspection of all pipework and fittings to ensure that:

- they are fully supported, including cisterns and hot water cylinders
- they are free from jointing compound and flux
- all connections are tight
- terminal valves (sink taps etc.) are closed
- in-line valves are closed to allow stage filling
- the storage cistern is clean and free of any unwanted material (bits of plastic from where the holes were cut).

It is useful at this stage to advise the customer and/or other site workers that soundness testing is about to commence.

Testing for leaks

1 Slowly turn on the stop tap to the rising main.

2 Slowly fill, in stages, to the various service valves, and inspect for leaks on each section of pipework, including fittings.

3 Open the service valves to appliances, fill each appliance and visually test for leaks again.

4 Make sure the cistern water levels are correct.

5 Make sure the system is vented to remove any air pockets prior to pressure testing.

Pressure testing

Pressure testing of installations in buildings is done using hydraulic equipment (see Figure 5.46). BS 6700 has separate procedures for testing rigid metal (such as copper) and plastic pipes – which in turn have two procedures (A and B).

Metal pipes
- Make sure any open-ended pipes such as vent pipes are sealed.
- Once the system has been filled, leave it to stand for 30 minutes to allow the water temperature to stabilise.
- Pressurise the system using the hydraulic testing equipment to a pressure of 1½ times the system maximum operating pressure.
- Leave it to stand for 1 hour.
- Check for visible leakage and loss of pressure. If sound, the test has been satisfactory.
- If not sound, repeat the test after locating and repairing any leaks.
- Complete a test certificate.

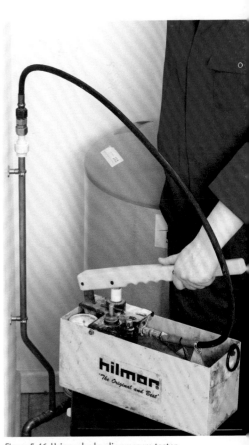

Figure 5.46: Using a hydraulic pressure tester on copper pipe

Plastic pipes

Test A

- Apply test pressure (1½ times the maximum working pressure), which is maintained by pumping for a period of 30 minutes. Visually inspect for leakage.
- Reduce pressure by bleeding water from the system to 0.33 times the maximum working pressure. Close the bleed valve.
- Visually check and monitor for 90 minutes. If the pressure remains at or above 0.33 times the working pressure, the system can be regarded as satisfactory.

Test B

- Apply test pressure and maintain by pumping for a period of 30 minutes.
- Note the pressure and inspect visually for leakage.
- Note the pressure after a further 30 minutes. If the pressure drop is less than 60 kPa (0.6 bar), the system can be considered to have no obvious leakage.
- Check and monitor for 120 minutes. If the pressure drop is less than 20 kPa (0.2 bar), the system can be regarded as satisfactory.

Final system checks

After you have completed the system tests, the system should be thoroughly flushed out, as described on page 220, to remove any debris or swarf before you carry out a final visual check for leaks. Also, make sure that all taps, valves, etc. are working. Advise the customer and/or other site workers that testing is complete. At this point a performance test of the system is conducted, and a handover and documentation are completed. This will be covered at Level 3.

Working practice

Nick qualified as a plumber several years ago. He has just installed a hot and cold water supply and a hot water heating system in a new-build property. He completed the installation very quickly using plastic tubes and fittings.

Nick tests the system using a hydraulic pressure testing kit. However, his machine has a leak so he is unable to gain a pressure at 1.5 times the system pressure. He is also in a hurry so, after checking the building for evidence of leaks, which he does not find, he completes the test certificate and leaves, ensuring the mains supply is turned off.

Several months later another plumber is asked to commission the system. This plumber knows that a test certificate was issued and so carries out the commissioning, checks round for visual leaks and hands back the keys. The same afternoon the owners move in. During the night a fitting on the cold water supply under the floor in the bathroom 'blows off', damaging carpets, the kitchen ceiling and causing water damage to the brand new kitchen units.

- Who is at fault? What went wrong? Why did the fitting not blow off during the day after commissioning and when Nick initially carried out his hydraulic test?
- Who would be responsible for the insurance claim? Would the insurance company pay, bearing in mind a certificate was issued? What is the probable cause(s) of the plastic fitting being able to blow off?

Describe commissioning checks for cold water systems

Figure 5.47 outlines the commissioning checks for both hot and cold water installations.

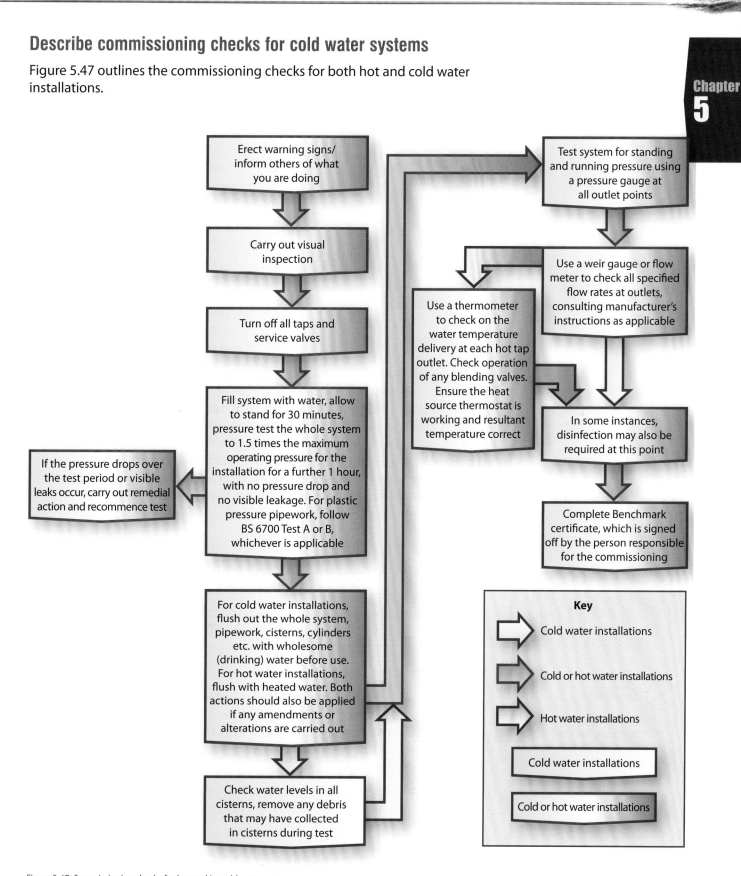

Figure 5.47: Commissioning checks for hot and/or cold water systems

Figure 5.48: Taking a water pressure reading

Figure 5.49: Keep the weir cup level when taking a flow rate reading.

System pressure and flow rate

The incoming water pressure and flow rate supplied via the mains have a key bearing on the size of pipework and fittings used in the system. Pipe sizes are covered in more detail at Level 3. However, at Level 2, you may be required to assist with taking readings of incoming water pressure and flow rate as part of the installation process.

- Water pressure is measured with a pressure gauge, as shown in Figure 5.48. The gauge usually reads in bar pressure.
- Flow rate is measured using a device called a weir cup, as shown in Figure 5.49. The reading is usually given in litres per second (l/s).

A common fault with domestic systems installation is that water pressure readings are not taken at the right stage in the job. They should be taken at the design or pricing stage. Insufficient supply pressure and flow rate can result in big problems. For example, if the wrong system components (such as combination boilers) are installed, they will not work correctly. If the pressure to an existing dwelling is poor, check first that there is no burst on the service from the external stop tap. You can usually tell by putting your ear against the stop tap key while it is on the stop tap and listening for a hissing noise.

Describe the differences between permanently and temporarily decommissioning a cold water system

Decommissioning domestic systems means turning off the supply to the system, removing system pipework and components and making sure that the hot or cold supplies are sealed or left so that they cannot be turned back on. The correct term for this is 'permanent decommissioning'.

This might be necessary if:

- an old system is to be completely stripped out of a domestic property and replaced with a new or alternative system (for example, a direct hot water system being replaced with an indirect system)
- a system is to be stripped out permanently (for example, prior to the demolition of a building).

Temporary decommissioning is the process of taking a system out of action for a while so that work can be carried out on it, for example maintenance.

Before commencing work, make sure that:

- anyone who may be affected by your actions is aware of what is happening
- items such as washing machines and dishwashers are not operating.

> **Remember**
>
> When decommissioning, you will need to dispose of redundant materials safely and correctly.

 Safe working

The Water Regulations state that, when a system or pipework length is permanently decommissioned, no dead legs should be left in the pipework system. This means removing any branched tee connections from an existing pipe run.

When decommissioning part of a system:

- Consider the use of other parts of the installation and the amount of time that disruption will occur.
- You may need to plan the work in order to minimise disruption to the customer, or you may be able to cut into pipes and fit valves so you can work without affecting others for long periods.
- Make sure that you seal off all open ends of pipes, either permanently or temporarily.
- Remove any dead legs from the system.
- If required, you may need to put up warning signs advising people not to use items.
- If you have turned a valve off, you may be able to remove the handle to ensure that someone does not turn it back on.

Explain the method for draining cold water systems

Figure 5.50 shows the procedure for draining hot and cold water installations (refer to Figure 5.45 on page 220 for the key to the diagram).

Describe the negative impact of dead legs in systems

Dead legs are long lengths of pipe from hot water storage vessels to appliances. The Water Regulations define a dead leg as 'a length of distribution pipe without secondary circulation'. Dead legs are problematic as they may lead to:

- the growth of bacteria
- noise in the system.

Dead legs are illustrated on page 267 in Chapter 6: *Domestic hot water systems*.

Growth of bacteria

Water within dead legs may stagnate, allowing the growth of water-borne bacteria such as legionella, which causes legionnaires' disease. Legionella bacteria can grow in hot and cold water outlets, atomisers, air conditioning plants, whirlpool or hydrotherapy baths or in dead legs. The bacteria multiply at temperatures of 20–40°C and are transferred to humans via inhaled mist droplets (aerosols). The bacteria are dormant (inactive) below 20°C and cannot survive above 60°C.

> **Case study**
>
> Approximately 350–400 people in England and Wales contract legionnaires' disease each year, although about half of these cases catch the virus while abroad. Anyone can catch this potentially fatal disease, but it is more likely to affect men, smokers, people over 50 and those with a weak immune system. Legionnaires' disease cannot be passed from one person to another.
>
> In 2011 there were 239 reported cases, 172 men and 67 women; 20 of these cases (around 12 per cent) were fatal. In May 2012 over 100 people in Edinburgh contracted legionnaires' disease, and by September 2012 three people had died of the disease.
>
> Use the Internet to find out more about recent outbreaks of legionnaires' disease.

Erect warning signs/inform others of what you are doing
Notices should clearly state 'SYSTEM DRAINING'
Pay particular attention to warning the occupier
what is expected

Ensure boilers/immersion heaters are electrically isolated if they are connected to the hot water system. Allow water and heat source to cool slightly as the shock of cold water hitting a hot boiler could fracture it

Ensure boilers are isolated if they are connected to the cold water system

If the property is occupied, ask for movable objects that could get damaged to be removed from the areas of work Ask if the occupier would like to draw water for drinks, etc.

Protect the area to be worked in with suitable absorbent materials or use suitable receptacles to contain any leaking/excess water safely

Connect a hose pipe to the lowest drain valve attached to the hot water pipe connected to the kitchen sink

Connect a hose pipe to the lowest drain point, normally above the internal stop valve

Secure the hose using a hose clip, protect the area immediately around the drain valve with suitable absorbent material, take the other end of hose to a suitable drainage gully point outside building

Turn off the nearest isolation valve; this may be the internal stop valve

Open all cold water outlets in the building that have been isolated, including the opening of a hot water tap to lower the float-operated valve in the CWSC and flush any WCs. All these actions let air into the system, ensuring water is not held in vacuum. When using the procedure to drain hot water systems, open hot water tap outlets

Check the hose for evidence of any running water

NOTE – If an indirect cold water system is involved, repeat the procedure, finding the drain valve at the lowest point of that system
There will obviously be more water to drain because of the CWSC, but the majority of this may be drained through the appliance taps

Turn off the drain valve, disconnect the hose, remove the hose to outside the building carefully to ensure no damage from any spillage and stow away correctly

Return to all taps previously opened, to allow air into the system, and close them off

Remove dust sheets, tools and equipment from the property

If the system(s) are to be left drained down at this stage, fix warning signs to the internal stop valve and any hot water/central heating boiler source, clearly stating 'SYSTEM DRAINED' and a contact number

Finally, inform the owner/customer/person in charge of the building of the work you have carried out

Figure 5.50: Draining down procedure for hot and cold water systems

Noise

Causes of noise in cold water systems are covered in more detail later in this chapter, on pages 234–235.

Working practice

You have been asked to soundness test a cold water system that someone else has installed. After a visual inspection, you are concerned about the quality of the soldering on the end feed capillary fittings throughout the installation.

After pointing this out, the builder is worried about potential water damage to the recently decorated property.

What would be the solution to this problem? Describe how this would be carried out, including the identification of any individual leaking fittings.

Progress check 5.05

1 Define the term 'wholesome water'.

2 How is the potentially fatal legionnaires' disease contracted?

3 List the installation situations in which legionella bacteria are known to multiply.

4 At what temperatures do legionella bacteria multiply?

5 How should you test a cold water system for soundness if there is no mains water supply connected to the installation?

6 When commissioning a new cold water supply installation, what pressure should the installation be subjected to?

7 For how long should you run the soundness test when commissioning a new cold water supply installation?

8 Produce a list of bullet points to indicate the method of draining down a direct system of cold water supply.

9 When decommissioning or draining a cold water system, why is it important to open all tap outlets and make sure that float-operated valves are in the open position?

10 What equipment would you use to check that flow rates meet those specified by the manufacturer or system designer?

UNDERSTAND THE BASIC MAINTENANCE REQUIREMENTS OF DOMESTIC COLD WATER SYSTEMS

Maintenance is important to ensure that system components operate correctly and to reduce water and energy waste through, for example, dripping taps. Refer back to earlier in the chapter for aspects such as inspecting and testing. There are a few basic principles of maintenance.

- Check the manufacturer's instructions.
- Identify the fault and ensure that you have any parts required for replacement or repair.
- Inform the customer before starting any work.

Remember

An important aspect of maintenance, which applies to all the procedures covered in the section, is liaison with the customer.

- Isolate the supply.
- Strip the components.
- Repair or replace the defective part.
- Reassemble the components.
- Turn on the supply and test the component for correct operation.
- Report back and inform the customer that the fault has been repaired.

Identify common defects found in cold water components

Most faults occur on taps and will depend on the type of tap. Faults tend to be:

- dripping taps
- leaking taps (glands)
- tap heads that are difficult to turn
- noise when the cold tap is turned on.

These may be caused by:

- worn/broken washers
- defective tap seats
- jammed headgear
- ceramic disc failure.

In some properties, servicing valves may have been fitted to the hot and cold supply under the appliance. This is good practice but not a requirement of the Water Regulations. If servicing valves are fitted, isolate the supply by turning them off. The checklist below assumes servicing valves are not fitted.

Checklist

Isolate the supply

On a direct cold water system, turn off the cold water supply to the system and drain through the sink tap.

Turn off the gate valve located on the cold feed from the CWSC to the domestic hot water cylinder for hot tap repairs.

On mains-fed hot water systems, close the service valve on or near the unit.

Drain down the system pipework through the relevant tap.

Repair or replace the defective part

This procedure will depend on the type of taps fitted. Most taps fitted to appliances are chrome-plated brass but some may be plastic.

Remove the cover from the tap head assembly; this will usually be screwed through the top of the tap cover and concealed behind a chrome or plastic cap. Lift the cap with a small flat-headed screwdriver. You can then remove the cover screw (which is usually cross-headed).

Continued ▼

- You are then left with a headgear similar to the stop tap (see pages 232–233).

- The process for rewashering the tap is the same as for the stop tap (see page 233).

Reassemble the component

This is the same procedure as for the stop tap (see page 233), but you will also need to replace the cover.

Turn on and test

- Make sure the hot and cold taps to the appliance are turned off.

- Turn on the stop tap if it is a direct system.

- Turn on the service valve on the cold water distribution pipe if it is an indirect system.

- Turn on the gate valve to the hot water storage cylinder.

- Return to the appliance taps and make sure the supply pressure to the cold water tap is satisfactory (direct) and that the cold water (indirect) and hot water supplies are flowing smoothly.

- Test the operation of the taps.

- Inform the customer that the fault has been repaired.

Float-operated valve

There are three types of float-operated valve:

- Portsmouth (see Figure 5.51)
- diaphragm (see Figure 5.52)
- equilibrium.

The trend for new installations is to use plastic diaphragm float-operated valves in both CWSCs and WC cisterns. In maintenance work, however, you could come across a range of different types. The Water Regulations require that servicing valves are fitted as close as possible to the cistern in order to isolate the float-operated valve for maintenance work.

> **Remember**
>
> When working on an appliance, it is a good idea to push the appliance plug into the waste hole so that you don't drop any tap parts into the waste system. However, remember to remove the plug before turning the water supply back on.

Figure 5.51: Portsmouth float-operated valve

Figure 5.52: Diaphragm float -operated valve

Remember

When working on a Portsmouth or a diaphragm ball valve, it is easier to 'split' the valve at the front between the body nut and the body. This enables you to take the valve out of the cistern and work on it in plenty of space.

Common faults

- Probably the most common fault is a running warning or overflow pipe. This is caused by the float-operated valve not closing off the supply properly.
- There is no water in the cistern, the toilet will not flush or there is no hot water. These faults are usually because the hole in the seating is blocked with tiny bits of debris that got into the system when it was installed.
- There is excessive noise in the system, particularly when the WC cistern is flushed or hot/cold water is drained off from the cold water supply cistern.

Checklist

1 Isolate the supply.

2 If a service valve is fitted, simply turn it off. If a service valve is not fitted, you will have to turn off the supply at the stop tap.

3 If you are working on a CWSC, run the level down to enable you to work easily on the float-operated valve. If it is a WC cistern, pull the flush to allow the float to drop. You might think this wastes water, but you need to check the correct operation of the float valve by watching it working as it fills up again.

4 Repair or replace the defective part (when dealing with valves creating overflows).

5 If the float valve is plastic, then it is either the washer in a Portsmouth valve or the diaphragm in a diaphragm ball valve that is defective (see Figure 5.53).

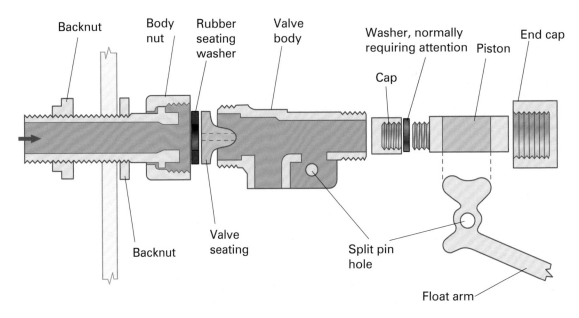

Figure 5.53: Section through a Portsmouth type float-operated valve

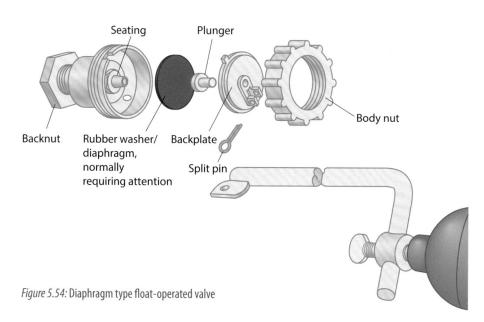

Figure 5.54: Diaphragm type float-operated valve

Activity 5.4

Look at the diaphragm valve in Figure 5.54. List the sequence for changing the washer on the valve and reassembling the components. Try to practise on a real valve too.

Defective washers

Most problems with float-operated valves are caused by defective washers. Others are caused by oxidisation or a build-up of scale between the lever arm and piston.

Checklist

Once you have split the body from the body nut, follow this sequence.

1 Remove the end cap.

2 Remove the split pin from the lever arm to free the piston.

3 Remove the piston cap, take out the defective washer and replace it.

Reassemble the components

☐ Refit the cap.

☐ Make sure the piston, and the end of the lever arm that locates the piston, are cleaned thoroughly.

☐ Replace the piston, relocate the arm and split pin.

☐ Replace the end cap.

☐ If a service valve is fitted, turn it on. If not, turn on at the stop tap.

☐ For both WC cisterns and CWSCs, allow the cistern to fill naturally. Make any final adjustments using the float adjustment.

☐ Service valves should be fitted in installations where they are not already present.

☐ Report back to the person who asked for the work to be carried out, informing them that the fault has been repaired.

Chapter
5

Internal stop tap

Common faults

- The stop tap won't turn off the supply properly. This is usually caused by a worn-out washer or a defective seating.
- The stop tap is stiff and hard to turn. This is the result of infrequent use and lack of regular maintenance. The spindle and packing-gland nut oxidise over time, which 'bonds' the two together.
- The stop tap is leaking. If the stop tap is a compression type, the nut and O ring joint could be leaking. Similarly, with a capillary fitting, the solder joint could be leaking. It is more likely, however, that the **packing-gland nut** is leaking. Another possibility is that condensation has been mistaken for a leak. The incoming main is very cold, and the stop tap is usually located in a kitchen, where moisture levels are high. If the pipe is not insulated, the moisture will condense on the cold surface of the pipe and trickle down it to the stop tap, giving the impression of a leak.

Key term

Packing-gland nut – a nut used to compress packing to make valve spindles watertight.

Checklist

Once you have split the body from the body nut, follow this sequence.

Isolate the supply

- Make sure you have a stop tap key ready.

- Locate the external stop tap. It should be at the boundary of the property (footpath at the front of the house). On other property types, it may not be easy to locate. In some cases, you may need assistance from the water undertaker to isolate the supply.

- Once you have located the external stop tap, lift the cover and locate the external stop tap head.

- Be prepared for the stop tap box to be filled with silt or other rubbish. You may have to clear that first to find the head.

- Once you have located the head, use the stop tap key to turn off the supply (clockwise). If it is stiff, do not force it – you may snap the head. Try twisting it both ways to get some movement. Again, if it is impossible to move it, you may have to call the water undertaker.

- Return to the property and drain down the cold water from the lowest point (usually the kitchen sink). Hopefully, there will be a drain-off tap close to the stop tap. In older properties, drain taps are not always fitted.

- Open all cold tap outlets and run a hot tap for a short period. This will lower the CWSC float-operated valve. Flush any WCs for the same reason: these actions let air into the system, removing the vacuum which is formed, and will reduce the amount of water present when the work is carried out.

Strip the component

- You will need to remove the headgear from the body. This is not always easy, as it may not have been stripped for years. It often helps

Continued ▼

to apply heat to the joint between the headgear and body. Remove the headgear using adjustable grips or spanners. If a drain-off tap was not fitted, get ready for a rush of water from the pipework above the stop tap.

Repair the component

- Strip the headgear by removing the **crutch head** and packing-gland nut.

- Clean the rising spindle and the inside of the packing-gland nut using fine emery cloth (or a file in some cases).

- Replace the defective washer.

- Check the seating for defects. If the seating is severely pitted, it might be worth replacing the stop tap. Alternatively, you could reseat the stop tap using a reseating tool that 'grinds' the seating flush again (see Figure 5.55).

Reassemble the component

- Fit all the component parts together, making sure that the spindle and the inside of the packing-gland nut are lubricated.

- Repack the gland nut using PTFE tape.

- Refit the crutch head.

- Before refitting the headgear, make sure the washer between the headgear and body is in place and intact.

- Refit the headgear and tighten it.

- If not already fitted, install a drain-off tap to bring the installation up to requirements.

Turn on and test

- Make sure the internal stop tap is turned off, as well as the drain tap.

- Turn on the external stop tap.

- Return to the building and make sure the cold tap on the sink is open. Close any other taps you have previously opened.

- Slowly open the stop tap and allow water to flow from the sink tap. This will help to remove any bits of debris that may have entered the pipework.

- Open the stop tap fully and assess the pressure at the sink tap.

- Turn off the sink tap and allow the supply to charge the system.

- Check that all the appliances fed by the supply are operating properly. Check the cistern water levels.

- Inform the customer that the fault has been repaired.

Key term
Crutch head – the handle of a stop tap.

Figure 5.55: Tap reseating kit

Identify the sources of noise in the system

Besides being annoying, in severe cases noise in plumbing systems can damage pipework and fittings, and eventually cause leaks. Figure 5.56 outlines the main sources of noise in installations.

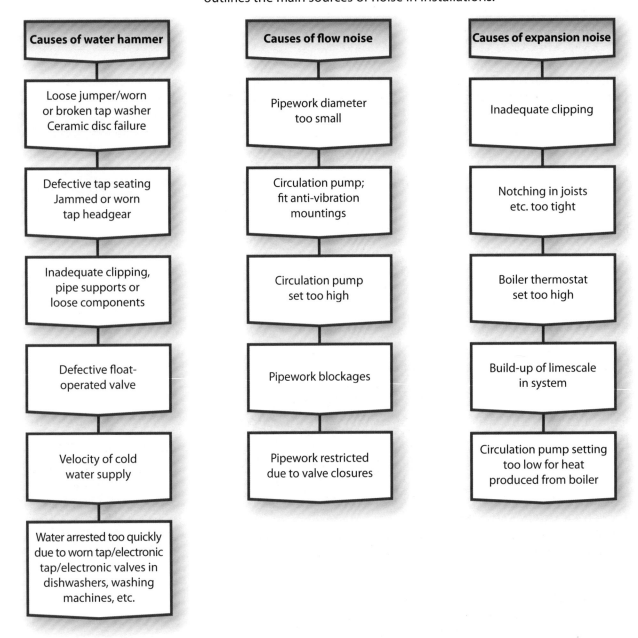

Causes of water hammer

Loose jumper/worn or broken tap washer Ceramic disc failure

Defective tap seating Jammed or worn tap headgear

Inadequate clipping, pipe supports or loose components

Defective float-operated valve

Velocity of cold water supply

Water arrested too quickly due to worn tap/electronic tap/electronic valves in dishwashers, washing machines, etc.

Causes of flow noise

Pipework diameter too small

Circulation pump; fit anti-vibration mountings

Circulation pump set too high

Pipework blockages

Pipework restricted due to valve closures

Causes of expansion noise

Inadequate clipping

Notching in joists etc. too tight

Boiler thermostat set too high

Build-up of limescale in system

Circulation pump setting too low for heat produced from boiler

Figure 5.56: Sources/causes of noise in plumbing systems

Water hammer

- This is probably the most common cause of complaint from customers.
- When a valve is closed suddenly, shock waves are transmitted along the pipework, making a loud hammering noise.
- It can be worse where cold water supply pipework is not clipped properly.
- It can be remedied by the regular maintenance of float valves or tap washers.

- The velocity of the cold water supply will further affect the problem.
- Installing a **water governor** to reduce the flow rate can help reduce the risk of water hammer.

Flow noise

- Pipework noise becomes significant at velocities of over 3 m/s. Therefore, the system should be designed to operate below 3 m/s, even if this means increasing the diameter of the pipe or installing a water governor.
- Flow noise is also sometimes heard from cisterns. Splashing noises are caused by incoming water hitting the water surface as the cistern fills.
- Silencer tubes on float valves were once used to cure this, but are no longer allowed, except for the collapsible type.

Expansion noise

Expansion noise usually occurs in hot water pipework. As the system expands and contracts, it causes creaks and cracking sounds. The use of relevant pipe clips, brackets or pads between pipes, fittings and pipework surfaces should help deal with expansion and contraction.

Sometimes you may find that noise in the system is caused by a lack of pipe support in the form of brackets or clips. In most cases, extra brackets can be added without the need to isolate the system.

Describe reasons for inadequate water supply

Underground bursts

If the leak is on underground pipework, try to locate it as follows:

- Shut off the stop tap at the property boundary and dig up the ground where the water is showing. In some cases, this may not be at the point of the leak. It is advisable to close the main stop tap in the property before turning on the boundary stop valve. This will allow you to control the water entering the property following the repair.
- After repairing the leak, leave any excavations open until tested.
- Open the kitchen tap and then slowly open the stop tap so that any dirt that may have entered the pipe is flushed out, rather than getting into the system and damaging float-operated valves.

Blocked/partially blocked components

- Check new components for blockages such as casting sand from the manufacturing process (see the case study on page 236).
- The more complex the component, the more chance there is a blockage. This is why all unvented systems connected to the cold water supply are protected by an in-line strainer. This prevents debris entering the delicate components and creating blockages.
- Take care when commissioning systems. Use service valves to remove float-operated valves when turning on the supply for the first time to prevent debris from blocking the valve outlets.
- Check tubes before installation: badly stored/transported tubes may have become blocked by their ends being caught in soil etc. on site.
- In large-diameter pipes visually check for vermin and/or spoil etc.

- Check that tubes are burr-free. If not, the slight indentation caused by tube cutters is an ideal point for debris, scale, etc. to build up, creating partial and eventually complete blockage.

Case study

The manufacture of brassware (taps etc.) involves pouring molten brass into a cast, which is formed using fine black sand (casting sand). When checking the inside of a new tap, the plumber found that some of the casting sand had not been successfully removed during manufacture. This could result in the tap being blocked from new, so always check.

Incorrectly sized components

- Ensure that all plumbing fittings and components comply with the *Water Fittings and Materials Directory* (published by WRAS). Materials and components not in the guide may be of poor quality and not sized or designed for use in the UK.
- If a component is designed with a smaller diameter than is required, the flow to appliances will be much restricted.
- Oversizing of components leads to higher installation costs and potential wastage of water.

Partially closed valves

Partially closed valves impede water flow – not water pressure. This is a common misunderstanding among both consumers and, on occasion, installers. Reducing the flow of water in this way may have operational effects on the equipment or components that the valve serves. This practice may also increase system noise, so is not recommended.

Airlocks to low pressure systems

Airlocks are one of the most common causes of problems in low-pressure hot and cold water systems. They are usually caused when pipework from CWSCs, hot water storage vessels and boilers is not installed correctly. Figure 5.57 shows two installation details that highlight the problem. Pipework running horizontally should be level or at an appropriate fall to allow air to escape the system. Horizontal runs of primary gravity circulation pipework should also be installed at the appropriate fall.

Activity 5.5

What are the main causes of airlocks in a hot water system? Create a table listing the problems that may be caused by airlocks and the ways in which airlocks can be avoided.

 **Safe working**

On the cold feed pipework from cisterns, always use purpose-made bends of slightly less than 90° to help avoid airlocks.

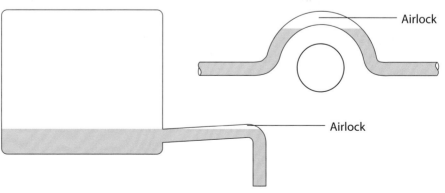

Figure 5.57: Horizontal pipework should be level or at an appropriate fall to prevent airlocks

Low incoming water pressure

Table 5.06 outlines causes of low water pressure and the action to be taken.

Cause of low water pressure	Action to be taken
Peak time – breakfast, lunch or teatime	Consider system type change
Burst pipe to adjoining property	Check adjoining properties; arrange access to vacant properties
Underground burst pipe	Locate and repair
Incoming main less than minimum 25 mm internal diameter	Consider replacement if smaller than 25 mm internal diameter
Incoming service pipe shared by adjoining property	Identify and inform Water Authority of the issue
Water Authority turned off supply to area while repair work was carried out	Report/enquire at Water Authority
Corrosion/partial blockage	Identify, locate and rectify
Location of property and length of underground service pipe	Consider increasing internal diameter of service pipe Notify Water Undertaker
Incorrect choice of cold water system/ alteration carried out to existing system	Consider system type change or modification

Table 5.06: Causes of low incoming water pressure and how to deal with them

State the procedure for leak identification

One common sign of leaks on accessible copper pipe or joints is that the water turns the fitting or pipe green and leads to a build-up of limescale. If the pipes are hidden, people may be unaware of the problem until some damage is caused. Visually inspect the system and report your findings to the customer before carrying out any work.

Explain the procedure for repairing leaks on cold water components

Figure 5.58 outlines this procedure (refer to Figure 5.45 on page 220, for the key).

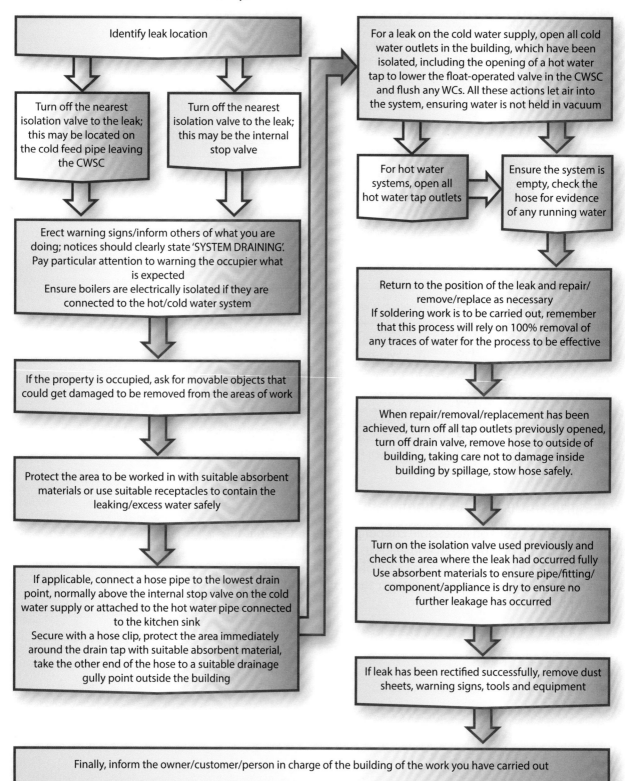

Identify leak location

Turn off the nearest isolation valve to the leak; this may be located on the cold feed pipe leaving the CWSC

Turn off the nearest isolation valve to the leak; this may be the internal stop valve

Erect warning signs/inform others of what you are doing; notices should clearly state 'SYSTEM DRAINING'. Pay particular attention to warning the occupier what is expected
Ensure boilers are electrically isolated if they are connected to the hot/cold water system

If the property is occupied, ask for movable objects that could get damaged to be removed from the areas of work

Protect the area to be worked in with suitable absorbent materials or use suitable receptacles to contain the leaking/excess water safely

If applicable, connect a hose pipe to the lowest drain point, normally above the internal stop valve on the cold water supply or attached to the hot water pipe connected to the kitchen sink
Secure with a hose clip, protect the area immediately around the drain tap with suitable absorbent material, take the other end of the hose to a suitable drainage gully point outside the building

For a leak on the cold water supply, open all cold water outlets in the building, which have been isolated, including the opening of a hot water tap to lower the float-operated valve in the CWSC and flush any WCs. All these actions let air into the system, ensuring water is not held in vacuum

For hot water systems, open all hot water tap outlets

Ensure the system is empty, check the hose for evidence of any running water

Return to the position of the leak and repair/remove/replace as necessary
If soldering work is to be carried out, remember that this process will rely on 100% removal of any traces of water for the process to be effective

When repair/removal/replacement has been achieved, turn off all tap outlets previously opened, turn off drain valve, remove hose to outside of building, taking care not to damage inside building by spillage, stow hose safely.

Turn on the isolation valve used previously and check the area where the leak had occurred fully
Use absorbent materials to ensure pipe/fitting/component/appliance is dry to ensure no further leakage has occurred

If leak has been rectified successfully, remove dust sheets, warning signs, tools and equipment

Finally, inform the owner/customer/person in charge of the building of the work you have carried out

Figure 5.58: How to deal with leaks

Working practice

A customer called you out, saying that the overflow from the CWSC in the loft overflowed through the warning pipe at night. You rewashered the float-operated valve, checked the water level and left.

The next day you are called back for the same problem. You decide to replace the valve completely to save another recall. After checking for leaks on the new connection and ensuring the water level in the cistern is correct, you leave once again. However, you are called back again the next day as the same problem is still occurring.

- Why is the cistern only overflowing at night?
- What action could you take to prevent further problems with the valve?

Progress check 5.06

1 List some common faults that might affect taps.

2 Name three types of float-operated valve that may be found in use.

3 Produce a list of bullet points identifying the possible causes of water hammer.

4 Which other two types of noise, not related to water hammer, might be experienced in plumbing systems?

5 List possible causes of low incoming water pressure to a property.

6 How can a leak be identified when there is no visual confirmation?

7 A float-operated valve has no water passing through it and it is confirmed that water is available up to the valve. What is the likely cause of the blockage?

8 When commissioning an indirect cold water system it is found that water from the CWSC is not reaching the bathroom appliances. Other than a closed isolation valve or blockage, what could the problem be?

9 While rewashering a dripping tap, you notice that the washer is badly worn. However, after replacing it with a new washer, the tap still drips badly. What could be the cause of this problem?

10 What should be fitted with a float-operated valve in a plastic CWSC? What is the purpose of this component?

Knowledge check

1 According to the Water Regulations, 'fluid that represents a significant health hazard' is identified as which fluid category?

  a 2   b 3   c 4   d 5

2 What type of air gap is built into the design of a pillar tap to prevent backflow?

  a AUK1
  b AUK2
  c AUK3
  d AUK4

3 Which source of water is not categorised as 'wholesome water' before treatment?

  a spring water
  b stored rainwater
  c deep well water
  d upland surface water

4 What is the purpose of 'snaking'?

  a to reduce the mains pressure at non-peak times
  b to use more pipe than is required
  c to reduce noise in the pipework
  d to allow for lineal movement if ground conditions change

5 How is the potentially fatal legionnaires' disease contracted?

  a passed from person to person orally
  b through drinking water
  c by breathing droplets of contaminated water
  d by close contact with an infected person

6 Screening, coagulation, sedimentation and filtration are all terms associated with what?

  a testing for legionnaires' disease
  b commissioning a cold water system
  c processes in the treatment of water
  d ways in which underground water reserves are found

7 Where is a service pipe located?

  a connecting the water main to the external stop valve
  b connecting the external stop valve to the internal stop valve
  c linking two cold water storage cisterns together
  d rising vertically through the building

8 Where is a service valve located?

  a immediately above the internal stop valve
  b connecting several appliances together
  c on top of the water main
  d adjacent to a float-operated valve

9 Which part of the Building Regulations covers cold water systems?

  a E   b F   c G   d H

10 In indirect cold water systems, which of these appliances is the only one connected directly to the rising main?

  a sink   b bath   c washbasin   d WC

11 Which of the following is **not** a suitable use for water which has been collected by rainwater or greywater harvesting?

  a flushing toilets
  b washing cars
  c watering the garden
  d drinking

12 How far should the rising main inside the building be from the outside wall of a building if the pipe in the duct is uninsulated?

  a 650 mm
  b 750 mm
  c 1000 mm
  d 1350 mm

13 What is the minimum size of a connection pipe when two cisterns are linked?

  a 15 mm
  b 22 mm
  c 28 mm
  d 35 mm

14 Which of the following alloys contains zinc?

  a brass
  b bronze
  c solder
  d stainless steel

15 What is the purpose of the stainless steel screen found in warning pipe connectors and cistern vents?

  a to prevent insects/vermin entering the cistern
  b to prevent filtered air entering into the cistern
  c to prevent dead insects leaving the cistern in case of overflow
  d to provide frost protection

16 How far should a warning pipe connection be made above the static water level in a CWSC?

  a 15 mm
  b 20 mm
  c 25 mm
  d 30 mm

17 In an indirect cold water system, how far above the cold water distribution pipe should the DHWS cold feed be positioned?

  a 20 mm
  b 25 mm
  c 35 mm
  d 50 mm

18 In an indirect cold water system, why should the DHWS cold feed be positioned above the cold water distribution pipe?

  a to prevent cross flow
  b to allow more pressure in the cold pipework than the hot
  c to prevent corrosion
  d to prevent scalding should the supply be interrupted

19 In the permanent decommissioning of a cold water supply, why should you remove all dead legs?

  a to prevent stagnation of the water left in the pipes
  b to have more scrap to recycle
  c to prevent corrosion
  d to make the remaining system more noisy

20 When fault-finding in an occupied building, which source of information should you seek first?

  a contacting the original installer
  b consulting the manufacturer's instructions
  c asking the occupier of the building
  d using the Internet to find similar issues experienced by plumbers

Domestic hot water systems

This chapter will cover the following learning outcomes:

- Know the types of domestic hot water systems
- Know the components used in domestic hot water systems
- Understand the installation requirements of domestic hot water plumbing systems
- Know the design features of showers
- Understand the basic maintenance requirements of hot water systems
- Understand the key requirements of testing and decommissioning of domestic hot water systems

Introduction

Perhaps surprisingly, the supply of hot water to domestic dwellings only became commonplace after the Second World War. Since then, the design of hot water systems and installations has improved enormously. This chapter looks at methods of heating water and at how hot water is then stored and distributed for various uses. Much of the information in this chapter is common to Chapter 5: *Cold water systems*, so you should refer back when necessary.

KNOW THE TYPES OF DOMESTIC HOT WATER SYSTEMS

There are basically two types of hot water system: storage and non-storage. However, within these two types there are many variations. Hot water storage was the original form of hot water provision and is still required as an alternative to non-storage (instantaneous) systems. Designers and installers must consider the end user in the selection of a suitable hot water system.

Identify the key sources of information related to the installation of hot water systems

As with cold water systems, the main information sources are:

- the Water Regulations (plus a guide to the Water Regulations which simplifies the content of the original document)
- BS 6700 – services supplying water for domestic use
- the Building Regulations – Parts G and L1 set out the legal requirements for the energy-efficient installation of hot water systems; Part G deals with installation requirements and water efficiency.

These details are covered at some length in the *Domestic Heating Compliance Guide* and *The Water Efficiency Calculator for New Buildings* published by the government's Department for Communities and Local Government.

You should also consult appliance and component instructions and manufacturers' literature during the installation, maintenance or servicing of equipment. You must leave these documents with the consumer when you hand over the installation or complete any maintenance or servicing work.

Identify the main types of hot water systems

There are many types of hot water system. However, all types may basically be categorised as storage or non-storage systems. The systems will be covered in more detail later in the chapter.

- **Storage systems:** Water is heated and stored in a vessel until needed. This type can be vented or unvented.
- **Non-storage (instantaneous) systems:** These supply hot water instantly. They can be simple **single-point** systems or more complex **multipoint** systems.

Key terms

Single-point – an instantaneous water heater supplying one outlet.

Multipoint – an instantaneous water heater that can supply water to more than one outlet.

The following factors should be considered in system selection and design:

- the amount of hot water required
- the temperature during storage and at outlets
- installation and maintenance costs
- fuel energy requirements and running costs
- any wastage of water and energy
- safety for the user.

Choosing a system

Systems range from simple single-point arrangements to complex centralised boiler systems supplying hot water to a number of outlets. Figure 6.01 shows ways of supplying hot water as set out by BS 6700. It is divided into centralised and localised systems.

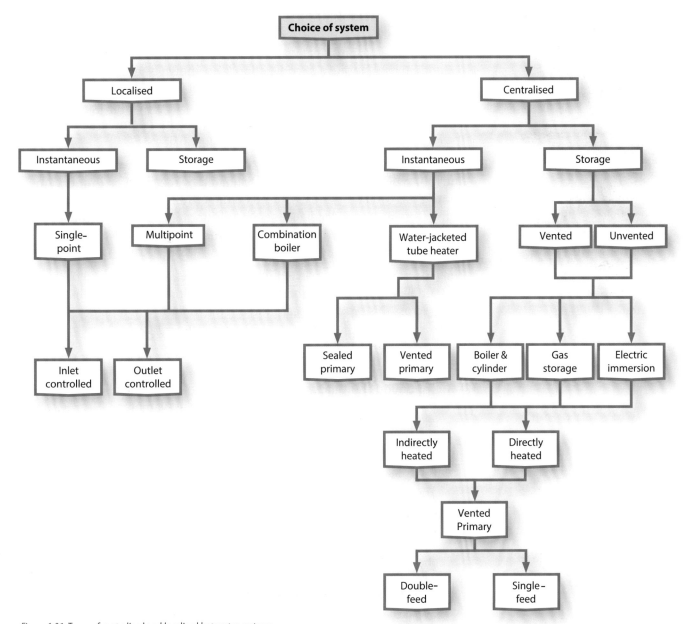

Figure 6.01: Types of centralised and localised hot water systems

Centralised system

- Water is heated and can be stored centrally in a building.
- A thermostat can control the heating of the water.
- A system of pipework supplies the heated water to various draw-off points.

Localised system

- Water is heated where it is needed, such as a single-point water heater sited over a sink.
- Localised systems are often used in situations where long distribution pipe runs would involve a waste of water and energy.

Describe the operating principles of basic hot water storage systems in domestic dwellings

Direct systems (vented)

Direct hot water storage systems are fed via a cold water storage cistern (CWSC). These systems heat water in a boiler (by gas, oil or solid fuel). The water then rises because of convection (also known as **gravity circulation**) through the primary flow pipe and into the hot water storage vessel, thus heating the contents of the vessel directly. The hot water from the boiler is replaced by the cooler and heavier water moving in the primary return from the lower area of the storage vessel (see Figure 6.02).

These systems do not always have to be heated by a boiler via primary flow and return pipes. They can be directly heated by means of an **immersion heater**. Gas circulators are also used on direct systems, connected directly to the storage vessel. Appliances such as instantaneous water heaters, when supplied directly, should be fitted with a servicing valve (ball type) as close to the appliance as possible.

The type of system shown in Figure 6.02 is no longer widely used because of recent energy efficiency requirements. Therefore, you are only likely to come across this type when carrying out maintenance or repairs.

Key points about direct systems using gravity circulation

- Minimum pipe sizes for primary circuits to hot water storage vessels are: 22 mm for short pipe runs; 28 mm for longer pipe runs (or from continuous burning appliances).
- Vent pipes should not be less than 22 mm in diameter.
- All pipes must be laid to falls to prevent airlocks and to help systems drain down.
- The vent route from the boiler, primary flow and open vent should not be valved.
- The cold feed pipe (normally sized at 22 mm in a small domestic property) should be sized in accordance with BS 6700. The cold feed is the key route by which expanding water is taken up from the cylinder when it is heated; the heated water from the cylinder moves through the cold feed pipe and the water level rises in the storage cistern.
- Any pipe feeding a hot water system (cold feed) must be a minimum of 25 mm above the cold water distribution pipe (if applicable) to avoid scalding situations should the supply be interrupted.

> **Key terms**
>
> *Gravity circulation* – cold water is heavier than hot water. This means that gravity exerts a stronger pull on cold water, drawing it down and allowing the hot water to rise through the system.
>
> *Immersion heater* – an electric element fitted inside a hot water storage vessel. It can be controlled by a switch and thermostat.

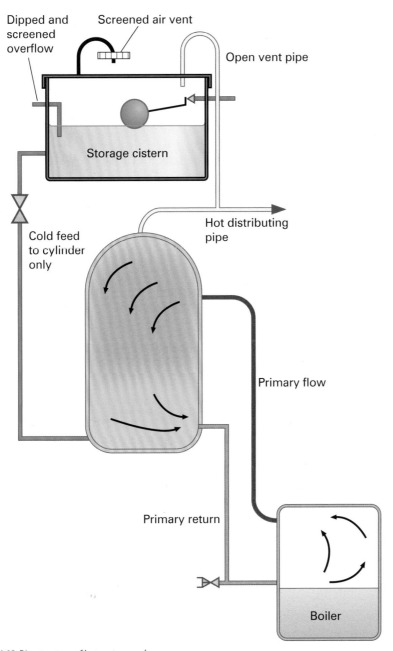

Figure 6.02: Direct system of hot water supply

- An isolation valve, a full-way gate valve or lever-type spherical valve should be located on the cold feed pipe for maintenance purposes, as near to the CWSC as practicable. In the case of indirect storage cylinders, a drain valve should be fitted to the cold feed at its lowest point, normally adjacent to its connection into the storage cylinder.
- No other pipework should be connected into the cold feed pipe.
- The open vent pipe cannot be taken directly from the top of the hot water storage vessel. This is to prevent one-pipe circulation (parasitic circulation).

- The hot draw-off pipe should rise slightly away from the storage cylinder to the open vent connection to aid the removal of any air in the system.
- The hot water draw-off pipe should also incorporate a 450 mm offset between the storage vessel and its point of connection to the open vent pipe to prevent one-pipe circulation.
- The vent pipe must be terminated at the correct height above the CWSC. In domestic situations, this is 150 mm plus 40 mm per metre of the system height (head).
- Corrosion inhibitors should not be used because the water in the boiler is fed directly to the appliances.
- No other supplies or draw-offs should be connected to the cold feed.

Figure 6.03 shows a direct system heated by an immersion heater. The immersion heater should be controlled by a thermostat.

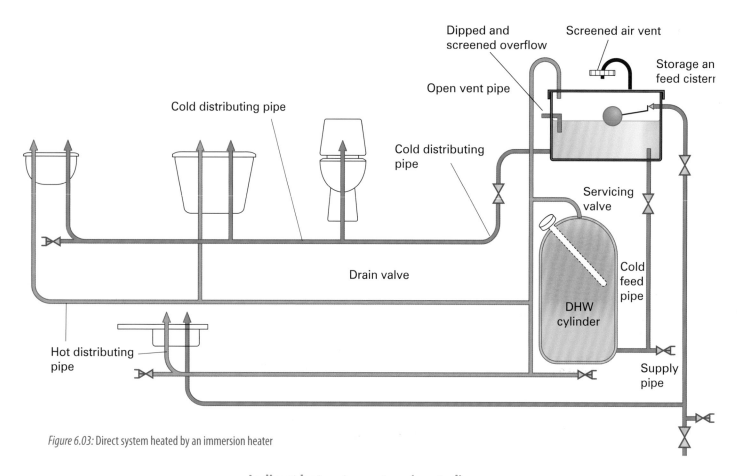

Figure 6.03: Direct system heated by an immersion heater

Indirect hot water system (vented)

This is the most common form of vented domestic hot water system and allows the boiler to be used for the central heating circuit. The system permits the use of a variety of different metals because the primary circuit is totally separate from the secondary circuit. The system is called indirect because the water in the storage vessel is heated indirectly through a heat exchanger. Figure 6.04 shows a vented double-feed indirect system.

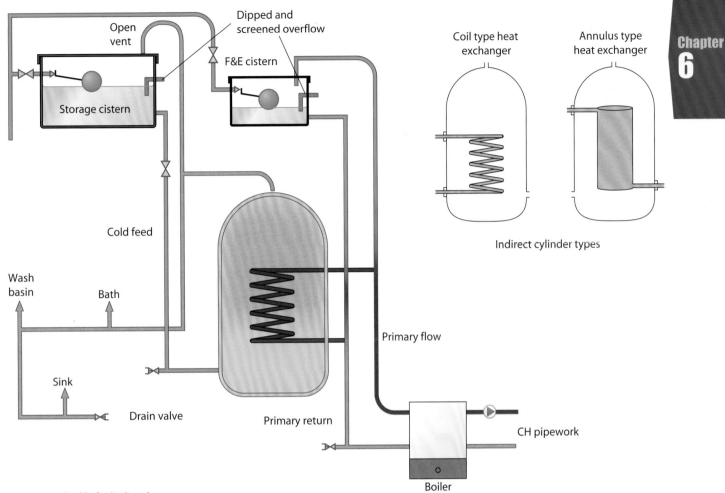

Figure 6.04: Double-feed indirect hot water system

Key points about double-feed cylinders:

- The open vent and cold feed pipes may be connected to the primary flow and return pipes, as shown in Figure 6.04, or fed separately into the boiler.
- Where the vent pipe is not connected to the highest point in a primary circuit, an air release valve should be fitted to aid the removal of air, but the open vent must also be retained.
- A separate feed and expansion cistern needs to be provided to feed the primary circuit. This ensures that where a double-feed cylinder is used, the primary water is kept totally separate from the secondary hot water.

Indirect single-feed system

This is also known as a primatic cylinder system (see Figure 6.05 on page 248). It uses a self-venting cylinder and does not require a separate feed and expansion cistern. The water in the primary and secondary circuits is separated by an air bubble. The cylinder must be carefully installed, in accordance with the manufacturer's requirements, to ensure that the air bubble is not dislodged in the cylinder, which would allow the two waters to mix. It is for this reason that this type of cylinder is not widely used in modern vented types of hot water system.

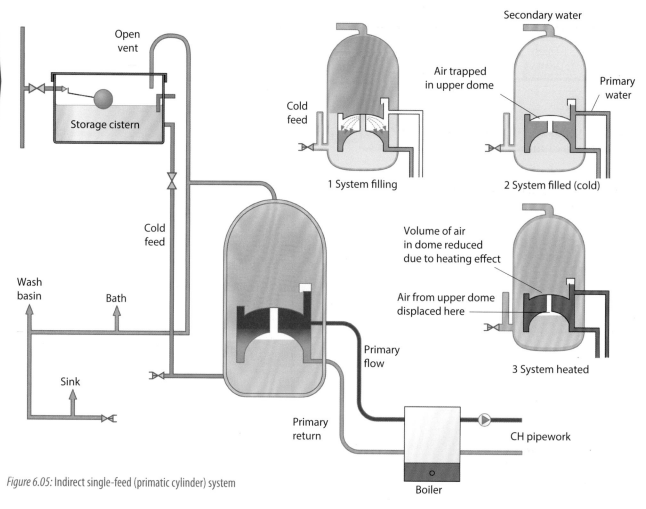

Figure 6.05: Indirect single-feed (primatic cylinder) system

Key points about single-feed cylinders

- Water enters the primary circuit via a number of holes at the top of the vertical pipe immediately under the upper dome.
- The system is self-venting through the air vent pipe while the primary circuit is filling.
- Once the primary circuit is filled, the filling of the secondary supply (the one that feeds the appliances) continues.
- When the secondary supply is full, two air seals are formed and a seal is maintained. However, this seal could be lost if the water is heated too much and expands through this route (solid fuel appliances) or if the system volume is exceeded.
- Once the water is heated, expansion of the water in the primary circuit is taken up by forcing the air from the upper dome. This air is displaced down the vertical pipe to form another air seal in the lower dome. It is not lost.

Vented cylinders

Vented hot water storage cylinders are available off the shelf in a number of sizes to suit particular circumstances, such as airing cupboard dimensions:

- Standard heights vary from 675 mm to 1800 mm.
- Standard widths vary from 300 mm to 600 mm.

The most common cylinder size for a standard one-bathroom property is 900 × 450 mm, storing approximately 120 litres of water. Some manufacturers also produce custom-made cylinders to the required dimensions for a job.

Cylinders are available in three grades, based on the overall head of water generated by the cold water storage cistern on the base of the hot water storage cylinder:

- Grade 3: maximum head 10 m (used in most domestic properties)
- Grade 2: maximum head 15 m
- Grade 1: maximum head 25 m.

Heat sources (boilers) connected to hot water storage systems (except solid fuel appliances) require a thermostat, which is set by the consumer. In existing systems, this is the minimum control for hot water storage. In systems complying with current regulations, much more control is required. These points are covered later in this chapter (see pages 263–264) and in Chapter 8: *Central heating systems*.

An immersion heater fitted in the cylinder must now be a dual thermostat control type. The first thermostat can be manually set by the installer, normally to 60–65°C. The second thermostat, known as the 'high limit', will be preset and will cut the system off until it is manually reset. However, if this does operate, it will usually indicate that the normal control thermostat has failed and may need replacing.

Recently there were changes to the Building Regulations and document G3, which now covers vented hot water cylinders as well as unvented hot water. G3 states that a hot water supply for any fixed bath must be designed and installed so that the temperature of the water cannot rise above 48°C. This requirement applies to new-build and conversion properties and is intended to prevent scalding. See the section on blending valves (thermostatic mixing valves) on pages 263–264.

Specifications

Any new or replacement vented hot water storage cylinders must comply with *Domestic Heating Compliance Guide* requirements. The main points are:

- A direct, indirect, single- or double-feed cylinder must be manufactured to BS 1566 standards.
- The cylinder must be labelled as Building Regulations Part L compliant and show:
  - the type of vessel
  - the nominal capacity in litres
  - the standing heat loss in kWh/day
  - the heat exchanger performance in kW
  - compliance with BS 1566.

BS 1566 lays down minimum requirements for the insulation applied by manufacturers to vented cylinders, as well as for the surface area of the heat exchanger (coil) in an indirect cylinder. The greater the coil surface area, the more energy efficient the cylinder will be.

The *Domestic Heating Compliance Guide* requires that all the pipework connecting to a hot water cylinder should be insulated:

- to a point not less than 1 m from the cylinder, or
- to the point at which the pipework becomes concealed.

Did you know?

Setting the maximum water temperature to 60–65°C helps prevent limescale deposits in the system.

Did you know?

'Super-duty' type cylinders have a higher performance rating than BS 1566 and so are more energy efficient.

The storage capacity of a hot water cylinder is based on the consumption of hot water in the property. The recovery rate should also be considered. According to BS 6700, the minimum storage capacity in a small property should be 100 litres.

Combination storage systems (vented)

Under the requirements of the *Domestic Heating Compliance Guide*, combination-type hot water cylinders should be manufactured to BS 3198. The main feature of a combination-type hot water storage cylinder is that it contains an integral cold water storage cistern (see Figure 6.06). The base of the cistern must therefore be positioned at a higher level than the level of the highest water outlet (tap) in the property. This height (head) must also ensure that sufficient water flow rate is available at the outlets. These systems are not widely used in newer installations as the relatively low head of water generated may not provide the required water flow rate at the outlets unless the system is boosted with a water pump.

Advantages of these systems

- They have low installation costs.
- They can be useful in flats, provided minimum water flow rates are achieved.

Disadvantages

- They cannot normally be used with gravity-fed showers (unless the system is pumped).
- They tend to generate lower pressure and hence lower flow rate at outlets.
- The cold water storage cistern is often of limited capacity.

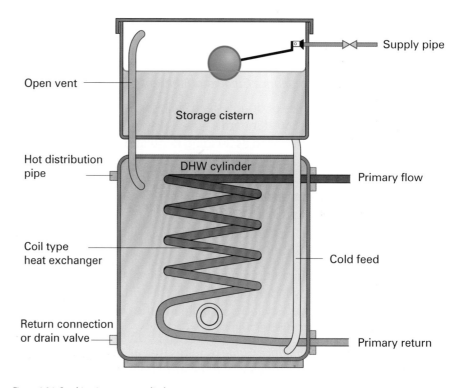

Figure 6.06: Combination storage cylinder

Unvented hot water storage systems

This type of system is mainly covered at Level 3, and to install or maintain these systems you must hold an unvented hot water certificate/card. An overview of the system is given below.

Figure 6.07 shows the typical layout of an unvented hot water storage cylinder. Table 6.01 outlines the purpose of the functional control devices fitted to the cylinder illustrated (numbers in brackets in the table).

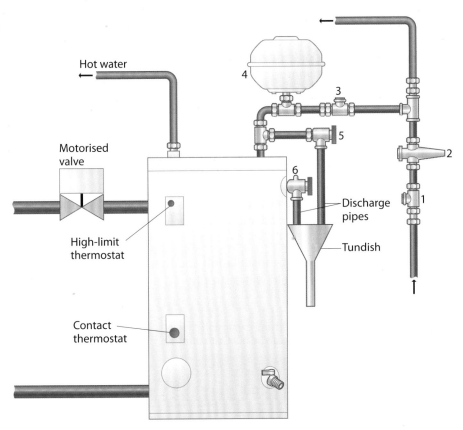

Hot water

Motorised valve

High-limit thermostat

Contact thermostat

Discharge pipes

Tundish

Figure 6.07: Typical unvented hot water storage cylinder

Functional control device	Tasks performed
Line strainer (1)	Prevents grit and debris entering the system from the water supply (which would cause the controls to malfunction)
Pressure-reducing valve; on older systems this may be a pressure-limiting valve (2)	Gives a fixed maximum water temperature
Single-check valve (3)	Prevents stored hot water from entering the cold water supply pipe (a contamination risk)
Expansion vessel or cylinder air gap (4)	Takes up the increase in water volume in the system caused by the heating process
Expansion valve (5)	Operates if the pressure in the system rises above the design limits of the expansion device (i.e. if the cylinder air gap or the expansion vessel fail)
Temperature relief valve (6)	Protects against excess temperature and pressure
Isolating (stop) valve (not shown in Figure 6.07)	Isolates the water supply from the system, for maintenance

Table 6.01: Functional control devices fitted to unvented hot water storage cylinder

This system does not include an open vent pipe or a cold water storage cistern. Essentially, it operates under the influence of mains/boosted cold water pressure, therefore providing higher water flow rates at taps (outlets). As the hot water storage cylinder operates as a closed vessel (at high internal pressure), the Building Regulations state the following requirements.

- The cylinder must be fitted with a range of safety (temperature) controls to ensure that the stored hot water temperature never exceeds 100°C.
- The discharge pipes from the cylinder must be located in a safe place where they are not going to scald anyone.
- The cylinder must only be installed/maintained by operatives who have passed an unvented systems competency test
- The cylinder installation must be notified to the local authority before installation or be self-certified by a Competent Persons Scheme member company.

An unvented hot water storage cylinder must be manufactured to BS EN 12897 requirements. This standard also details insulation and heat exchanger performance requirements for the cylinder to comply with Part L1 of the Building Regulations. Under the Water Regulations requirements, the cylinder installation must include a range of functional controls preventing:

- over-pressurisation of the installation and therefore wastage of water
- contamination of the cold water supply.

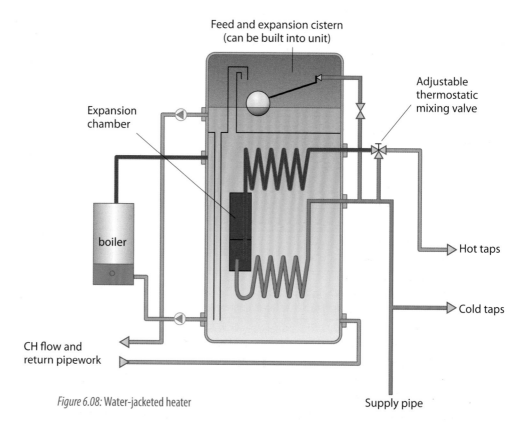

Figure 6.08: Water-jacketed heater

Water-jacketed heater

This is also known as a thermal storage system (see Figure 6.08). When the hot tap is turned on, cold water from the mains or a storage cistern passes through a heat exchanger, which is in a heat store of primary hot water. The size of this heat store is based on the volume and rate of flow that can be delivered without an unacceptable drop in temperature. The cylinder thermostat programmes the primary water flow from the boiler. Hot water is pumped to the radiator heating circuit and is returned to the heat store. The cooler water from the heat store is then returned to the boiler, where it is reheated. This is similar to an indirect domestic hot water system but in reverse.

Describe the operating principles of basic hot water non-storage systems

Non-storage (instantaneous) systems work by passing cold water from the service pipe through a heat source, which heats the water before it emerges at the appliance outlet. The heat source can be gas, oil or electric, and the system can have single, multipoint or combination boilers.

The rate at which the water can be heated is limited. Therefore, the flow rate needs to be controlled so the water can be heated properly. Because of the reduced flow rate, it is not possible to supply a large number of outlet points all at once, so these systems are not installed in situations where there is high demand. For example, you might find a multipoint boiler in a small property, or a single-point in an office kitchen area or WC.

Combination boilers

Combination (combi) boilers heat up a cold water supply instantaneously for domestic hot water, so they are classified as a hot water supply. They can also supply hot water for central heating systems.

Multipoint (gas-fired)

As shown in Figure 6.09, this consists of a gas burner sited beneath the heat exchanger. When the hot tap is opened, water passes through the heater. This causes the gas valve to open as a result of the drop in pressure in the **differential valve**. This drop in pressure is caused by water passing through the **venturi**, which creates a negative pressure as it sucks the water from the valve. The diaphragm is connected to a push rod and, as this lifts, it opens the gas line. The gas is then ignited by the pilot light. When the hot water tap is turned off, the pressure in the differential valve is equalised, the diaphragm closes and the gas supply is turned off.

> **Key terms**
>
> **Differential valve** – a valve that opens and closes automatically in response to the flow of water passing through the unit.
>
> **Venturi** – a constricted piece of tube that has the effect of reducing fluid or liquid pressure when a fluid flows through it.

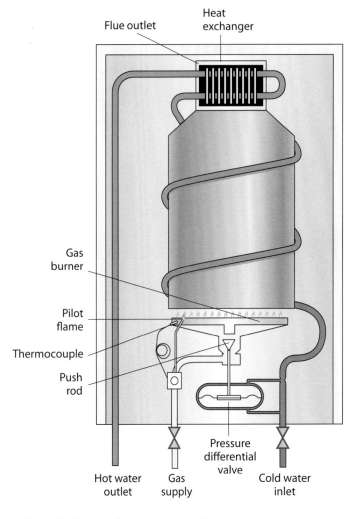

Figure 6.09: Typical gas-fired multipoint water heater

Single-point instantaneous water heater

This type of heater uses electricity or gas. Sited directly above the appliance, it is usually inlet controlled, with the hot water delivered via a swivel spout. The electric multipoint is a small tank of water with an electric heating element inside. Because of the low volume of water, it quickly heats up as it is drawn through the heater. The temperature at the outlet depends on the water flow rate and the kW rating of the heater. Gas single-point heaters are now less common, but you may cover them within the gas qualification.

Single-point heaters are used where a small number of hot water draw-offs are fed by individual heaters in a non-domestic type building and where the use of a centralised hot water system would be uneconomical. You might find them in the WC of small cafes, for example.

Storage heaters

Outlet controlled

These are more common in large domestic or small commercial/industrial buildings. Figure 6.10 shows a typical storage heater. In this case, it is heated by gas, but electric storage heaters are also available. A gas storage heater is basically a self-contained boiler and storage system. This system also includes an open flue, which must be terminated externally. Often referred to as pressure-controlled water heaters, they are usually designed to be fed by a cistern (indirect) or mains (direct) supply, in which case they would be included in an unvented system. This type of storage heater is classified as outlet controlled because the supply is controlled at the appliance outlet, such as a hot tap. It will also serve multiple outlets.

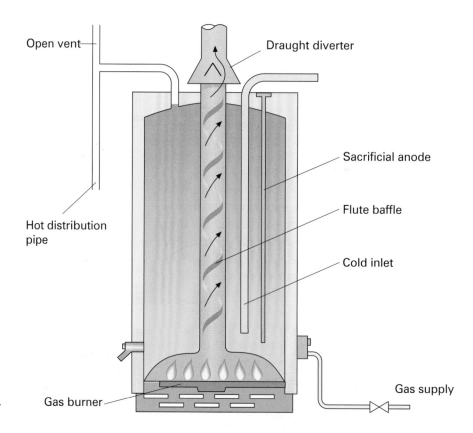

Figure 6.10: Outlet controlled gas storage heater

Inlet controlled

As Figure 6.11 shows, inlet controlled storage heaters are generally single-point heaters, fitted either above the appliance with a swivel outlet spout, or under the appliance. The heater is fed from the supply pipe, which has an inlet control. The outlet, and any connections made to it, must not be obstructed, as the open outlet allows for expansion of the water on heating. If an under-sink model is used, a special tap is required to allow venting of the water heater. They can be heated by gas or electricity.

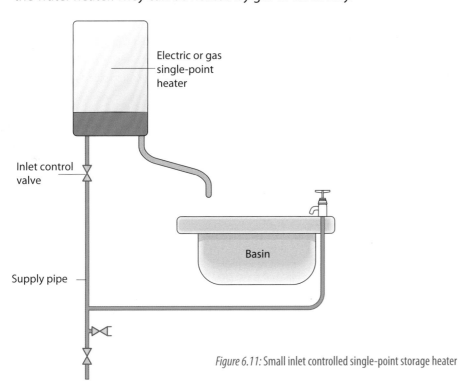

Electric or gas single-point heater

Inlet control valve

Supply pipe

Basin

Figure 6.11: Small inlet controlled single-point storage heater

Working practice

A couple lives in a two-bedroom flat, which is heated by night storage heaters. A combination storage cylinder (vented) served by an immersion heater supplies the hot water.

The couple have called Adam to the flat as the hot water storage cylinder is leaking. Adam tells them that the cylinder is not repairable as its base has started to corrode. He offers the customers an estimate to replace the faulty cylinder. They ask if there is an alternative way of providing hot water, as they are at work all day and the heated water tends to get wasted as the immersion has to be on but they never get the benefit. They consider this a waste of money and energy.

Consider the following:

- For hot water demand, the flat has a kitchen and bathroom.
- A wet central heating provision will not be considered.
- Gas is available in the dwelling.
- This is a ground-floor flat.
- The cylinder cupboard is in the bathroom on an outside wall.
- There is no shower as the system head is currently too low.
- Water pressure and flow are good.

What are the advantages and disadvantages of storage and instantaneous hot water in this case? What type of system or appliance would you recommend, taking into account the clients' wishes? What type of fuel should be used and be most convenient? How many hours would the work take?

Progress check 6.01

1 What is a centralised storage system of hot water?

2 What is a localised storage system of hot water?

3 What is an instantaneous hot water system?

4 What is a storage system of hot water?

5 Explain the term 'gravity circulation'.

6 How can an immersion heater be controlled?

7 What is the minimum horizontal length of the draw-off pipe from the top of a storage vessel, to prevent one-pipe circulation?

8 What is the maximum temperature that the thermostat controlling water in any hot water storage cylinder should be set to?

9 Which grade of hot water storage cylinder is most common in domestic properties?

10 What information would you find in the *Domestic Heating Compliance Guide*?

Identify the fuel types used with direct and indirect hot water storage systems

The following energy sources or fuels are used for heating hot water:

Electricity
- immersion heater
- instantaneous heater
- storage heater

Gas
- boiler
- water circulator
- instantaneous heater
- storage heater

Solid fuel
- boiler
- combined cooker and boiler

Oil
- boiler
- combined cooker and boiler

Solar thermal

Solar water heating

Solar water heating systems use heat from the sun to work alongside a conventional water heater. They collect heat from the sun's radiation via a flat plate system or an evacuated glass tube system before being stored in a hot water cylinder. The system is not usually pressurised and often uses 'drain back' technology: the pipework slopes between items and so does not have water in it once it is used. This means that there is no need for insulation.

When a solar water heating and hot water central heating system are used in conjunction, solar heat will be concentrated inside a pre-heating cylinder that feeds into the cylinder heated by the central heating. Alternatively, the solar heat exchanger will replace the lower heating element. The upper element will remain in place to provide for any heating that solar energy cannot provide.

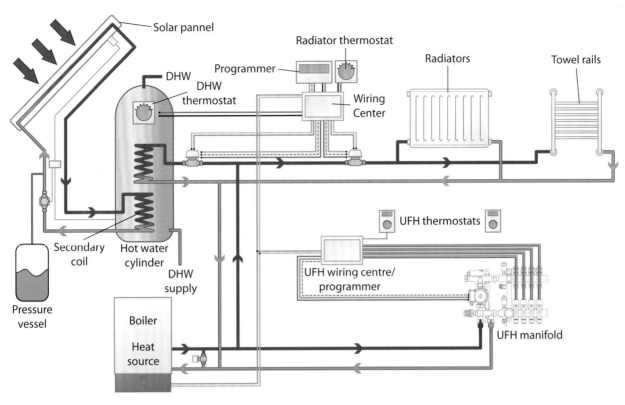

Figure 6.12: Solar thermal store system

However, we usually use our central heating at night and in winter when solar gain is lower. Solar water heating for washing and bathing is often a better option than for central heating, because supply and demand are better matched (see Figure 6.13).

Advantages

- It is environmentally friendly.
- No fuel deliveries are needed.
- It is low maintenance.

Disadvantages

The main disadvantage is that the system has a relatively high set-up cost.

Did you know?

In many climates, solar-powered systems can provide up to 85 per cent of domestic hot water. In many northern European countries, combined hot water and space heating systems (solar combi systems) are used to provide 15–25 per cent of home heating energy.

Figure 6.13: Solar water heating system

Did you know?

Heat pump capacities range from a few kW to many hundreds of kW.

Geothermal

Ground source heat pumps (GSHPs) take low-level heat that occurs naturally underground and convert it to high-grade heat using an electrically driven or gas-powered heat pump. This heat can then be used to provide space heating for a building. GSHPs can also be driven in reverse to provide comfort cooling.

The heat is collected through a series of underground pipes laid about 1.5 m below the surface, or from a borehole system. In both of these options, water is recirculated in a closed loop underground and delivered to the heat pump, which is usually located inside the building.

The installation of GSHPs requires a large amount of civil engineering works, such as sinking boreholes (50 m+) or digging trenches 1–2 m deep to house the collector pipe. The feasibility of doing this will depend on the geological conditions at the site.

Connecting a GSHP into an existing system is often constrained by the requirements of the system to operate at temperatures above those delivered by the GSHP. This can often be overcome, but at extra cost. GSHPs are best suited to new-build projects, where they can be included in the building design.

Biomass

Biomass is biological material (derived from living or recently living organisms) that can be converted into electricity or clean-burning fuels in an environmentally friendly and sustainable manner. It is accepted as a renewable replacement for fossil fuels, as it can be replaced at the same rate as it is used. The essential difference between biomass and fossil fuels is the timescale involved.

Fossil fuels such as coal, oil and gas originally came from biological material, but that material absorbed CO_2 from the atmosphere many millions of years ago. Using these materials as fuel involves burning them, which leads to the carbon oxidising into carbon dioxide and the hydrogen becoming water vapour. Unless these combustion products are captured and stored, they are released back into the atmosphere.

Biomass, on the other hand, takes carbon out of the atmosphere while it is growing and returns it as it is burned. If it is managed on a sustainable basis, biomass can be harvested as part of a constantly replenished crop. This maintains a closed carbon cycle, with no net increase in atmospheric CO_2 levels – so it is 'carbon neutral'.

Biomass energy is derived from five distinct energy sources:

- virgin wood from forestry or wood processing
- energy crops (high-yield crops grown specifically for energy applications)
- agricultural residues (residues from agriculture, harvesting or processing)
- food waste from food and drink manufacture, preparation and processing and consumer waste
- industrial waste and **co-products** from manufacturing and industrial processes.

Key term

Co-products – by-products of some industrial processes which can be used as biomass fuel.

Wood energy is derived from the direct use of wood as a fuel to provide heat (for example, in domestic wood-burning stoves) or as a replacement for fossil fuel in a power station. Crops such as corn and sugar cane can be fermented to produce ethanol, which can be used as transportation fuel or as an addition to petrol. Biodiesel, another transportation fuel, can be produced from leftover food products such as vegetable oils and animal fats.

Advantages

- Biomass production is carbon neutral – it produces no more carbon dioxide than it absorbs.
- The materials used to fuel biomass energy are otherwise sent to landfill, so it is an excellent green use for waste.
- Biomass energy could realistically produce up to 80 per cent of a home's energy needs.

Disadvantages

- Although biomass energy (like most alternative fuels) is more cost-effective than fossil fuels, it does incur initial set-up costs and ongoing running costs.
- Storage space is needed and this requires ventilation and a dry environment.

Explain the advantages and disadvantages of hot water storage systems

Tables 6.02 and 6.03 (on page 260) outline the advantages and disadvantages of hot water storage systems and non-storage hot water systems.

Advantages	Disadvantages
Stored hot water is available in the event of non- or low-pressure supply.	The amount of water stored is restricted by the size of the cylinder.
Power showers can be used.	Heating of water needs to be planned.
The cylinder is usually located in a cupboard, giving a clothes airing space.	A CWSC is needed in the roof space (except in unvented systems); also increases the risk of frost damage.
Bulk hot water is available on demand.	More pipes, and larger pipe diameters, are needed than in non-storage systems.
They are relatively low maintenance.	Maintenance costs increase for unvented systems.
Good flow rates are possible, but these depend on head (height of storage cistern) or mains pressure (in unvented systems).	If stored hot water is used up, it will take time for the system to heat up again.
	Unvented systems rely on a suitable, uninterrupted mains water supply.
	Heated water will chill if not used, and must be reheated before further use.

Table 6.02: Advantages and disadvantages of hot water storage systems

Advantages	Disadvantages
No CWSC is required.	Breakdown of the heat source is a greater risk than in storage-type systems.
Mains pressure throughout the system means that shower installations are simplified.	There is no airing cupboard (though this may be seen as an advantage). A radiator may be placed in a cupboard as an alternative, if required by the client.
May use only one heat source – a combination boiler (heating and hot water).	May not cope with consumer demand in larger installations.
Water is heated instantaneously, on demand – no stored hot water, so no chilling factor.	Pressure of supply is easily interrupted when other appliances are used (e.g. if a WC is flushed). This depends on the available pressure from the water main.
Fewer pipes, and smaller pipe diameters, are needed so these systems are less labour-intensive and more cost-effective.	They are usually more expensive to repair and maintain.
There are no pipework components in the roof space (usually) so less risk of freezing.	Flow rates may be lower than in storage-type systems – it will take longer to fill appliances (e.g. baths).
No traditional airing space is required (though some customers may see this as a disadvantage).	

Table 6.03: Advantages and disadvantages of non-storage hot water systems

Working practice

You have been asked to advise a potential customer on the types of renewable energy sources and how they could be incorporated into providing hot water to an existing detached three-bedroom property.

Gas and electricity are available at the property, but the owner would like to find alternative renewable means to provide either full or supplementary energy.

Make up a table to include the following information.

- Identify the different renewable energy sources.
- Indicate the advantages and disadvantages of each energy source.
- Briefly outline how the energy can be used to provide a hot water supply. Sketches or images may help with this.

Progress check 6.02

1 Which British Standard deals with the services supplying water for domestic use?
2 What are the two main categories of hot water system?
3 Define the term 'stratification' in the context of a hot water supply.
4 Explain the difference between storage and non-storage systems.
5 Name four energy sources or fuels in common use (excluding renewable energy).
6 Identify five advantages of hot water storage systems.
7 Identify five disadvantages of hot water storage systems.

KNOW THE COMPONENTS USED IN DOMESTIC HOT WATER SYSTEMS

Identify the types of storage cylinder used in domestic hot water systems

The various types of cylinder have been covered earlier in this chapter. However, Table 6.04 summarises these types.

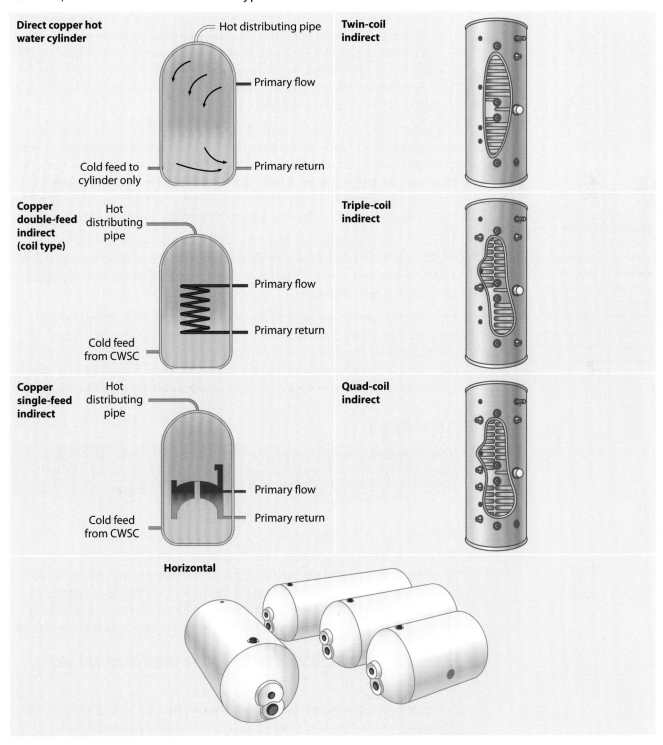

Table 6.04: Types of hot water storage cylinder

Chapter 6

Describe the operating principles of immersion heaters

Immersion heaters consist of an electric element that is controlled by a thermostat housed in the same unit. They are used either as a sole means of heating a hot water supply or combined with another heat source. Some consumers prefer to heat water this way during the summer to avoid using other fuel types (such as solid fuel appliances).

- Traditional immersion heaters are located in the top of the storage cylinder to give the maximum heating of the stored water. In this case, the element is long and reaches about two-thirds into the cylinder.
- Dual types have an additional shorter element. This type comes with a switch to enable the selection of either element: for example, 'Bath' for the long element, requiring a large amount of stored hot water, or 'Sink' for the shorter element, for tasks such as washing up, requiring a small amount of stored water.

Hot water cylinders solely heated by electricity are often designed to take advantage of low-tariff electricity at night. This means a much larger cylinder than normal is installed, together with two immersion heaters. The bottom heater works on the low electricity tariff and is used to heat the entire contents of the cylinder during the night. The top heater uses normal tariff electricity and provides daytime top-up if there is a shortage of hot water.

The thermostat switches on and off as the water temperature demands. Immersion heaters are best connected using a timer to enable economical selection of when they are used and also to ensure that the unit is not left on when it is not required.

Immersion heaters are normally electrically rated at 3 kW, which means they must be wired directly to the electrical consumer unit by a qualified electrician and should never be connected into the ring main circuit by any means. The cable connecting the power from the fused outlet to the immersion heater connections must be of a suitable heat-resistant quality.

Safe working

New immersion heaters are now supplied with a secondary thermostat or overheat thermostat to guard against overheating should the normal control thermostat fail. Following a number of scalding incidents with immersion heaters, it is now advisable to replace single-thermostat immersion heaters with the safer dual thermostat variety.

Activity 6.1

Visit the Health and Safety Executive (HSE) website and search for 'hot water systems'. Create a leaflet identifying the main health and safety risks associated with hot water systems and explaining how these risks can be reduced or avoided.

The purpose of the vent pipe

The vent pipe maintains atmospheric conditions in the pipework. It allows any air entering the system to escape and, should the water in the system become overheated, it allows it to expand up the vent pipe and discharge into the cistern. The use of better system controls has reduced the risk of overheating, but it can occur on direct systems, particularly those using solid fuel back boilers.

Explain the importance of temperature control on hot water systems

Some domestic hot water storage systems can exceed 80°C under normal operating conditions. These are vessels used as heat stores and those connected to solar heat collectors or solid fuel boilers. The outlet from these vessels should be fitted with a device such as an in-line hot water tempering valve in accordance with BS EN 15092:2008. This will ensure that the temperature supplied to the domestic hot water distribution system does not exceed 60°C.

Building Regulations document G3 3.65 states that hot water to a bath should be supplied at a maximum of 48°C by the use of an in-line blending valve or other appropriate temperature control device, with a maximum temperature stop and a suitable arrangement of pipework.

Safe working

The thermostat controlling the water temperature in any type of hot water storage cylinder should normally be adjusted to a maximum of 60–65ºC.

Although the temperature of water delivered to a bath outlet must not exceed 48°C, it is strongly recommended that temperatures are set lower.

Case study

Hot water is responsible for the highest number of fatal and severe scald injuries in the home. Every year around 20 people die as a result of scalds caused by hot bath water, and a further 570 suffer serious injuries. Young children and older people are most at risk because their skin is thinner.

Identify methods of controlling temperature

Table 6.05 outlines some temperature-related control devices and their operation.

Safety device	How this controls temperature
Control thermostat	Maintains the water temperature in the cylinder between 60 and 65°C
High limit thermostat (energy cut-out device)	A non-self-resetting device that isolates the heat source at a temperature of around 80–85°C
Temperature relief valve	Discharges water from the cylinder at a temperature of 90–95°C (water is dumped from the system and replaced with cooler water to prevent boiling water)

Table 6.05: Temperature control safety devices

Describe the operating principles of blending valves

Blending valves are also known as thermostatic mixing valves (TMVs). They blend hot and cold water to a set temperature to ensure constant, safe outlet temperatures on appliances such as showers and baths to prevent scalding. They are used in most public buildings such as hospitals and schools. In sheltered accommodation, all showers and hot water outlets must be controlled by TMVs. They are also becoming more common in domestic properties (see Figure 6.14).

There are many serious cases of scalding, mainly of young children and elderly people, some of which have been fatal. Document G of the current Building Regulations states that all new-build work and any properties involved in renovation/modernisation must have a blending valve installed in the pipework to a bath so that a temperature of 48°C is not exceeded.

Figure 6.14: Blending/thermostatic mixing valve

① Safe working

Mixing valves to limit the temperature of hot water outlets should not be easy for the end user to alter.

Valves are now available that have been certified to the new BuildCert TMV2 scheme, which recommends maximum hot water outlet temperatures for use in all premises. These valves maintain the preset temperature, even if the water pressure varies when other appliances are in use. The TMV3 valve is made to a different standard for NHS applications.

The hot and cold supplies to the valve normally have similar pressure, although because of the requirement for non-return valves to be fitted to prevent cross-contamination, this is not so important. If the pressures are considerably different, it may be harder to set up the blending valve. You will need to check the manufacturer's specification before you install the valve to ensure it will operate with unequal pressure supplies.

A valve can supply one or more fittings but, to prevent legionnaires' disease, the blended water pipe run should not be more than 2 m (see page 265–266 and Chapter 5, page 225).

Working practice

You have been asked to carry out the labour-only plumbing work for the conversion of a large house into a care home for the elderly. This is a change of use conversion.

There are nine en-suite bathrooms, a hot and cold water supply connected to an existing storage cylinder (of adequate size) and soil and waste pipework.

A friend (a retired plumber) has advised the owner on the sanitary ware, fittings, pipework and components required for the job. These have been purchased and are on site awaiting installation.

You notice that the purchased goods do not include blending valves. You ask the owner whether they have been ordered but he looks confused and says that he has never heard of them. You explain that they are expensive but essential. However, the owner tells you to carry on without them.

- How many blending valves are needed?
- Which appliances should they be connected to?
- What is the maximum temperature that the blending valve should be factory-set at?
- Is the owner within his rights to tell you to continue without fitting the blending valves?
- Who (if anyone) should you report this matter to?

Identify types of showers used in domestic water systems

The three main types of shower – gravity, pumped and electric – are covered in detail on pages 269–272.

UNDERSTAND THE INSTALLATION REQUIREMENTS OF DOMESTIC HOT WATER PLUMBING SYSTEMS

Installation requirements for hot water systems are very similar to those for cold water systems. Refer to the section covering this in Chapter 5 on pages 215–219 for more information.

Identify the key installation features of hot water storage cylinders

In all domestic hot water cylinder installations, it is good practice to sit the base of the cylinder on timber blocks to aid air circulation under the cylinder and to help prevent corrosion. Wherever possible, place the cylinder centrally in the identified position to enable ease of access for routine maintenance or replacement work.

All new domestic hot water storage cylinders must now comply with Part L1 of the Building Regulations. Other key installation features, such as temperature control and avoiding parasitic circulation, are covered elsewhere in this chapter, on pages 244–250.

Stratification

In a hot water storage vessel, stratification is where layers of water form from the top of the vessel (where it is hottest) to the base of the vessel (where it is coolest) – see Figure 6.15.

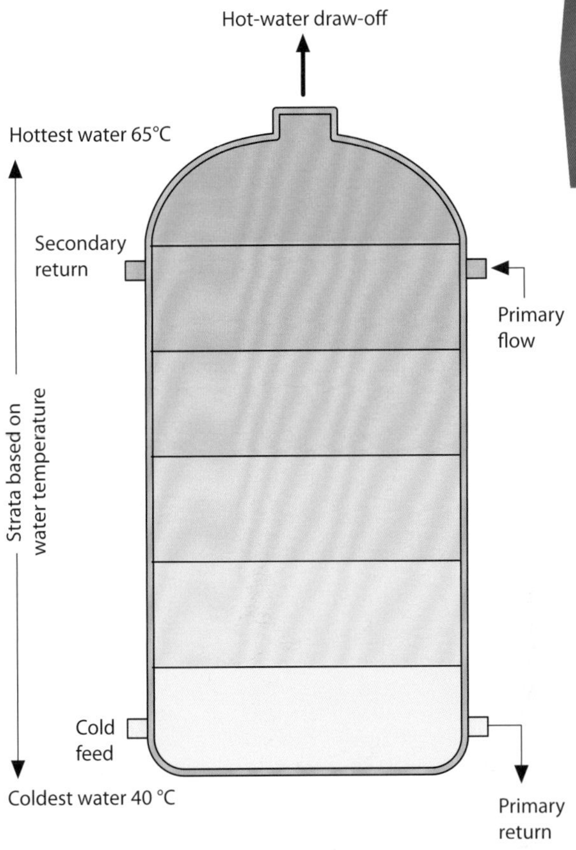

Stratification must take place for the vessel to function to its maximum efficiency. Manufacturers of storage vessels build stratification into the design. The following design rules enable stratification to take place:

- The vessel should be cylindrical.
- The cylinder vessel should be installed vertically rather than horizontally. If the cylinder must be installed horizontally, a specialist cylinder will be needed (see Table 6.04 on page 261).
- The cold feed connection should be in a horizontal position.

Figure 6.15: Stratification in a hot water cylinder

Describe requirements for minimising the installation of long hot water draw-offs

These include the need for energy conservation and to avoid the risk of legionella, which could be caused by dead legs. Dead legs are covered in Chapter 5 on page 225 (though the term is used slightly differently in hot and cold water systems) and should be avoided for two reasons:

1 They waste water because the user has to run off cold water before it turns hot.

2 Energy is wasted heating up the volume of water contained in the dead leg, which then cools.

One method of overcoming the problem of dead legs is secondary circulation. This can be demonstrated by the typical installation layout using a pumped secondary circulation system, shown in Figure 6.16. However, you will not be able to use secondary circulation if you have installed a combination boiler. Trace heating can be used as an alternative to secondary circulation (see page 266).

A flow-and-return loop is installed, which feeds all the appliances. The water is kept circulating, either by gravity or by a non-corrosive circulating

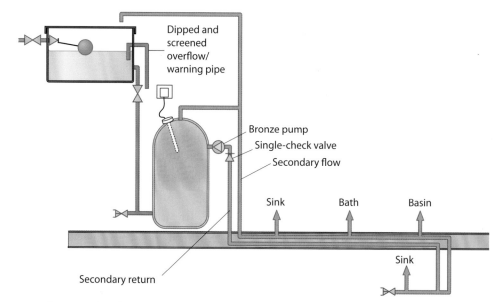

Figure 6.16: Secondary circulation system to overcome dead legs

pump (made of bronze). The return pipe is connected in the top third of the cylinder. This prevents the cooler water lower down the cylinder from mixing with the hot water. The return leg of the cylinder normally includes a single-check valve fitted in the secondary return before the circulation pump to prevent reversal of the water flow in the return pipe when a tap is opened. The operation of the pump is normally controlled by a simple time clock, which is set to turn off the pump and prevent wasteful circulation of heated water during periods when the building is not in use.

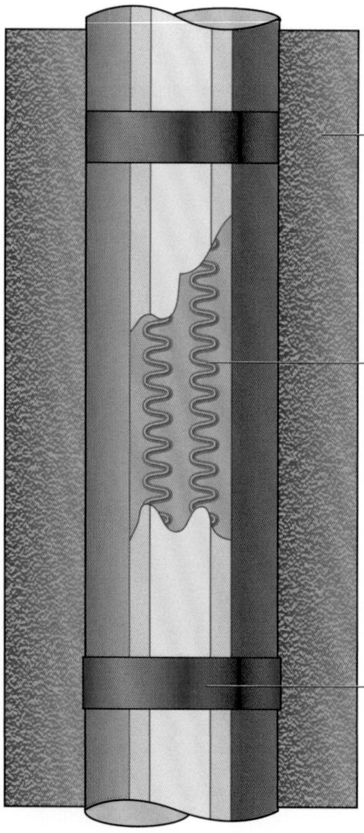

List the methods of optimising the length of hot water draw-offs

Trace heating

Trace heating involves attaching a low temperature heating element to the outside of the pipe, controlled by a thermostat which activates the heater when the temperature is low, thereby preventing the pipe from freezing (see Figure 6.17). It is more common on industrial installations, but there are domestic products available.

Trace heating is also sometimes used instead of secondary circulation, which can be very costly due to the extra pipework and the bronze pump. Instead, low voltage tape is fitted to the hot water draw-off pipework along the dead legs of the system. When activated, this applies a gentle heat along the pipe, reducing the chilling factor.

Other methods include using pumped secondary circulation systems, centralised direct heat sources and centralised hot water storage cylinders, all of which are covered elsewhere in this chapter.

Identify the considerations when installing pipework in relation to hot water pipework and cold water pipework

When designing a hot water system, you need to take into account the length of pipe run from the cylinder to the individual outlets, avoiding excessive lengths of dead legs (see Figure 6.18 on page 267). The Water Regulations recommend that uncirculated hot water distribution pipes

Figure 6.17: A trace heating element protects pipework from frost or may be used in place of a secondary circulation system

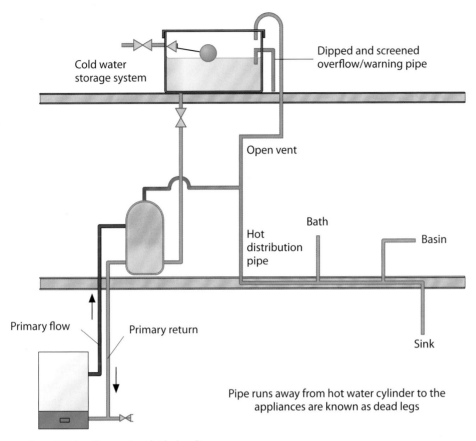

Cold water
storage system

Dipped and screened
overflow/warning pipe

Open vent

Bath

Basin

Hot
distribution
pipe

Sink

Primary flow

Primary return

Pipe runs away from hot water cylinder to the
appliances are known as dead legs

Figure 6.18: Dead leg runs to individual appliances

should be kept as short as possible and, if uninsulated, should not exceed the maximum length stated.

In Table 6.06, the right-hand column gives the approximate length of time it would take to draw off the cool water based on the draw-off rate of a washbasin tap with a flow rate of 0.15 l/s. The Health and Safety Legionella Code L8 states that the maximum draw-off period for hot water to reach its correct temperature should be 60 seconds.

Care should be taken in positioning hot and cold pipework for two reasons.

- Hot water pipes may transfer their heat to cold water pipes if too close (heating pipes close to cold water pipes obviously have similar consequences), which will raise the temperature of the cold water. This may even be detectable by touch or taste, which could mean a change from fluid risk category 1 to 2.
- When installing hot and cold water pipework together horizontally, the hot pipe should be *above* the cold pipe. This is because the heat lost by the hot water pipe would rise and form condensation on the cold pipe.

Pipe OD (outside diameter)	Length	Seconds
<12 mm	20 m	11
<22 mm	12 m	25
<28 mm	8 m	26
>28 mm	3 m	>15

Table 6.06: Maximum lengths of uninsulated pipes

A plumber was called to a property during particularly cold weather. The owner had reported a leak under the kitchen sink. The plumber found a lot of water nearby but no leak. On further investigation, he found that the cold pipework had been fitted above the hot pipework. This caused condensation to form, giving the appearance of a leak.

The installation was not new and the problem would have been present throughout the life of the installation.

- Why was the problem more prominent at the time the customer reported the leak?
- Why had the condensation problem not been noticed before, bearing in mind that condensation would have been forming?
- What could be done to alleviate the problem?

Identify the reasons for insulating hot water system pipework and components

The importance of insulation for frost protection, prevention of heat loss and energy conservation is covered in Chapter 5, so refer back to pages 215–216 as necessary.

The key requirements in the *Domestic Heating Compliance Guide* relating to insulation to conserve energy are as follows:

- Primary circulation pipework should be insulated wherever it passes outside the heated living space or through voids in a building, which are ventilated from unheated spaces.
- Primary circulation pipes for hot water services should be insulated throughout their length, except where they need to penetrate joists or building structural elements.
- The pipes connected to a hot water storage cylinder should be insulated up to 1 m from the cylinder or to the point where they become concealed.
- If a secondary circulation system is used, the pipework included in this circuit should be fully insulated.
- Pipework in insulated areas should comply with the requirements of BS 5422.
- Insulation that complies with the *Domestic Heating Compliance Guide* must not have a greater heat loss than the figures quoted in Table 6.07.

Pipe diameter (mm)	Maximum permissible heat loss (W/m)
15	7.89
22	9.12
28	10.07

Table 6.07: Maximum permitted heat loss for insulation

Describe the fluid categories and methods of backflow prevention

The fluid risk categories and devices and methods for backflow prevention have been identified and discussed in Chapter 5. They also apply to hot water supply systems. Refer to pages 217–218 to remind yourself of how these categories are identified and what risks they carry; methods of backflow prevention and the devices that may be used.

Identify the key contamination issues in plumbing systems

As with cold water supply systems, there is a risk of the hot water supply becoming contaminated with legionella bacteria. Look back at pages 265–267 and Chapter 5 page 225 to remind yourself how legionella can affect hot and cold water supplies and to identify the steps that should be taken to reduce the risks.

Progress check 6.03

1 What are the specific requirements of the cold feed pipe from its connection in the CWSC to its termination into the storage cylinder?

2 What criteria or requirements should you adopt when installing a hot water storage vessel (cylinder) in a plumbing system?

3 Produce a table to show the maximum permissible length of dead legs in a hot water supply system.

4 Where should a secondary return connection be located on a domestic hot water storage vessel (cylinder)?

5 When installing hot and cold pipework horizontally at 50 mm centres, which pipe should be above the other? Give a reason for your answer.

6 Why must hot water pipework and components be insulated?

7 Which fluid category represents a slight health hazard'?

8 Explain the term 'point of use backflow protection'.

KNOW THE DESIGN FEATURES OF SHOWERS

This section looks at the layout and connection requirements of shower mixing valves. A shower can be fed by mains pressure (high pressure) or it can be storage fed by gravity (low pressure) or by using booster pumps.

Identify the pipework configurations of hot water showers

Mains-fed showers (high pressure)

The electric shower shown in Figure 6.19 is designed for mains connection, although some can be fed indirectly. The electrical rating can be over 10 kW, so it is important that the supply is adequate and wired directly from the mains distribution unit (MDU). For a rating of 9.6 kW, the circuit protection device requirements are 45 amps and the cable 10 mm^2. The shower should also be isolated with a switch, which should be within easy access outside the shower room. Any pressure variations in the cold water supply to the shower are handled by the flow governor. Most electrical instantaneous showers are fitted with a flexible hose outlet – they need a check valve as required by the Water Regulations. This will usually be provided by the manufacturer.

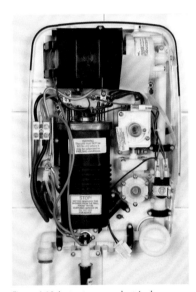

Figure 6.19: Instantaneous electric shower

The operation is simple: when the control knob is turned on, an electrical element located inside a heating chamber full of water is activated. The temperature of water requested by the user on the control knob simply reduces the water flowing across the element for hotter water and increases the water flow if cooler temperatures are requested.

The Water Regulations require that provision is made to ensure that backflow cannot occur. This is done by installing a double-check valve or by using a rigid connection to the showerhead. Alternatively, if a sliding rail is used, the flexible hose will pass through some form of device designed to prevent it from being submerged in water (a flexible hose retaining ring, attached to the slide rail); this will prevent backflow. Mains connection can only be used on thermostatic mixing valves, using the manufacturer's valves designed specifically for this purpose.

Figure 6.20: Shower mixer

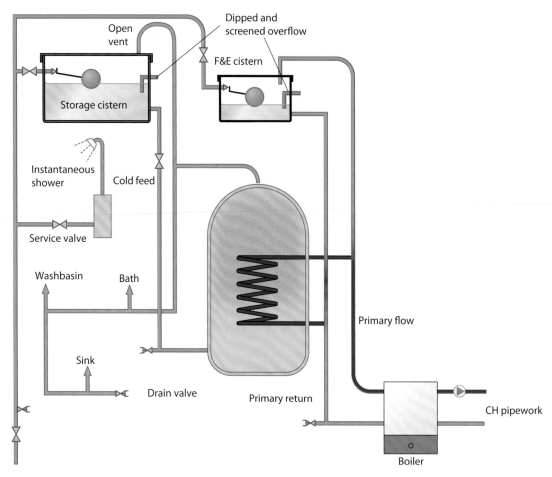

Figure 6.21: Layout of mains-fed instantaneous shower

Working practice

Neeta is a final year apprentice. During alterations to a plumbing system on a house extension, the customer gives her an ordinary shower mixer for fitting, even though the cold water is mains fed.

Neeta explains to the customer that the shower should not be fitted directly to the mains because it is not good practice and would be against Water Regulations. The customer accepts her explanation. However, he goes on to say that a thermostatic shower is three times the price and, since it is his house, Neeta should just fit the shower as he has asked.

- What should Neeta do now?
- Who could she contact for expert advice?

Safe working

For non-thermostatic shower mixers, the cold feed connection to the hot water storage vessel in the CWSC should be above the cold water supply connection to the shower. This is to prevent scalding should the water supply to the CWSC be turned off accidentally (or interrupted) and the content of the cistern allowed to drain down.

Storage-fed gravity showers (low pressure)

The hot and cold supplies need to be of equal pressure to ensure the correct mix ratio of the hot and cold water. There are two types of mixer, manual or thermostatic (Figure 6.20). Thermostatic is the safest option. Thermostatic units mix the two waters to the desired temperature set by the user. Manual types are set by the user but there is no compensation as the water from the hot water storage cylinder will eventually start to cool down the longer the shower is used. The showerhead needs to be at least 1 m below the bottom of the CWSC to ensure adequate pressure from the showerhead (see Figure 6.22 on page 271).

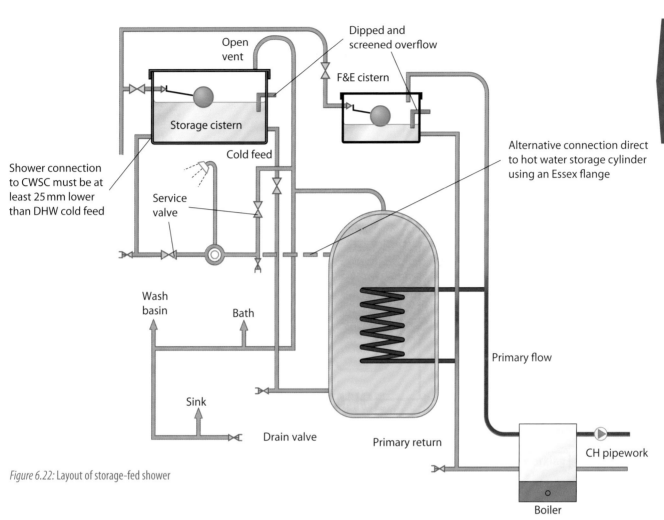

Shower connection to CWSC must be at least 25 mm lower than DHW cold feed

Open vent

Storage cistern

Dipped and screened overflow

F&E cistern

Cold feed

Service valve

Alternative connection direct to hot water storage cylinder using an Essex flange

Wash basin

Bath

Primary flow

Sink

Drain valve

Primary return

CH pipework

Boiler

Figure 6.22: Layout of storage-fed shower

Pumped (boosted supply) showers

A booster pump may be used to increase the water pressure. Figure 6.23 shows a single-impeller booster pump and Figure 6.24 shows a double-impeller booster pump. The pump increases the pressure, which means that

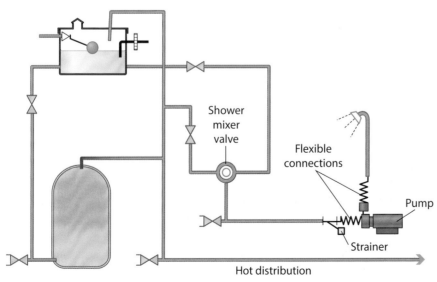

Shower mixer valve

Flexible connections

Pump

Strainer

Hot distribution

Figure 6.23: Single-impeller pump installation

Indicator	Fault	Solution
Unvented hot water systems		
No flow of hot water	• Mains supply too low or off • Blocked strainer • Pressure-reducing valve incorrectly installed or faulty	• Check availability and condition of supply • Remove and clean strainer • Check and refit pressure-reducing valve as required
Primary water mixing with secondary water (lack of hot water, discolouration)	• Fractured coil in double-feed indirect cylinder or indirect coil(s)	• Replace damaged cylinder, conforming to current regulations
Vented and unvented hot water systems		
Noise	• See Chapter 5, pages 234–235	• See Chapter 5, pages 234–235
Immersion heater	• Over time, the element sheath in the unit may deteriorate, causing the safety fuse to blow or trip	• Replace immersion heater, taking care not to rupture the cylinder when loosening and tightening
Blockage	• Foreign matter in system • Limescale build-up	• Ensure heat sources (including immersion heater) are set at 60–65°C • Clean or replace affected pipework
Temperature of water is poor	• Heat source or cylinder thermostat set too low • Faulty heat source or cylinder thermostat • Programmer or timer incorrectly set or faulty	• Adjust thermostat to desired temperature • Replace faulty thermostat • Adjust settings • Replace faulty programmer or timer
Excessive cool water must be run off before hot water is delivered	• Water has too far to travel before delivery • Pipe runs are too long, creating dead legs	• Shorten dead leg zones or install a secondary return pipe
Central heating is working but there is no hot water supply	• Three-port valve or two-port valve not working correctly • Cylinder thermostat faulty or incorrectly set • Programmer faulty or incorrectly set	• Change electronic actuator head • Check valves for freedom of movement using spring return lever • Replace or reset cylinder thermostat • Replace or reset programmer
Continuous water discharge	• Pressure-reducing valve on main not working correctly • Faulty pressure or temperature relief valve • Faulty expansion relief valve	• Replace valve if pressure exceeds 2.1 bar • Use manufacturer's instructions to identify faults, then repair or replace as necessary
Water boiling in system	• Faulty immersion heater thermostat and no overload button • Faulty heat source control thermostat • Solid fuel appliances loaded with too much fuel (vented system only)	• Replace immersion heater with a dual thermostat type • Replace faulty boiler thermostat • Educate user regarding over-fuelling of solid fuel appliances. Consider enlarging system to incorporate heat leaks (e.g. radiators)
Vented, unvented and instantaneous systems		
Poor flow rate through shower mixer valve or shower rose (outlet)	• Limescale formation, causing partial blockage • Mesh screen filters in shower valve blocked with foreign matter or limescale deposits	• Inspect inlets and mesh filters in valve • De-scale shower rose outlets (simply rub with thumb on modern outlets but older types may need a replacement rose outlet)

Table 6.08: Indicators of and solutions to faults in domestic hot water systems (continued)

Explain the reasons for the build-up of limescale in hot water systems

Limescale caused by hard water can cause problems in plumbing systems. In extreme cases, if limescale build-up is undetected or steps are not taken to reduce or stop its formation, it will completely block pipes, cylinder coils and appliances, as shown in Figure 6.25.

One key method of ensuring that limescale deposits are kept to a minimum in plumbing pipework and systems is to ensure that the maximum temperature of the heated water is set between 60°C and 65°C.

An important property of water is its ability to dissolve gases and solids to form solutions. This is referred to as its solvent power and has a bearing on how soft or hard the water eventually becomes. Water hardness and limescale are covered in Chapter 3 on scientific principles (page 105) and Chapter 5 on cold water systems (page 184), so refer back as necessary.

Figure 6.25: Limescale can clog pipes and other components

Identify the methods of removing limescale in hot water systems

There are various methods of treating water in domestic systems, not only to soften hard water to prevent limescale but also to remove any limescale build-up.

Water filters

A water filter removes impurities from water using a fine physical barrier, a chemical process or a biological process. Filters cleanse water to various extents depending on the type of filter. A simple line filter, which contains a fine wire gauze, reduces the amount of solid impurities like sand or grit (see Figure 6.26). This can usually be removed, washed out and replaced as required. This type of filter can be found on the cold supply to an unvented hot water cylinder.

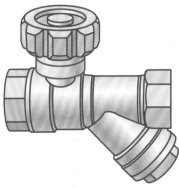

Figure 6.26: Simple line filter

A cartridge filter helps reduce particles and substances in the water, such as chlorine, to give it a pleasanter taste (see Figure 6.27).

Water conditioners

A water conditioner installed in the cold water supply softens the water before it is heated, which reduces scale formation. The process that softens the water is called 'ion exchange'. There are three main methods of treating water hardness:

- base exchange softeners
- scale reducers
- magnetic water conditioners.

Base exchange softeners

- These pass hard water through a tank containing resin particles. The resin attracts and absorbs the hardness salts – mainly calcium and magnesium – from the water and replaces them with sodium.

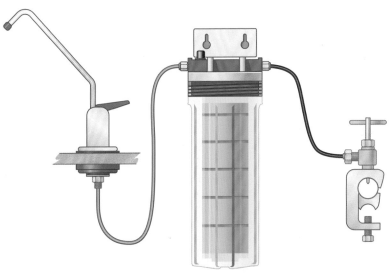

Figure 6.27: Typical cartridge filter

- The **resin particles** become saturated with hardness salts and needs to be regenerated, using salt solution to put sodium back into the resin.
- The hardness salts are released from the resin and washed down the drain.

The unit requires regular maintenance and checking, and an annual service. A check on water hardness also needs to be carried out by a service engineer. The salt can usually be topped up by the end user. BS 6700 makes recommendations for the installation of base exchange water softeners. Its main consideration is preventing **backflow** and contamination of the water supply. As there is a backflow risk associated with the use of a base exchange water softener, where salt additives are put into the cold water system, the installation must be protected using a single-check valve.

When installing a base exchange water softener, at least one tap (normally the kitchen sink) is not connected via the softener. A base exchange water softener is shown in Figure 6.28 below.

Scale reducers

Scale reducers are chemical conditioners that reduce limescale in water. They dispense chemical additives into the system to treat the water. The

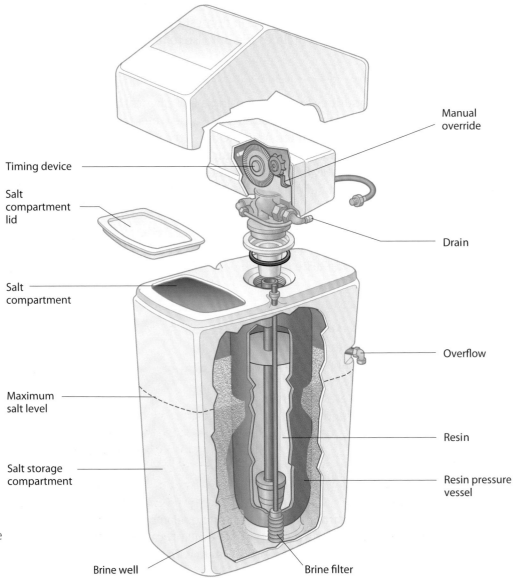

Timing device

Salt compartment lid

Salt compartment

Maximum salt level

Salt storage compartment

Manual override

Drain

Overflow

Resin

Resin pressure vessel

Brine well

Brine filter

Figure 6.28: Section through a base exchange water softener

chemicals alter the molecular structure of the calcium salts in the water, preventing them from bonding. Scale reducers are commonly used to provide protection to individual system components such as multipoint water heaters or combination boilers (see Figure 6.29).

Magnetic conditioners

These are clamped to the pipework and use a small electrical field to change the molecular structure of the salts in the water. There are many variations: some simply comprise a strong magnet; others are connected to the mains via a transformer to generate the electric current (see Figure 6.30).

Figure 6.29: Scale reducers must be replenished with chemical additives

| Progress check 6.06 |

1   Produce a simple chart to identify five indicators of typical faults associated with hot water supply. The chart should include:
   • the fault
   • the probable cause of the fault
   • the action required to repair the fault
   • the type of system the fault could be found in.

2   If limescale deposits remain undetected in a hot water system, what will happen to the pipework or, in some cases, the components?

3   Which water type is most likely to create limescale?

4   What measures can you take to reduce or prevent the formation of limescale in plumbing systems?

5   If long runs of horizontal pipework are installed on low-pressure systems and water does not flow, what is the likely cause?

| Working practice |

A customer asks you to investigate a combination boiler that has become very noisy. Also, the hot water flow is very poor. The customer admits that the problem has been getting worse over a prolonged period and is not something recent. The water in the area is known to be of a temporary hard quality.

• What is the likely cause of the problem?
• How serious could the issue be (bearing in mind it has been going on for a long time)?
• What advice will you give the customer to prevent this happening again?

Figure 6.30: Magnetic conditioners are attached to pipes

UNDERSTAND THE KEY REQUIREMENTS OF TESTING AND DECOMMISSIONING OF DOMESTIC HOT WATER SYSTEMS

With some minor changes regarding the use of different valves and components, the testing and decommissioning of hot water supplies is very similar to the requirements for cold water supplies. Look back at Chapter 5, Figure 5.47 on page 223 (flow chart of commissioning checks) and Figure 5.50 on page 226 (draining down procedure). These flow charts identify the differences in the procedures for hot and cold water systems. These flow charts and other areas of Chapter 5 cover the assessment criteria for this learning outcome. See Table 6.09 for cross-references to common areas.

Commonality between hot and cold water supply	Cross-reference to Chapter 5
Testing hot water pipework	
Hydraulic test equipment	Pages 219–222
Soundness testing	
System flushing	Page 220
Commissioning checks	Page 223
Permanent and temporary decommissioning of hot water system	Pages 224–225
Draining hot water systems	Page 226

Table 6.09: Commonality between hot and cold systems

Describe methods of testing hot water pipework systems

Soundness testing of both hot and cold water systems includes:

- visual inspection
- testing for leaks
- pressure testing
- final checks.

These topics have been described in detail in Chapter 5, pages 221–222

Describe the requirements for flushing a system

Following any work on a system, you should flush it to get rid of any particles or foreign matter that might cause damage. This is described in more detail in Chapter 5, page 220.

Describe commissioning checks for hot water systems

These are covered by Figure 5.47 on page 223 of Chapter 5.

Describe the differences between permanently and temporarily decommissioning a hot water system

This is covered in Chapter 5, pages 224–225.

Describe the method for draining hot water systems

You must follow the correct procedures when draining down a system, in order to prevent damage to the system or the property it is in. These steps are covered in detail in Figure 5.50 on page 226 of Chapter 5.

Working practice

Kasha was called to a house where the solid fuel central heating and hot water system had stopped providing hot water, although the heating system still worked. The hot water supply worked by gravity circulation.

Kasha found that heated water from the solid fuel boiler was gravitating up the primary flow to just behind the storage cylinder. A heavyweight check valve had been fitted on the primary flow, but this had seized up in the closed position and needed to be removed.

Kasha decided to move to another job, giving time for the solid fuel to burn out and the water in the system to cool down. When she returned two hours later, the house owner said that he had drained the system to save her time. Kasha thanked him and got to work.

First, she removed the cylinder so that she could access the primary return pipe and the check valve. Then she removed the valve. As she did so, she heard rumbling and steam started to emerge from the open-ended pipe. She didn't want to make a mess, so she held a cloth over the end of the pipe to try to stop any water escaping. Her hand was so badly scalded that she needed hospital treatment and a week off work to allow the skin to heal.

She later discovered that the owner of the house had had his elderly father staying with him. Noticing that the solid fuel burner was going out, his father had stoked it back up. Because the system had been drained, there was very little water in the pipework and what was left boiled rapidly, creating steam.

- List the errors Kasha made when draining the system.
- Why had a heavyweight check valve been fitted to the primary return?
- Is it feasible to fit a valve on the primary circuit? Explain your answer.
- What should Kasha have done immediately after she injured herself?
- Should she report the incident to anyone? If so, who? Explain your answer.

 Safe working

When working with hot water, there is a risk of burns and scalds. It is a good idea to be prepared with clean, cold (running, if possible) water available close by. If you do burn yourself, place the burn in cold water as soon as possible. This will reduce the amount of blistering to the area. In addition, wherever possible, you should let the system cool down before you begin work.

Progress check 6.07

1 Three standard pressure tests are applied to new pipework installations. Describe the application of each one.

2 What is the name of the piece of equipment used to carry out pressure tests?

3 Write a simple bulleted list to indicate the stages required to flush a cistern-fed hot water supply system.

4 Which two documents should you refer to in order to find out the test pressure and length of time a test should be carried out?

5 When commissioning a hot water supply system, what equipment would you need to test for:

   a standard and running pressure

   b flow rates at outlets

   c temperature delivery
     at outlets?

6 Explain the following terms:

   a permanent decommissioning

   b temporary decommissioning.

7 Why is it good practice to let the water and heat source cool slightly before draining a hot water system?

Check your knowledge

1 Which of the following British Standards provides the standards for soundness testing of a hot water supply installation?

a BS 1192

b BS 7593

c BS 6700

d BS 3402

2 Which of the following is not a type of hot water system?

a direct

b indirect

c non-storage

d primatic

3 What is the purpose of a vent pipe?

a to fill the system

b to maintain atmospheric conditions

c to reduce atmospheric conditions

d to circulate the water, preventing dead legs

4 What is the main reason for installing a secondary circulation supply to a hot water tap?

a to improve pressure

b to avoid fitting a pump

c to prevent dead legs

d to reduce pressure to the taps

5 According to the Water Regulations, what is the maximum length of 15 mm uninsulated pipework in a hot water system?

a 10 metres

b 11 metres

c 12 metres

d 13 metres

6 How far above the cold distribution connection should the DHWS cold feed connection be installed, in order to prevent scalding if the water supply is cut off or reduced?

a 15 mm

b 25 mm

c 30 mm

d 40 mm

7 Mains fed, storage fed and boosted supply are all types of what?

a cold water supply

b hot water supply

c central heating system

d shower

8 When installing a water softener, which of the following outlets should be left connected directly to the incoming main (un-softened water).

a bath

b kitchen sink

c wash basin

d mains fed shower

9 What is a weir cap used to measure?

a low pressure

b high pressure

c system pressure

d flow rate

10 After the commissioning of a hot water supply system is completed, which of the following documents should be completed?

a SEDBUK

b ROSPA

c Benchmark

d FENSA

Chapter 7

Sanitation

This chapter will cover the following learning outcomes:

- **Know the appliances and associated components used in sanitary installation**
- **Know the requirements for installing sanitary appliances**

Introduction

In this chapter you will gain the knowledge, understanding and skills involved in installing common sanitation systems. The chapter will cover a range of appliances, such as baths, WCs and showers, and their associated fittings and installation methods. You will also learn how to carry out basic installation tasks commonly used in plumbing.

KNOW THE APPLIANCES AND ASSOCIATED COMPONENTS USED IN SANITARY INSTALLATION

The appliances and components covered in this section include: toilets (WCs), bidets and urinals; sinks, baths and showers.

Describe the working principles of different types of toilets

Working principles of sanitary appliances

The main working principles that you need to consider regarding sanitary appliances apply, on the whole, to WC pans and cisterns. Some WC pans clear their waste by the wash-down action of the water as it is released into the bowl.

Other designs use siphonic action: siphonic WC pans, siphons in cisterns and automatic siphons. The process of siphonage and its use is covered in Chapter 3: *Scientific principles for domestic, industrial and commercial plumbing*, on pages 111–112.

There are many types and designs of water closets (WCs), and they are categorised as:

- back-to-the-wall
- close-coupled
- low-level
- high-level
- concealed.

WCs are usually manufactured from **vitreous china** conforming to BS 3402. Vitreous china is made from a mixture of white burning clays and finely ground materials. These are fired at high temperatures, and even before glazing the material cannot be contaminated by bacteria, so it is totally hygienic.

Vitreous china is coated with an impervious non-crazing glaze in either a white or coloured finish. The material is stain proof, burn proof, rot proof, rust-free, non-fading and resistant to acids and alkalis.

WCs used in public places may also be manufactured from stainless steel. This material is more resistant to vandalism than vitreous china.

Key term

Vitreous china – a material produced from a solution called slip, or casting clay. Slip has the consistency of pouring cream and contains ball clay, china clay, sand, fixing agent and water. Vitreous products have a glasslike appearance.

WC cisterns

Prior to 1993, the capacity of a WC flushing cistern was 9 litres. The Water Regulations brought this down to 7.5 litres and, from January 2001, reduced it further to 6 litres. In addition, the Water Regulations also permit the use of dual-flush cisterns, which deliver 6 litres for a full flush and 4 litres for a lesser flush. The dual-flush cistern is now specified more often than the single-flush. To meet the requirements of the Water Regulations, a dual-flush cistern should have a label on it, stating that it is a dual-flush cistern.

There are commonly two types of cistern flushing mechanism:

1 siphonic

2 dual-flush valve (often known as a drop valve).

With the siphon type (see Figure 7.01), when the lever is pressed, a disc (or diaphragm) lifts the water in the bell of the siphon (1) up and over into the leg of the siphon (2). This creates the siphonic effect, which continues until the water in the cistern has dropped to a level that allows air to enter the bell.

> **Remember**
>
> WC pans and cisterns are manufactured as a single unit and therefore replacement cisterns to existing pans must always be the same capacity as the original. It is not possible to 'mix and match' as each component has been designed carefully in conjunction with the others to operate properly as a 'suite'.

> **Activity 7.1**
>
> Examine the WC cistern in your house and see if you can find out what its flushing capacity is. What is the purpose of dual-flush cisterns?

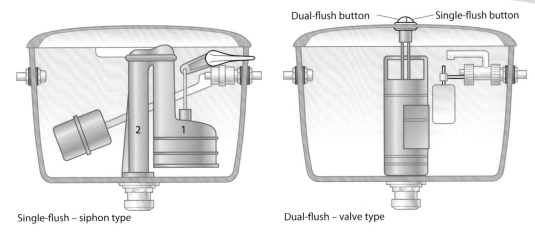

Single-flush – siphon type

Dual-flush – valve type

Figure 7.01: Different types of cistern flushing mechanism

The dual-flush valve-type cistern (see Figure 7.01) is operated by pressing either the full flush or lesser flush buttons, which have to be clearly marked on the cistern. These operate a valve that releases the water into the WC pan.

Another type of fitting (see Figure 7.02) is also available for 6-litre cisterns. This works by lifting a hinged flap as a flushing valve and is suitable for full flush only.

Cisterns of 9 litres and 7.5 litres are still available. The Water Regulations permit these to be installed as replacements for existing cisterns where the WC pan was designed to work on those capacities.

The overflow pipe can now discharge into the WC pan using an integral overflow, eliminating the need to provide an external overflow pipe. This pattern of cistern is now supplied with most WCs.

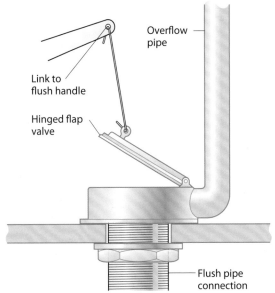

Figure 7.02: Cistern with hinged flap mechanism

> **Safe working**
>
> Always use suitable PPE when handling used sanitary appliances (see page 299).

WC pans

There are a number of designs for WC pans, which come under two main types: wash-down and siphonic. The wash-down pan (see Figure 7.03), which tends to be the most common type, uses the force of the water from the cistern to clear the bowl.

The siphonic pan (see Figure 7.04) creates a negative pressure below the trap seal. With the single-trap siphonic pan, this is done by restricting the flow from the cistern and is achieved by the design of the pan. The double-trap close-coupled pan uses a pressure-reducing device between the cistern and the pan. As the water is released into the second trap, it has the effect of drawing air from the void between the two traps, and siphons the content from the bowl.

Figure 7.03 label: Flush pipe

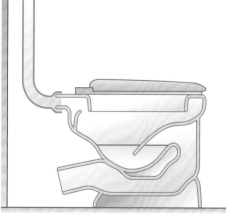

Figure 7.03: Wash-down pan

Figure 7.04: Single- and double-trap siphonic pan

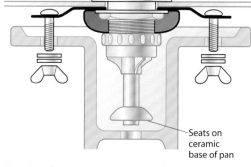

Figure 7.05 label: Seats on ceramic base of pan

Figure 7.05: Typical pressure-reducing valve between cistern and pan

Figure 7.05 shows what the pressure-reducing valve looks like between the cistern and pan.

Popular WC types include:

- concealed – where the cistern is hidden (see Figure 7.06)
- close-coupled – where cistern and pan are joined in one unit (see Figure 7.07)

Figure 7.06: Concealed WC

Figure 7.07: Close-coupled WC

- low-level – where the cistern is no more than 1 metre above the pan
- high-level – where the cistern is more than 1 metre above the pan, necessitating flushing by a chain (see Figure 7.08).

The choice of WC suite depends on a number of factors:

- cost – the customer's budget
- location – for public toilets, factors such as durability and ease of cleaning are a consideration. For example, stainless steel units are often used because of their resistance to vandalism.
- aesthetics (how good it looks) – Victorian versions of the high-level cistern are now considered very desirable.

On larger housing contracts, or in public buildings, the choice of WC may already have been made by the client's architect as part of the contract specification document.

Most WC pans manufactured today are P trap and use either a straight or bent pan connector, depending on the soil stack location.

Some cisterns are reversible. This means that the handles, overflow outlet and water inlets can be either left- or right-handed. In some cases, the cistern is plastic rather than vitreous china.

Kits are usually supplied for the soil outlets, and will be either a P trap or an S trap.

List different types of urinal

Urinals are specifically for use by men and are normally installed in non-domestic buildings, such as public toilets and schools. The water supply to urinals is controlled by an automatic flushing cistern or a pressure-flushing valve. Urinals can be:

- single or multiple bowl
- trough or slab type
- made from vitreous china, plastic, stainless steel or fire clay.

Figure 7.09 shows a typical urinal set-up with individual wall-fixed bowls made of vitreous china. The diagram also shows the cistern and flush pipe arrangements.

Figure 7.08: High-level WC

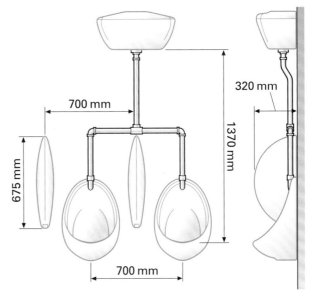

Figure 7.09: Typical two-bowl urinal layout

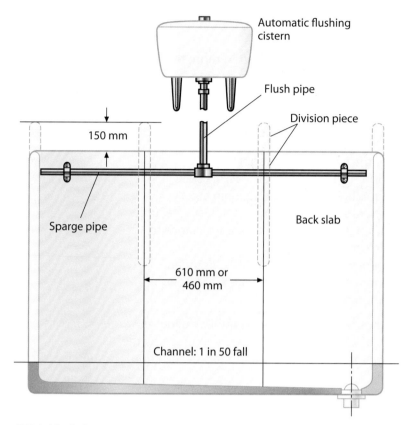

Figure 7.10: A slab urinal

Key term

Sparge pipe – a horizontal pipe that connects to the flush pipe. It is mounted on the face of the urinal and has pre-drilled holes (urinal spreaders), which are used to wash the face of the slab and channel.

Slab urinals are usually found in public buildings. The flush pipe is connected to a horizontal pipe called a **sparge pipe**. This is often mounted on the face of the urinal and has a number of pre-drilled holes in it (see Figure 7.10).

Trough urinals are used in toilets that may have a high risk of vandalism. The trough is sized for the maximum number of people that will be using it, and can be various lengths. The outlet is positioned at one end of the trough, which has a slight fall across the base (see Figure 7.11).

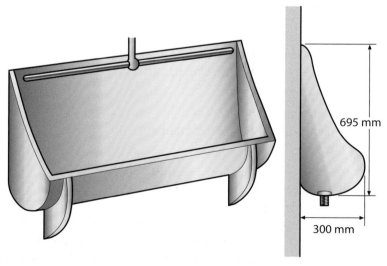

Figure 7.11: A trough urinal

Explain the working principles of an automatic flushing cistern

As mentioned on page 285, urinals are fitted with an automatic flushing cistern (see Figure 7.12). Its automatic siphon operation is very simple.

- As the cistern fills and the water rises, the air inside the dome of the automatic siphon is compressed.
- The increased pressure forces water out of the U tube, which reduces the pressure in the dome.
- The reduction in pressure causes siphonic action to take place, flushing the cistern.
- When the cistern has emptied, the water in the upper well is siphoned into the lower well.

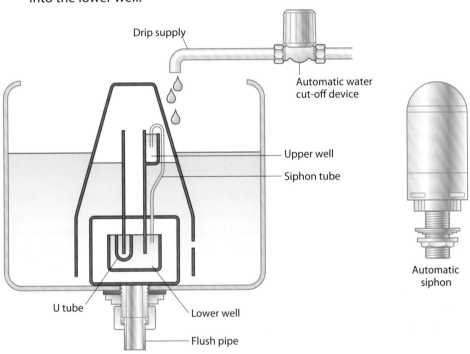

Figure 7.12: An automatic flushing cistern

The Water Regulations state that auto-flushing cisterns must not exceed the following maximum volumes:

- 10 litres per hour for a single urinal bowl or stall
- 7.5 litres per hour, per urinal position, for a cistern servicing two or more urinal bowls or stalls, or per 700 mm of slab.

The flow rate can be achieved using urinal flush control valves, which allow a small amount of water to pass into the system (see Figure 7.13).

Timed flow control valves can also be used. These have the additional advantage of being switched off when the building is not in use (for example, evenings, weekends, factory shutdowns or school holidays). They can be set for 'one-off' hygiene flushes, such as once every 24 hours. Many public buildings now control the supply of water to the automatic flushing cistern using infrared sensors that detect motion (similar to some burglar alarms). When motion is detected, a solenoid valve is opened, permitting a flow of water to the automatic flushing cistern.

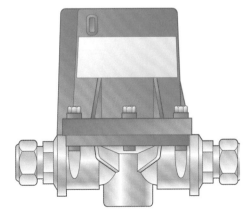

Figure 7.13: Hydraulic flush control valve

Activity 7.2

Visit the WRAS website and watch the short film about the Water Regulations. Write a brief summary of the key pieces of information in the video.

Progress check 7.01

1 According to the current Water Regulations, what should the maximum capacity of a WC flushing cistern be?

2 Why is it still permitted to buy and install WC cisterns with the older regulation capacities of 7.5 and 9 litres?

3 How is the water supply to a urinal automatic flushing cistern controlled?

4 What is the maximum volume of water that the Water Regulations allow for an automatic flushing cistern for a single urinal bowl or stall?

5 What is a sparge pipe?

Identify different types of bidet

Bidets are made from vitreous china and come in two types:

- ascending spray (see Figure 7.14).
- over-the-rim supply.

The ascending spray bidet has to be piped up correctly to avoid the risk of contamination to the supply. It cannot be used on the mains supply, such as with combination boilers and unvented hot water systems. This is a requirement of the Water Regulations. The bidet can be supplied for use with pillar taps, **monobloc** fittings and pop-up waste.

An over-the-rim bidet presents less of a water contamination risk and may be connected to supply pipework fed directly from the mains. However, an AUK2 tap gap (non-mechanical) must be maintained above the spillover level of the appliance (see Figure 7.15). See Chapter 5: *Cold water systems*, page 218 for more on air gaps.

If the over-the-rim bidet includes a flexible hose, it should be treated as an ascending spray type. This is because the flexible hose and spray head could drop into the contaminated water in the bidet bowl.

Key term

Monobloc – a mixer tap with a single mounting hole.

Safe working

To avoid potentially serious contamination of the water supply, an ascending spray bidet or an over-the-rim bidet with a flexible hose must be fed from storage and must have its own dedicated piping system.

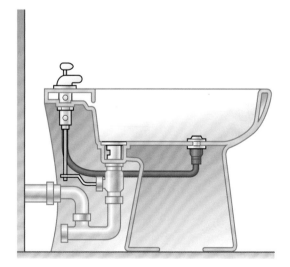

Figure 7.14: Ascending spray bidet

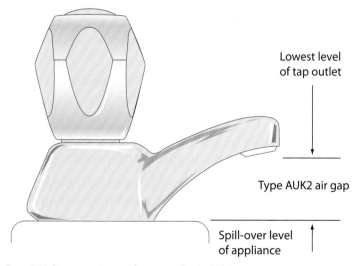

Lowest level of tap outlet

Type AUK2 air gap

Spill-over level of appliance

Figure 7.15: Air gap requirement for an over-the-rim bidet

Identify different types of washbasin

Washbasins can be:

- pedestal
- wall-hung
- countersunk/countertop
- in-wall (recessed).

All basins are available with either single tap holes for a monobloc mixer tap, or dual tap holes for pillar taps. Basins are also available with three holes for fittings with an independent spout.

Pedestal

The pedestal (freestanding) type is probably the most popular choice in homes. Appearance is the main factor, as all the pipework and fittings are concealed within the hollow pedestal (see Figure 7.16).

Wall-hung

These are fixed to the wall with brackets supplied by the manufacturer or may be of the towel rail type. Larger washbasins, secured using brackets, are mostly found in non-domestic installations.

Countersunk/countertop

These come in three types:

- countertop – sits proud of the work surface
- semi-countertop – used where space is tight as the front of the basin projects clear of the top
- under the countertop – has the flange surface on top of the basin, secured under the countertop using fixings supplied by the manufacturer.

Just as the name suggests, countertops sit on top of the counter and can also be recessed. They are becoming increasing popular and come in many stylish designs and shapes. They are installed in all sorts of buildings, from homes to public houses and restaurants. They are manufactured from vitreous china, stainless steel or a high-impact plastic called acetyl. Figure 7.17 shows a modern example.

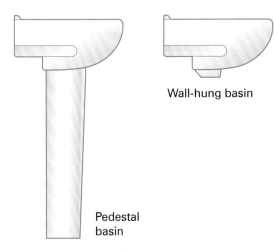

Wall-hung basin

Pedestal basin

Figure 7.16: Fixed to the wall basins

Figure 7.17: Modern countersunk basins

Figure 7.18: Hand-rinse basins are useful in cloakrooms and small spaces

Did you know?

A slotted waste fitting is required with a vitreous china sanitary appliance (such as a basin) that has an integral waste fitting. The slot allows the overflow water to discharge to waste.

The basin waste outlet is designed to take a 32 mm (1¼ inch) slotted waste fitting. Waste fittings can be the standard slotted type with a plug and chain or a pop-up type operated by lifting and pressing a knob on the top of the basin.

In-wall (recessed)

This type of basin is set back in the wall (recessed). This helps reduce the amount of room needed, as it does not stick out as far as a wall-mounted or pedestal washbasin. These are popular in cloakrooms and other limited spaces.

Hand-rinse basins

Called hand-rinse basins due to their size, these are used in cloakrooms with WCs. They can be of traditional or corner design. Options include pedestal, bracket-mounted or one of three countertops:

- standard, which fits into a pre-cut hole; brackets secure the underside of the basin to the countertop and the joint is sealed with waterproof sealant
- semi-countertop, which can be made from vitreous china or acetyl
- under countertop, which is made from acetyl.

Identify factors to consider when selecting taps for installing with a bath and basin

Different types of draw-off taps include pillar, mixer and globe taps. These are covered in detail in Chapter 5, on pages 207–211, so refer back as necessary.

Fitting the taps is a relatively simple process. Tap manufacturers' designs have to comply with regulations, so your installation will also satisfy the regulations. When choosing taps for a bath or basin, you should consider the following factors:

- size (¾ and ½ inch (19 and 13 mm)), depending on the appliance
- location
- appearance
- type
- operation
- design of basin or bath.

- **Location:** Ensure there will be enough room to operate the taps properly.
- **Appearance:** Don't damage the taps when fitting them. Hold the taps firmly when tightening the backnut or use soft jaw grips and a cloth to make sure you don't mark them.
- **Type:** When considering the type of tap to fit, you should think of the job the tap has to perform. They could be rising or non-rising spindle pillar taps, monobloc mixer or bi-flow mixer. Some mixer taps may include a shower attachment. In this case special consideration of their water supply is needed.
- **Operation:** Taps that are likely to be used, for example, by elderly people or young children should be easy to operate. Quarter-turn taps are more appropriate than conventional taps, which require a firm grip to operate.
- **Design of basin or bath:** Consider the correct design of tap to ensure comfort of use and choose a style appropriate to the appliance, room and age of the building.

Did you know?

The correct tap conventions are: hot on the left, cold on the right.

Pillar tap to a basin

The grip washer goes on the underside of the tap (see Figure 7.19). Sometimes a thicker washer is also supplied, which goes between the backnut and the underside of the basin. Tighten the backnut using a purpose-made basin wrench. You can test if it is tight enough by checking by hand that the tap does not move. Make sure both taps are pointing outwards in parallel.

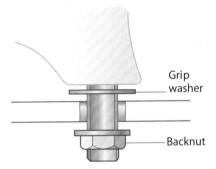
Grip washer

Backnut

Figure 7.19: Seal to basin tap

Monobloc taps

These are often supplied with an O ring, which should be fitted between the base of the tap and the basin. The washer goes underneath, as shown in Figure 7.20. The assembly is completed using a metal washer held in place by a fixing nut.

Tap and spout assembly: three-hole

These are installed in three-hole basins and ascending spray bidets, as follows.

- Remove all the parts that go above the basin.
- Fit the spout with its washers.
- Loosely connect the tubes between the body and side valves.
- Raise the assembly into position from below, making sure the sealing washers are in place.
- Fit the body seal, washer and lock nut loosely, then fit the valve flanges on top of the basin, aligning them with the top of the side valve, as in Figure 7.22 on page 293.
- Carefully tighten all the components, and then finish by fitting the valve headgear, any shields, drive inserts and hand wheels.

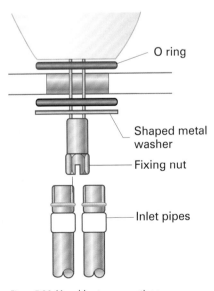
O ring

Shaped metal washer

Fixing nut

Inlet pipes

Figure 7.20: Monobloc tap connections

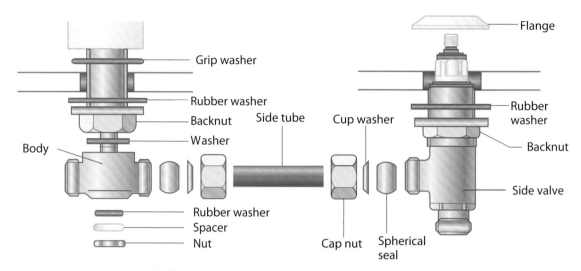

Figure 7.21: Three-hole basin tap

Identify different types of sinks

Belfast and London sinks

Belfast sinks are fairly large but are used in both domestic and commercial situations. They are set below work surfaces and are usually fed by bib taps, but can also be supplied by pillar taps or monobloc taps in domestic installations.

London sinks are similar but without an overflow. Both types are made of heavy-duty fire clay, which is manufactured in a similar way to vitreous china. They are available in white only and are designed to take a 38 mm (1½ inch) threaded waste.

Kitchen sinks

A vast range of designs and materials are available. Designs include:

- single drainer – either on the left or right
- single with basket – suitable for connection to a small waste disposal unit
- double drainer – where space is not an issue
- double sink with single drainer
- corner – versatile and useful in irregular shaped kitchens.

These are designed to take a 38 mm (1½ inch) slotted waste fitting for use with a sink overflow. However, most types of kitchen sink use a combination waste outlet with overflow, which is often supplied with the sink top. Taps can be monobloc or pillar.

Materials include:

- stainless steel
- plastic-coated pressed steel
- fire clay
- acetyl.

Safe working

Take great care when handling stainless steel sinks, as their edges are razor sharp.

Describe methods of connecting waste fittings to sanitary appliances

For washbasins, the range of waste fittings includes slotted, moulded seals and pop-up wastes.

Slotted waste fittings

These can be sealed using non-setting plumbing compound or silicone sealant as follows.

> **Safe working**
>
> Never use linseed oil based jointing compounds on plastic waste fittings as they degrade the plastic, causing it to fail. It can also ruin the bath and shower tray.

1 Mould a thin strip of compound or sealant around the flange of the waste.

2 Place the waste in the basin, making sure that the slots in the waste line up with the overflow slots of the basin. Apply more compound or sealant around the bottom of the basin and the thread of the waste.

3 Press a plastic washer into the compound and tighten the back nut to complete the joint. Clean off excess sealant or compound with a soft dry cloth.

Also check to see that the compound or silicone has not obstructed the overflow slots in the basin/waste fitting. This is the internal overflow that was moulded into the basin when it was manufactured. The slots on the waste allow any overflowing water in the basin to travel through into the internal overflow and safely down the waste pipe through the trap seal.

Figure 7.22: Standard 32 mm slotted basin waste, plug and chain

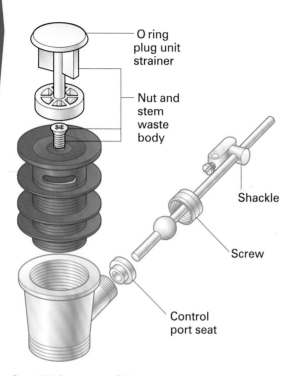

O ring
plug unit
strainer

Nut and
stem
waste
body

Shackle

Screw

Control
port seat

Figure 7.23: Pop-up waste fitting

The overflow level on any sanitary appliance is designed to carry away any water that has reached a certain point before it spills over the edge of the appliance. An overflow can be internal or external to the appliance. External overflows are commonly used on baths, kitchen sinks and in some designs of countertop washbasins. This type is called a combined waste and overflow.

Pop-up wastes

Pop-up wastes are available for washbasins, baths and bidets. On a bath, the plug unit is raised by twisting the circular control knob anticlockwise; this serves as both an overflow outlet and pop-up waste control. The plug unit is lowered by twisting the control knob clockwise, which allows the plug to drop into the waste by gravity (see Figure 7.23).

Moulded seals

These are supplied for pop-up wastes and combined wastes and overflow, as used on baths, plastic washbasins and sinks. The moulded seal fits between the waste flange and the appliance. For pedestal and plastic basins, a shaped seal and O ring are used together to make the joint for the shaped metal connecting washer or the metal pedestal fixing bracket (see Figure 7.24).

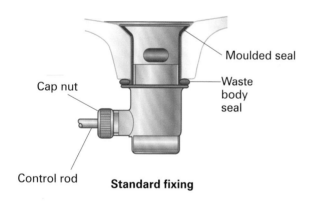

Moulded seal

Waste
body
seal

Cap nut

Control rod

Standard fixing

Plastic basin with overflow fitment

Pedestal basin fixing

Figure 7.24: Moulded seals

Where the appliance requires an overflow assembly, you should fit a rubber ring between the overflow flange and the back of the appliance. When fitting this set-up:

- do not cross-thread the overflow grill and the fitting to the rear of the appliance
- do not overtighten the fitting
- make sure you do not damage the flexible overflow tube when tightening the overflow fitting.

> **Remember**
>
> When tightening any backnuts:
> - do not overtighten
> - do not cross-thread.

Fixing a washbasin pop-up waste

- Fit the waste body into the basin outlet.
- Remove the cap nut and control rod.
- Position the appropriate seals, then couple the pop-up body onto the waste tail thread.
- Tighten with the control port correctly aligned with the back of the basin.
- Slacken the lock nut on the stem of the pop-up plug and fit into the waste body.
- Position the control rod in the control port so that the end of the rod is in the hole at the end of the plug stem.
- Assemble the cap nut.
- Check the operation of the plug. It should be flush with the waste flange and should lift about 10 mm when operated.
- To reduce the lift, turn the plug clockwise. To increase the lift, turn it anticlockwise.
- Remove the cap nut and control rod and take out the plug.
- Tighten the lock nut against the strainer stem of the plug and then refit the plug and control rod and tighten the cap nut.
- Fit the lift rod through the fitting and locate with the eyelet on the shackle.
- Adjust so that the left knob is just clear of the fitting when the plug is fully open.
- Tighten the screws on the shackle and check for smooth operation of the assembly.

Identify alternative means of waste disposal

Macerator

A macerator is a unit installed behind a WC pan that collects waste from the WC and reduces it to a 'liquid' state (see Figure 7.25). It is then pumped via a 22 mm internal bore pipe into a discharge stack. There are a few things to remember about macerators.

- A drain-off valve should be fitted at the base of a vertical rise.
- The macerator can only be fitted when there is another conventional WC in the dwelling.
- The electrical supply to the unit must be via an unswitched fused spur connection.

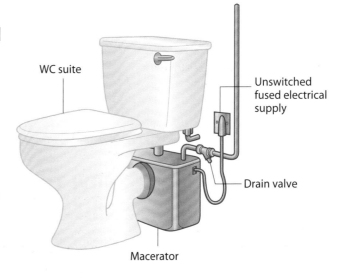

WC suite

Unswitched fused electrical supply

Drain valve

Macerator

Figure 7.25: WC macerator unit

- The unit can discharge its contents to a maximum of 4 m vertically upwards and 50 m horizontally, but not at the same time.
- Horizontal pipework must be laid to a minimum fall of 10 mm per metre.
- Waste pipework is typically 22 mm diameter for short pipe runs. With longer pipe runs, the diameter of pipework will need to be increased in accordance with manufacturer's requirements. Simple charts are included in the manufacturer's installation instructions to make the pipe sizing easy based on the overall length of the pipe run.

Sink waste disposal unit

These are usually installed under kitchen sinks to dispose of refuse such as vegetable peelings, teabags and leftover food. They require a larger waste outlet than normal and manufacturers produce sink tops to suit this. The unit should:

- be connected to the electrical supply via a fused spur outlet
- be connected to a waste trap – this must be a tubular 'P' trap as for all kitchen sinks
- be supplied with a special tool to free the grinding blades should they become blocked (keep your fingers clear)
- have some form of cut-out device that will turn the unit off if it jams.

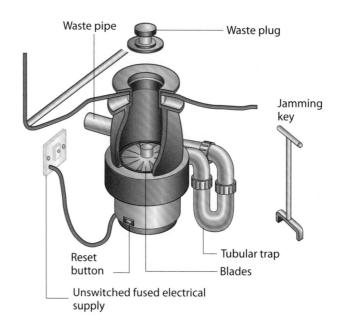

Figure 7.26: Waste disposal unit

KNOW THE REQUIREMENTS FOR INSTALLING SANITARY APPLIANCES

This learning outcome will introduce you to the methods and practices involved with the installation of common sanitary components and appliances. It includes areas such as storage and handling, tools and equipment and checking installations.

Identify safe storage methods for sanitary appliances and why they are important

Sanitary appliances are expensive so it is important that they are stored correctly to prevent damage or theft. This will help you to avoid replacement costs and to keep customers happy.

Storage of materials on site is particularly important on larger jobs where you might take delivery of a number of items of sanitary ware. However, some aspects also apply to one-off jobs.

Checklist

Safe storage

- Before a delivery takes place, make sure there is somewhere suitable to store the appliances – a lockable materials cabin, with adequate room, is ideal. The storage surface should be raised off the ground of the cabin (use pallets or similar) and it should be clean and dry.

- Before accepting delivery, check the appliance(s) carefully for signs of damage but without removing the wrappings. These should be left on as long as possible to protect the appliances. Production defects are rare, but keep an eye out for these as well.

- Check that all traps, taps, wastes and plugs, brackets and seats are with the appliances. There is nothing worse than starting a job and then finding that a waste fitting or bracket is missing.

- When you are happy that everything is complete, check off all the items ordered against the delivery note. If you have the authority, sign for the delivery. You will receive a customer's copy of the delivery note. If anything is not correct, note the missing items and contact the supplier immediately. If you need the delivered items urgently, sign for the items and get the person making the delivery to countersign that some items are missing.

- Take care when handling the material. Some sanitary items are heavy: use the lifting technique described on page 45 of Chapter 1. Most pieces of sanitary ware are fragile and can break into sharp pieces, so always wear gloves, especially when cleaning up accidents.

- Store the materials in the materials cabin, which should be clean and dry. The appliances should be stored away from plaster and concrete, and should be kept away from areas where other materials (such as brick stacks) could fall onto or into them.

The ways of storing appliances shown in Figure 7.27 on page 298 are simple but effective. WC pans should be stacked no more than four high. Make sure the storage surfaces (battens/pallets) are clean. Battens are used for all sorts of things on site, and if they are covered in grit, storing materials on them will be no better than storing them on the ground.

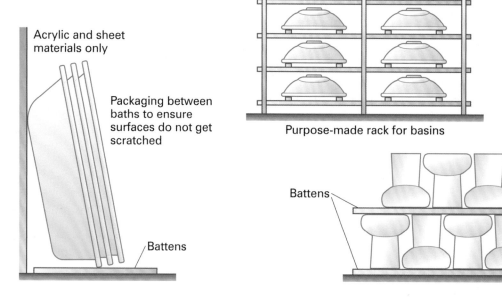

Acrylic and sheet materials only

Packaging between baths to ensure surfaces do not get scratched

Battens

Purpose-made rack for basins

Battens

Figure 7.27: Safe storage patterns for baths, WCs and basins

Checklist

Work site

- Check access to the work area, and check that the work area itself is free of obstructions and potential hazards. This should minimise the risk of accidents.

- Make sure you have any job details or specifications and drawings showing the fixing positions of the sanitary appliances and any specific fixing requirements. You will also need the manufacturer's fixing instructions.

- If you find any installation problems or defects in the appliance as you are working, report them to your supervisor.

- On some jobs you will cut holes or chases or install ducts for pipework. On other jobs this preparatory work is done by other trades in advance, so you will need to check that it has been done. If not, report back to your boss or the site supervisor.

- When you are at the stage of fixing the appliances, the first fix pipework must be installed. This may have been done by you, particularly on a single dwelling job, or it could have been done by another plumber on a large housing contract.

- Make sure servicing valves have been installed where required. Good practice involves putting them on every appliance that requires regular service and maintenance.

- On first fix jobs, open-ended pipes should have been capped. This is to prevent debris getting into the system.

- On jobs where the appliances are being fixed against tiles, baths and showers should be installed before tiling, and concealed bidets, washbasins and WCs after tiling. This makes the job easier for the tiler as they won't have to cut around the shape of the appliance.

- Edges of baths should be set into plaster or plasterboard. Careful planning and liaison with other trades can ensure that plasterwork is not taken to full floor level on the bath edge wall. In the case of stud-partitioned walls, extra struts can be fixed to support the edges of the plasterboard.

- Finally, as you are carrying out the work, avoid standing on any of the appliances. This is sometimes unavoidable with showers or baths with a shower appliance installed above them. If you do have to stand on a shower or bath, make sure the surface of the appliance is fully protected with dust sheets and cardboard packaging.

Identify safe handling of sanitary appliances

The manual handling or lifting of objects causes more injuries on work sites than any other factor. Back strains and associated injuries are the main reason for lost hours in the building services industries. Manual handling may involve pushing, pulling or lifting and lowering heavy loads such as sanitary appliances.

The movement of loads requires careful planning in order to identify potential hazards before they cause injuries. You should follow the correct safety precautions and codes of practice at all times.

Before moving a heavy load, you should carry out the basic risk assessment in Table 7.01.

Safe working

Always use the correct lifting and carrying techniques when moving materials from the store to the job. Ensure that your route is planned and safe.

Consider the task	*Consider the load*
Does it involve: • stooping • twisting • excessive lifting or lowering distances • excessive carrying distances • excessive pushing or pulling distances • frequent or prolonged physical effort • the risk of the load moving suddenly?	Is it: • heavy • bulky or unwieldy • difficult to grasp • unstable or with contents that are likely to shift • sharp, hot or otherwise potentially damaging?
Consider the environment	*Consider yourself*
Does it have: • space constraints • slippery or unstable floors • variation in levels • poor lighting • hot, cold or humid conditions?	Do you have: • any restriction in your physical capabilities • the knowledge and training for manual handling?

Table 7.01: Risk assessment for moving a heavy load

Full details of correct manual handling procedures are described in Chapter 1 on pages 44–45.

You should also wear all the necessary personal protective equipment (PPE) before handling any sanitary appliances.

- When carrying out work operations, you may need safety shoes, knee pads and suitable workwear.
- When removing old sanitary appliances, you should wear suitable gloves to help prevent contamination and the risk of disease. Disposable gloves are the most suitable, but they must be disposed of correctly so they cannot be handled at a later date.

Describe the different fixing methods required for installing sanitary appliances

Different types of appliance require different fixing methods so you should always refer to the manufacturer's installation instructions. In general, most appliances require fixing to the wall and floor in some way. This can normally be done with the correct size of plastic wall plug and screw for masonry walls or with a suitable cavity fixing for hollow walls. The various types of fixings are covered in detail in on pages 138–143 of Chapter 4.

Freestanding appliances

Freestanding appliances are not fixed in any way: the weight of the appliance is enough to prevent it from moving. The older cast iron baths with claw feet for decoration are a good example of these types of appliances. Some types of washbasin are freestanding, as they simply rest on a countertop with the tap outlet entering the basin over the side of it.

Wall-hung appliances

Wall-hung appliances require a support bracket, which is usually supplied by the manufacturer. It should be fixed to the wall and possibly also the floor. You should give careful thought to the weight the appliance will have to deal with when in use. You should use zinc-plated screws, as the support bracket will be boxed in for aesthetic reasons and any screws that become corroded will be difficult to replace later.

In-built appliances

These appliances cannot be moved easily as they are designed to work with the unit in which they are supplied. The manufacturer will supply details for the fixing of these units, which you should follow.

Describe the method of installing sanitary appliances

There is no fixed sequence for the installation of sanitary appliances. In general, plumbers will decide on the best sequence based on the position of the appliances and the size and layout of the bathroom. They will base their decision on what they think will be the quickest and easiest order of installation.

You should also consider the needs of the customer. For example, if you are replacing a bathroom suite in a dwelling with only one WC, you should install the WC suite first, to minimise the inconvenience to the occupants.

WC suites

We will assume that all the components have been installed in the cistern during **dressing** and the cistern has been fixed to the WC pan. The installation process is as follows.

- Offer the cistern to the wall, check the level of the cistern and pan and mark the holes through the back of the cistern. At the same time, mark the fixing holes through the base of the pan. If a soil pipe is already in place, either remove the pan connector or ease the pan outlet into the pan connector.
- Carefully take out the close-coupled suite, drill and plug the holes.
- Refit and screw back the cistern; screw down the WC pan. For the cistern, use soft washers and brass or alloy screws. Some manufacturers supply screws with caps that match the colour of the suite.

Working practice

Daniel, an apprentice, has been asked by his employer to assemble a bath and washbasin ready for installation later that day. Make a list, stating the order in which these two sanitary appliances should be assembled.

Key term

Dressing – plumbers often refer to 'dressing the suite'. It means installing the taps, wastes and – in the case of baths – the cradle frame or feet. It also includes installing float valves, overflows or flushing valves and the handle assembly to the WC cistern.

- Some plumbers, once they have positioned the pan, drill through the fixing holes rather than removing the close-coupled suite to do the drilling. This is acceptable, but take great care not to damage the suite with the drill, and also make sure the plug is fully located in the fixing holes.
- As previously mentioned, most modern WC cisterns do not need an external overflow pipe. However, if this is not the case and the hole for the overflow has not been prepared, mark this out and drill it before fixing the suite.
- Most float valves are plastic, so, when connecting the pipework with capillary fittings, do not solder the joint where it is connected to the plastic thread because it will melt and distort.
- Make sure a fibre washer is in place before tightening the tap connector.
- Remember, all float-operated valves must have a service valve fitted.
- Fit the overflow pipe, which is usually made from plastic.

Washbasins

Our sample bathroom specification includes a pedestal basin. The taps and waste are in, and the metal bracket to hold the basin to the pedestal has been fitted (see Figure 7.28).

To install the washbasin:

- Loosely attach the clamps using the pins and washers. Do not tighten them at this stage.
- Run sealing compound along the top edge of the pedestal; this will provide a better finish to the front.
- Place the basin on the pedestal. Adjust the clamps so that they rest against the precast shoulders of the pedestal.
- Check the basin is level and make any final adjustments.
- Carefully tighten the pins and then clean off any excess sealing compound.

The basin in this example has moulded screw holes at the back edges of the bowl for fixing it to the wall (see Figure 7.29), as follows:

- Offer the basin up to the wall; make sure it is level and in the correct position (you might have to pack the pedestal to achieve level). Make a mark on the wall through the holes.
- Remove the basin and pedestal and place them somewhere safe, preferably on a dust sheet.

> **⚠ Safe working**
>
> Close-coupled suites are heavy. Seek assistance if possible when handling the suite.

> **Remember**
>
> Do not overtighten or you will split the plastic. Watch out for cross-threading, too.

> **Did you know?**
>
> It is often a good idea with pedestal installations to fit the trap to the waste fitting before securing the basin to the pedestal; this is easier than when the basin is on the pedestal.

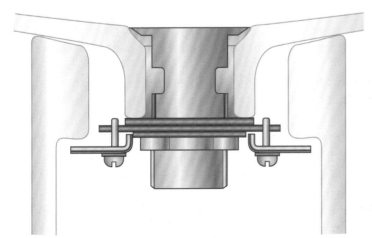

Figure 7.28: Basin fixing brackets in place

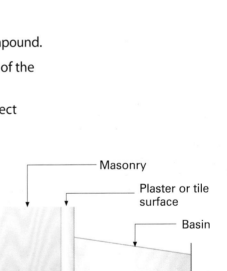

Masonry

Plaster or tile surface

Basin

Soft washer

Screw fixing aligned with hole in sanitary ware

Figure 7.29: Screw fitting of basin

- The moulded holes in the basin are angled. You need to estimate this angle and drill the hole accordingly. If you drill it level, as you screw the basin to the wall, the edge of the basin will crack off adjacent to the hole!
- Reposition the basin and pedestal and screw back to the wall using brass or alloy screws and soft washers. This will make the basin easier to remove for maintenance or replacement.
- If the pedestal has fixing holes in the base, you should mark the floor at the same time as the wall is marked for the basin. On wooden floors, drill pilot holes before refitting. On solid floors, place plugs in the drill holes before repositioning the pedestal.
- Basins supplied without fixing holes at the back of the bowl will have moulding holes on the underside edge.
- Fit the bracket to the basin and tighten so that it holds firmly in the hole.
- Offer the basin to the wall, then adjust the brackets so that they are flush with the wall. Tighten the fixing bolt into the basin, and then mark the position of the bracket holes.
- The pipework installation can be prefabricated so that it fits neatly behind the basin and the pedestal. Make sure you have fitted fibre washers before tightening tap connections for pillar taps. Monobloc taps usually come with compression fittings. Mark off the position of the joints between the pipework tabs and prefabricated pipe at the same time as you make the fixing holes.

Washbasins with brackets

A range of brackets are available, mainly for use in industrial or commercial situations (see Figure 7.30). When selecting a suitable bracket for the job, you should:

- make sure it is the correct pattern for the basin
- be confident that it will support the weight of the basin.

The normal fixing height of a basin is 800 mm from the floor to the front rim. The following is good practice when marking out for the brackets.

- Mark the centre line on the wall where the basin is to be fixed.
- Place the basin on the floor with its rim 800 mm from a wall.
- Position the brackets on the basin and measure, at the same time, the distance to the fixing holes from the wall and the centres of the fixing holes between the two brackets. Halve that distance to get the measurement from the centre line marked on the wall to the centre of each bracket.
- Transfer the measurements to the wall where the brackets are to be fixed. Position the brackets on the centre lines and mark the holes.
- Drill and fix the brackets, then place the basin in position and check height and level.

Alternatively, if you are working with a colleague, mark out the centre line for the basin and the rim height. Then, one person holds the basin in position, checking that it is in line and level. The other person puts the brackets in position and marks the holes.

> **Remember**
>
> Don't forget to use pilot holes on wooden floors. Do not fully tighten at this stage because you will need to adjust the bracket.

Figure 7.30: Selection of wall-mounted basin fixing devices

Countertop basins

These are usually supplied with a cutting template, which you must follow carefully when cutting out the hole. Use a jigsaw with fine teeth and a downwards cutting action to avoid damaging the worktop.

- Once you have cut the hole, seal the exposed edges with a waterproof varnish to prevent water/moisture damage to the chipboard.
- Seal the joint between the lips of the bowl with a recommended sealant, and then secure the bowl using the brackets supplied.
- The edge of the hole for an under-countertop basin is finished off with a strip of laminate in the same material and colour as the countertop.

Bidets

Bidets are screwed to the floor using brass or alloy screws. Marking and connection to supply pipework follow a similar pattern as for other appliances. The waste fitting is a 32 mm (1¼ inch) threaded unit, either slotted or solid depending on whether or not the bidet is fitted to an overflow; pop-up wastes are also available. Taps can be pillar or monobloc for over-the-rim applications. Figure 7.31 shows a typical arrangement for a douche type (under-rim) bidet.

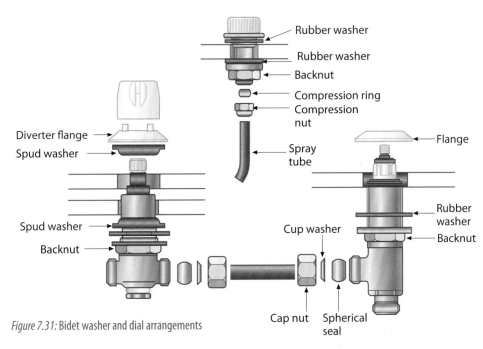

Figure 7.31: Bidet washer and dial arrangements

Urinals

Urinals are fixed on the type of bracket shown in Figure 7.32. The measurements can be set out in a similar way to those of washbasin brackets.

Wall urinals: These are usually installed at a height of 610 mm from floor level to the front lip of the bowl. When a number are installed, one or two might often be installed at about 510 mm for children or smaller adults. Dress the urinal bowls first, fitting the inlet spreader and waste.

Slab urinals: These are generally manufactured in fire clay, and can be supplied to any length on one or more walls, either in one piece or in smaller slabs. The floor channel is manufactured so that the internal surface is laid to a fall. This means that the actual channel block can be installed level.

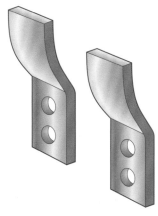

Figure 7.32: Urinal bowl fixing brackets

> ## Working practice
>
> Liam has been asked to install a range of wall urinals in a refurbishment job at a local swimming pool. A total of six wall urinals must be installed, fed by one cistern. He accepts delivery of the goods and checks them on delivery. However, when he opens the packaging, he notices that a 6-litre flushing cistern has been supplied, which is too small for the Water Regulations requirements, and that it has a siphon operated by a flushing handle.
>
> Discuss the following points:
>
> - What type of cistern and flushing arrangement should have been delivered?
> - Should all the urinals be hung at the same height? What is the reason for this?
> - What is the normal height of urinals from the floor to the rim of the appliance in each case?
> - What other specialist pipework and fitments should the delivery contain for the flushing of the urinals?
> - The pool is closed overnight for eight hours every day. What measures should be taken to prevent undue wastage of water in the urinals?
> - What do the Water Regulations specify about the flushing requirements of urinals? Will the installation provide enough volume for the required flushing arrangements?

Baths

- Place the bath on its feet and in position.
- Check the measurement from the top edge of the bath and/or bath panel to the floor. It should be the manufacturer's recommended distance.
- Check the bath is level across its length and width. You can make adjustments to level and height via the feet.
- Once the bath is level and at the correct height, tighten the locking nut on each foot.
- Mark the position of the fixing holes for the feet and the wall fixings. If the wall is plastered, and you have been instructed to let the bath into the plaster, mark it off for the full length of the bath. Take out the bath.
- If the floor is wood, drill pilot holes to receive the screws for the feet. This will make fixing much easier. If the floor is solid, drill and plug the floor. Drill holes for the wall fixings.
- Chase out the plaster for the wall brackets and, if letting the bath into the plaster, chase out the plaster where you marked the line of the bath. This enables the bath to be let into the wall and provides a good watertight seal once tiled.
- Refit the bath and screw the wall and floor brackets into position. If fixing to a timber floor, make sure the screws will not penetrate the underside of the floor.
- Now make the service and waste connections. It is best to prefabricate the tap connections if using soldered fittings to avoid using a blowtorch under the bath. Connect the trap and, if there is an external waste system, extend the waste pipe to the outside.
- If the bath is fixed flush to the plaster, fill the bath to one-third so that the weight causes it to settle. Then, after removing the protective film from the bath, seal the joint between bath and wall with silicone sealant.

It is sometimes advisable to fit wooden battens (75 × 50 mm) under the feet of the bath. This reduces the length of the feet adjustments and so makes the bath more secure. Battens are essential when installing onto composite flooring like chipboard, to stop the feet puncturing the surface when wet or damp.

Fitting bath panels

Methods for fitting bath panels vary, so follow the manufacturer's fixing instructions. The following steps should serve as a general guide.

- Check the measurement from the underside of the roll edge of the bath to the floor. Check for any adjustments needed if the floor is not level.
- If the panel has to be cut, support it firmly, cut it using a fine tooth saw and then finish off with a file.
- If panel support frames or bath leg clips are supplied, you should fit these.
- Use a level or plumb line to mark out for a fixing batten at floor level. Once you have made an allowance for any 'kicking space' on the panel, mark the position for the batten. If using self-adhesive pads, allow for their thickness when positioning the batten (four pads should be sufficient).
- Alternatively, you can use screws. Mirror screws provide an attractive finish, and some manufacturers provide coloured screwhead caps to match the bath panel. Pre-drill the panel to take the screws and then position the panel by inserting its top edge between the bath roll and the timber frame of the bath. Secure the panel at the base with the self-adhesive pads or screws.

Shower trays and enclosures

The waste is fitted to the tray during dressing. It will help with future cleaning and maintenance if a removable waste is fitted. The waste trap on a shower tray should always be accessible for maintenance (see Figure 7.33). Failure to leave access is a common mistake made by plumbers, which can cause real problems later. The waste requirements for shower trays are straightforward and are fitted using the washers supplied with a 38 mm (1½ inch) waste fitting. On fire clay trays, non-setting plumbing compound is applied in the same way as for a standard slotted washbasin waste installation (see page 293).

> **Remember**
>
> Bath panels should only be fitted once testing has been completed.

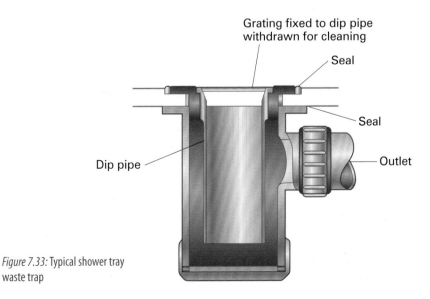

Grating fixed to dip pipe withdrawn for cleaning

Seal

Seal

Dip pipe

Outlet

Figure 7.33: Typical shower tray waste trap

Activity 7.3

Make a list of all the sanitary appliances in your home. Describe them in as much detail as you can.

Did you know?

Acrylic trays often have panels to allow access to the underside, so they do not need to be raised.

Safe working

To avoid the drill sliding over the smooth surface of a tile and causing damage to it (and possible injury to you), stick a piece of clear tape on the hole markings. This will provide a key for the drill bit as you start to drill the hole. You should manually rotate the chuck to break through the glaze and start the drill without hammer action.

The most important thing to ensure when fitting the tray is that it is level in all directions. If it is not, you will face big problems when fitting the enclosure.

Fixing a shower tray to a timber floor

- First, check whether the trap and waste pipework can be accessed under the floor. If not, you may have to raise the tray as for an installation on a solid floor (see below).
- If the pipework can be accessed, make sure the floor is clean and dry, then coat the underside of the tray with a recommended sealant. Alternatively, the tray can be bedded on the tile cement.
- Position the tray and press it firmly into place. Make sure it is level in all directions.
- Once the bed is dry or cured, you can seal the tray edges to the adjoining walls.

Fixing a shower tray to a solid floor

On a solid floor installation, it is likely that you will have to raise the shower tray to gain access to the trap and waste pipe. This is done by building a sub-frame out of external grade plywood and timber battens. The thickness of the battens will be governed by the amount of clearance you need, but they should be placed at 250 mm intervals. Once you have fixed the sub-frame, install the tray as you would for a timber floor.

Fitting a shower enclosure

Shower enclosures come in all shapes and sizes. The best advice here is to follow the manufacturer's instructions carefully, but here are a few general points.

- Check that the walls are plumb. Most manufacturers make allowances for the walls being slightly out of plumb. On older properties, if the wall is seriously out of plumb, some plastering work may be required before you start.
- It is almost certain that the frame of the enclosure will be fixed to a tiled surface, and this will involve drilling the tile.
- Using the manufacturer's measurements for the frame, mark out for the fixing holes. The best way is to mark the wall the full length of the frame, making sure it is plumb. Then offer the frame to the line and mark out for the fixing holes.
- The remainder of the process involves fitting the panels and the door. Make sure the doors open and close correctly. The installation should be completed by fitting any sealing strips supplied with the kit, and applying a recommended sealant as specified in the manufacturer's instructions.

State bespoke tools used for the installation of sanitary appliances

Table 7.02 describes the main tools you will need for installing sanitary appliances. No particular PPE is needed for these hand tools, but refer back to Table 4.01 on pages 124–135 of Chapter 4 on common plumbing processes for other general tools and equipment.

Tool	Care, operation and maintenance	Usage
Tap box spanner	Double-ended box spanner for quick assembly of tap backnuts on basins • Keep lightly oiled to prevent corrosion.	Used to tighten backnuts of taps on a washbasin, as it is easier to use than pump pliers or an adjustable spanner.
Telescopic basin wrench	Telescopic basin nut wrench with reversible jaw Spring-loaded jaws • The tension plug holds the head in any position up to a 90° angle. • Keep lightly oiled, jaw clean and free of debris to prevent slipping.	Used for accessing nuts on washbasins, baths, sinks or bidets.
Basin wrench	Basin wrench suitable for 15 mm and 22 mm nuts made from cast iron or aluminium • Keep lightly oiled to prevent corrosion.	Can be used in awkward places, to tighten backnuts fitted to taps behind sinks or baths.
Soft jaw plumbing pliers	• Keep the jaws free of dirt and grease. • Ensure the moving parts are kept lightly oiled. • Ensure the handles are in good condition.	Can be used to grip anything that has a decorative finish without risk of damage from the jaws, e.g. tap headgear when changing washers.
Tap reseating tool	• Keep the cutting teeth free from debris and dirt. • Keep all moving parts lightly oiled.	Used to reseat a tap so that the new washer has a good seal against it. Reseating will only work if the old seat is not too badly damaged; if it is, a new tap will be required.
Magnetic telescopic mirror	• Keep the mirror separate in your toolbox to prevent breakage. • Keep the telescopic sections lightly oiled.	The flexible head allows close inspection work and good visibility in hard to reach areas. It can also see around the back of pipework to ensure it is clean ready for soldering. The pickup tool is great for retrieving small ferrous objects that have been lost in confined spaces.
Toilet seat installation tool	• Ensure the end is not damaged, which may cause it to slip.	Used to tighten or slacken the nuts that secure the toilet seat to the pan.

Table 7.02: Main tools for installing sanitary appliances

Describe quality checks of sanitary appliances

Upon completion of the installation of sanitary appliances, and with the water off, you should do the following.

- Check that the appliances are level and have been secured correctly and firmly to the wall and/or floor to prevent them moving or even collapsing when in use.
- Make sure any locking nuts on the bath feet have been tightened so the legs won't come loose.
- Make sure the taps are secure and won't turn in the tap hole when operated.

You should then turn on the water to each appliance in turn at their individual isolation valves and make checks for leaks.

- Fit the plug in the waste outlet and run some water into the appliance. Then release it, checking the trap and waste for any leaks.
- Ensure that all the water runs away so that no water is left behind in the appliance.
- Any sealants such as silicone should be applied all around the edge of the appliance where it is in contact with the wall to prevent water from running down behind it and to provide a neat finish to the job.

Progress check 7.02

1 What term can be used to describe a WC that has its water cistern 1 m or more above the bowl?

2 The flushing capacity of WC cisterns has been reduced over recent years. What is the main reason for this change?

3 Which scientific principle explains the working principles of the automatic flushing device fitted to a urinal?

4 What is the purpose of an AUK2 air gap for taps?

5 If a plumber was 'dressing the suite', what would he be doing?

Knowledge check

1 The term 'back-to-the-wall' describes what type of sanitary appliance?

   a shower tray
   b bath
   c washbasin
   d WC

2 What is the maximum flushing capacity of a modern WC cistern?

   a 2 litres
   b 3 litres
   c 4 litres
   d 5 litres

3 What is the waste outlet on a bath called?

   a a combined waste and overflow
   b a slotted outlet
   c plug and chain
   d non-slotted

4 Currently, what is the required flushing capacity for a single urinal bowl?

   a 8 litres
   b 9 litres
   c 10 litres
   d 11 litres

5 On which of the following appliances could you find a pop-up waste outlet?

   a washbasin
   b shower tray
   c Belfast sink
   d urinal

6 What size is a basin waste outlet?

   a 32 mm
   b 40 mm
   c 50 mm
   d 65 mm

7 What size taps are fitted to a bath?

   a ¼ inch (6 mm)
   b ½ inch (13 mm)
   c ¾ inch (19 mm)
   d 1 inch (25 mm)

8 At what height should a wall-mounted washbasin be fixed?

   a 790 mm
   b 795 mm
   c 800 mm
   d 810 mm

9 At what height should a urinal bowl for children be fitted?

   a 500 mm
   b 510 mm
   c 520 mm
   d 530 mm

10 The term 'monobloc' describes which of the following?

   a single waste outlet
   b single hole fixing mixer tap
   c pair of pillar taps
   d single trap outlet

Central heating systems

This chapter will cover the following learning outcomes:

- **Understand the types of domestic central heating systems installed in domestic dwellings**
- **Know the different materials used to install domestic central heating pipework**
- **Understand heat emitters and their components**
- **Understand mechanical central heating controls**

Introduction

The idea of being able to heat all the rooms in a house was first contemplated in the 1950s. Up until then, which room was heated depended on which room was going to be used. The normal heat source was an open coal fire.

Over many years, central heating has developed into an efficient and sophisticated system that allows full control to suit individual lifestyles. This chapter looks at the requirements for central heating, from the beginnings of home heating to modern systems. In particular, it details the different pipework layouts and controls used for each type of system.

UNDERSTAND THE TYPES OF DOMESTIC CENTRAL HEATING SYSTEMS INSTALLED IN DOMESTIC DWELLINGS

Having central heating in dwellings is now considered part of normal life. However, a few decades ago it would have been considered a luxury.

State the purpose of central heating systems

The purpose of central heating is to:

- provide thermal comfort inside the building or dwelling
- be economical to operate and ensure maximum efficiency.

A newly installed central heating system must meet all the requirements for efficiency as set out in British Standard BS EN 14336:2004 *Heating systems in buildings: Installation and commissioning of water based heating systems.* This covers the installation, commissioning and testing of modern systems to ensure that they are working as efficiently as possible and that they maximise energy usage.

A modern, efficiently installed central heating system provides many benefits:

- Warmth will be provided throughout the dwelling.
- It allows us to make full use of all the rooms in a building or dwelling.
- A modern central heating system is fully controllable using programmers and thermostats.
- It helps prevent black mould growth and reduces condensation.
- It can help alleviate health problems such as asthma and bronchitis.

Identify the principle pipework systems

Gravity systems

Gravity heating systems work by having the pipework at the correct fall to aid circulation. Full-gravity systems are no longer installed, but you might come across them during maintenance work in older buildings such as village halls. They were sometimes referred to in domestic properties as background heating. One radiator, for example in the bathroom, was fed from the primary flow and return pipe from the boiler. Flow and return pipework to the radiator was usually 22 mm diameter.

Pumped heating with gravity hot water systems (semi-gravity)

These can be either one-pipe or two-pipe systems.

One-pipe system

This works using a one-pipe run (a complete ring of pipework) with a flow and return from the boiler (see Figure 8.01). The disadvantages of this system tend to outweigh the advantages, so they are no longer installed in domestic properties (see Table 8.01). However, as with full-gravity systems, you may come across them in older properties.

Advantages	Disadvantages
Lower installation cost compared to two-pipe system	The **heat emitters** (radiators) on the system pass cooler water back into the circuit, meaning that the emitters at the end of the system are cooler.
Quicker to install	The pump only forces water around the main circuit and not directly through the radiators. This means it is important to select radiators that allow minimum resistance to the flow of water.
Lower maintenance costs	The 'flow' side of the heat emitter (radiator) is usually installed at high level to improve circulation, creating additional unsightly pipework.

Table 8.01: Advantages and disadvantages of a one-pipe system

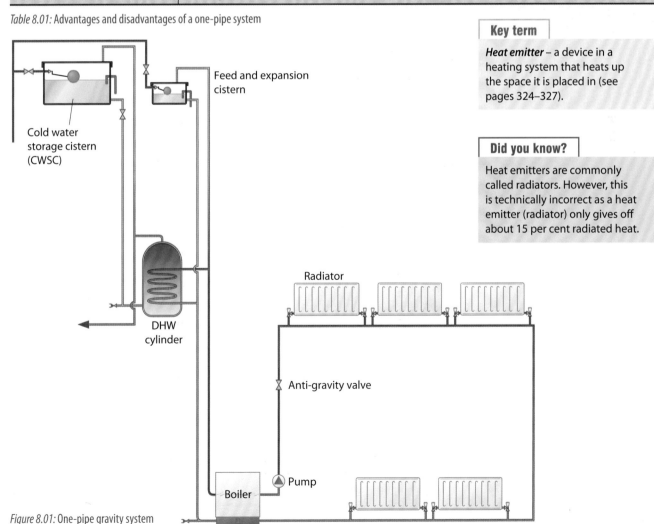

Feed and expansion cistern

Cold water storage cistern (CWSC)

DHW cylinder

Radiator

Anti-gravity valve

Boiler

Pump

Figure 8.01: One-pipe gravity system

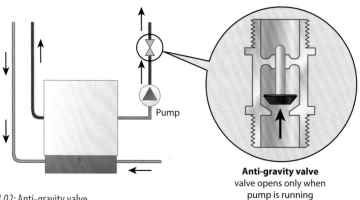

Figure 8.02: Anti-gravity valve

Anti-gravity valve
valve opens only when
pump is running

The one-pipe system features an anti-gravity valve to prevent the unwanted circulation of heated water by gravity when the central heating pump is not activated (see Figure 8.02).

Careful balancing and positioning of the radiators is important.

- For a given heat requirement in a room, the radiators at the end of the circuit must be larger than those at the beginning, due to the cooling effect of mixing return water into the flow pipe.
- To ensure correct operation, radiators must be as close to the pipework ring as possible. If you try to cut a couple of connections 100 mm apart into the pipework and run it for 5 m to the radiator, it will not work. The full ring has to be extended to run under the radiator with short tail connections onto it.

Remember

Feed and vent pipes are essential on many systems, but are often omitted on diagrams like these for clarity.

Two-pipe system

This system was particularly popular in the 1970s, but is no longer permitted on new properties (other than with solid fuel boilers) unless

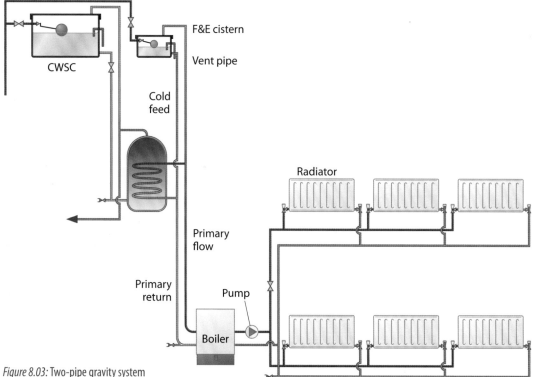

Figure 8.03: Two-pipe gravity system

additional controls are installed. This type of system no longer meets the requirements of the Building Regulations for extension or boiler replacements to existing oil- or gas-fired systems.

The two-pipe system is the most common method of feeding radiator circuits. Water is pumped around both the circuit and the radiators. This improves the speed at which radiators heat up. The system can be balanced easily by adjusting the **lockshield valve** on each radiator. See page 331 later in this chapter for more on lockshield valves.

Fully pumped system

In this system, the hot water and the heating circuits are operated completely by the pump. Because there is no requirement for gravity circulation, the boiler can be sited above the height of the cylinder, giving more design options (see Figure 8.04).

Installations are controlled by **motorised valves**. There are a number of system designs incorporating two-port zone valves or three-port valves (two-position and mid-position) that meet the requirements of the Building Regulations Approved Document L1. Motorised valves are covered later in this chapter (see page 335).

Key terms

Lockshield valve – used to balance the system when it is installed or maintained. It is also used as a service valve and is solely intended for use by the installing or maintenance plumber and not by the consumer.

Motorised valve – a valve in a water pipe that can be turned on or off by an electric motor. It is one of the heating controls in a system and a valve can be two- or three-port.

Remember

Any new system in a domestic property, other than one using a continuous-burning appliance, should be fully pumped.

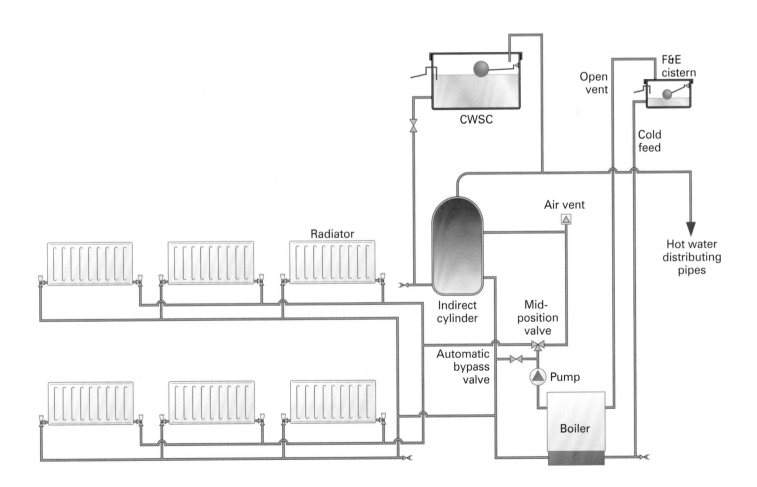

Figure 8.04: Fully pumped system

Compare the operating performance of principle pipework systems

Table 8.02 compares the main systems discussed on pages 310–313 and summarises their main features.

Pipework system	Features
Gravity No longer allowed to be installed under Part L of the Building Regulations	• Low installation cost • Very low operating efficiency • Poor general performance because it works using natural convection; long heat up times and long cool down times • Only used in a few rooms to provide background heat due to low operating temperatures • Large-diameter pipework required to ensure adequate flow to the heating surfaces • Low maintenance requirements; no external controls, pumps or motorised valves • Very limited temperature control – boiler thermostat only
Semi-gravity No longer allowed to be installed for gas- or oil-fired appliances under Part L of the Building Regulations	• More expensive to install than gravity systems • Increased efficiency as heating temperature now controlled by room thermostat, although hot water temperature still controlled by boiler thermostat • Cannot select central heating only as hot water is heated continually by natural convection, which reduces system efficiency • Higher maintenance cost because of greater use of mechanical and electric components • Increased performance of heating system; faster heating and cool down times and full heating to each room
Fully pumped Modern installation meets requirements of Part L of the Building Regulations	• The most expensive system, but greatly increased efficiency offsets installation cost • Separate control of heating and hot water with individual on/off time periods • Fast heating and cool down times of heating; fast recovery of hot water using high-efficiency cylinder • More maintenance needed because of control valves, thermostats, etc. • Boiler can be located higher than the cylinder as all circulation is now pumped • Full system interlock to ensure no energy consumption when not in use • Zoning possible for areas of irregular use

Table 8.02: Comparison of main central heating pipework systems

Explain the function of the pipework component parts

Feed and expansion (F&E) cistern

This is used on all open-vented central heating systems. Its main purpose is to allow water in the system to expand, while the cistern allows the system to be filled.

- The water level should be set low in the cistern when filling the system. The cold feed to the system in an average domestic property is usually 15 mm minimum, and this pipe must not include any valves.

This is to ensure that, in the event of overheating, there is a constant supply of cooler water to the system to prevent the dangerous condition of boiling.

- The servicing valve to the system should be located on the cold water inlet pipework to the cistern. If a valve were fitted in the cold feed, it coud be closed inadvertently; this could have disastrous consequences if the open vent also became blocked.
- The F&E cistern is located at the highest point in the system, and it must not be affected by the position and head of the circulating pump. To avoid any problems with gravity primaries, a minimum height can be obtained by dividing the maximum head developed by the pump by three.
- In fully pumped systems, the water level in the F&E cistern should be a minimum of 1 m above the pumped primary to the direct hot water storage cylinder.

Space for expansion of water

The system volume expands by about 4 per cent when heated, so a system containing 100 litres would expand by 4 litres in volume. Space must be allowed in the F&E cistern to take up the additional volume when heated.

Float-operated valve

The float-operated valve in the F&E cistern controls the flow of water into the cistern. It should be set low enough so that there is enough water to cover the 15 mm cold feed to the heating system. However, it must also allow for expansion of water in the system when it heats up. As indicated above, the expansion is 4 per cent of the system volume. In normal domestic installations, an 18-litre F&E cistern is adequate for this purpose.

The float-operated valve also allows cool water to enter the system should the boiler overheat because of thermostat failure. This will help prevent the water in the system from reaching boiling point, which would be very dangerous. Therefore, the cistern and float-operated valve must be able to withstand a temperature of 100°C. There is more on float-operated valves on pages 204–206 of Chapter 5.

Primary open safety vent

The purpose of the primary open safety vent is to:

- provide a safety outlet should the system overheat due to component failure
- ensure that the system is kept safely at atmospheric pressure.

The minimum diameter of the safety vent is 22 mm, and the pipe should never be valved. In a fully pumped system, the primary open safety vent should usually rise to a minimum height of 450 mm above the water level in the F&E cistern. This allows for any pressure surge effects created by the pump. The open safety vent also aids the removal of any air that the system might collect, particularly on commissioning or refilling (see Figure 8.05 on page 316).

Air separator

The air separator enables the cold feed and vent pipe to be joined closely together in the correct layout to serve the system. The grouping of the connections inside the air separator causes turbulent water flow in the separator, which in turn removes air from the system. This reduces noise in the system and lowers the risk of corrosion.

> **Remember**
>
> The Water Regulations require that all float-operated valves must be fitted with a service valve.

> **Did you know?**
>
> An 18-litre feed and expansion cistern is normally used with an open-vented central heating system in domestic properties.

Figure 8.06 shows a fully pumped system containing an air separator. This is also called a close-coupled method of feed and vent pipe connection, featuring only two pipes connected to the boiler.

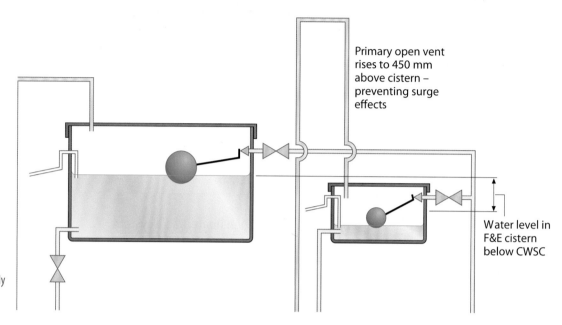

Primary open vent rises to 450 mm above cistern – preventing surge effects

Water level in F&E cistern below CWSC

Figure 8.05: Open safety vent to F&E cistern in a fully pumped system

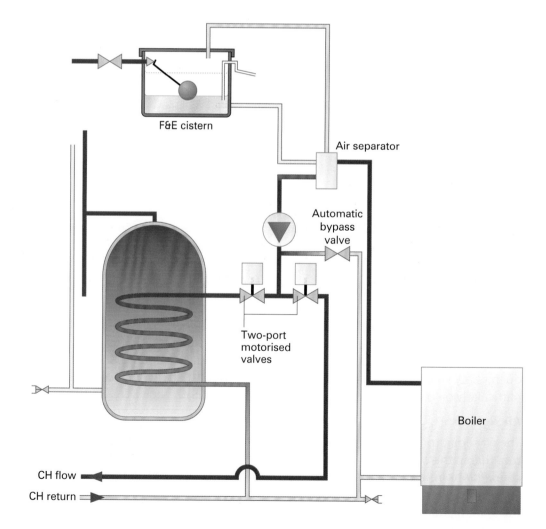

F&E cistern

Air separator

Automatic bypass valve

Two-port motorised valves

Boiler

CH flow

CH return

Figure 8.06: Air separator installed in a fully pumped heating system

Gate valve

A gate valve allows equipment to be serviced. When open, there is no restriction through the valve. The valve has a wheelhead attached to the spindle. When the head is turned anticlockwise, the threaded part of the spindle screws into the wedge-shaped gate, raising it towards the head (see Figure 5.24 on page 202). Gate valves are usually found in low-pressure pipelines, such as the cold feed from the cold water storage cistern (CWSC) to the hot water storage cylinder. You may also find them on supplies to shower valves where the shower is fed from low pressure via a CWSC.

Full-bore valve

A full-bore valve provides the same function as a gate valve but is easier to operate. It works in one quick movement by turning the handle a quarter of a turn (see Figure 8.07). A sphere with a hole through it inside the valve, much like a service valve, is turned as the valve opens and closes. Unlike a service valve, in which the bore is reduced considerably by the sphere, the bore of the pipe in a full-bore valve is maintained through the valve, just like a gate valve, and so provides minimal restriction to the water flow.

Drain-off valve

Also called a drain-off tap, this is covered on page 209 of Chapter 5.

Lockshield valve

These are fitted on heat emitters (radiators) and used to balance the system when it is installed or maintained. They are also used as service valves and are solely intended for use by the plumber, rather than the occupier of the building. They are covered in more detail on page 331.

Automatic bypass valve

These are used to make sure water can flow through the boiler to maintain a minimum water flow rate if thermostatic radiator valves (TRVs) or motorised valves close when there is no more demand for heat in the rooms or dwelling. Once set, the bypass valve will open automatically when the system pressure reaches a set point: a head of pressure will build up because of the still-running pump (see Figure 8.08).

Bypass valves reduce system noise and increase pump life by preventing it working against a '**dead head**'. The bypass should be installed between the flow and return. The valve has a direction arrow on it and should be installed with that arrow in the direction of the flow. If a higher capacity is required for large installations, two or more valves can be installed in parallel.

> **Key term**
>
> *Dead head* – a situation in which the pump is running but all heating controls are closed, preventing the flow of water around the heating circuit. This can damage the pump impeller and lead to pump failure.

Figure 8.07: A full-bore valve

Figure 8.08: Automatic bypass

Thermostatic radiator valve

Thermostatic radiator valves (TRVs) are a cost-effective way of controlling room temperature. The user adjusts them to a chosen temperature and the valve then works automatically to maintain that temperature. TRVs help lower system running costs by reducing demand on the boiler. This in turn helps the environment by reducing the amount of carbon dioxide (CO_2) released into the atmosphere.

Compare the different types of space heating systems

Central heating can be classed as full, background or selective.

- A **full** heating system heats all habitable rooms to the normal design temperatures of 21°C for the living room and bathrooms and 18°C for all other rooms. This is based on an outside temperature of –1°C.
- **Background** heating also heats the rooms but to lower temperatures.
- **Selective** heating allows a particular area or areas to be heated, usually by some form of control system.

Describe the configuration of space heating systems

The main types of heating systems – gravity and pumped – have been discussed on pages 310–314, so refer back as necessary.

C and C+ system

This semi-gravity system uses a two-port valve and provides independent temperature control of both the heating and hot water circuits in a pumped heating and gravity domestic hot water system (see Figure 8.09). The pump and boiler are switched off when space and hot water temperature requirements are met.

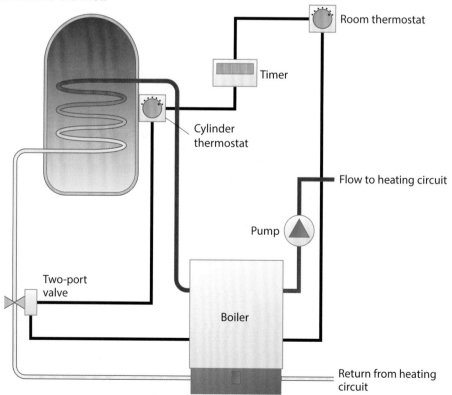

Figure 8.09: Semi-gravity system with two-port valve (simplified for clarity)

A zone valve can also be fitted to the heating flow pipe, enabling temperature control of the heating via a room thermostat. These modifications will ensure that the system meets the requirements of Part L of the Building Regulations with minimal cost. However, the best course of action would be to convert the system to fully pumped. TRVs can also be fitted to provide overriding temperature controls in individual rooms. Time control can be managed by either a time switch or a programmer.

Fully pumped system using a two-position three-port diverter valve (W plan)

This was one of the first fully pumped systems to be installed in domestic properties, but it is no longer widely used. It is designed to provide independent temperature control of the heating and hot water circuit in fully pumped heating systems. When used with a programmer, this design satisfies the Building Regulations.

The two-position three-port diverter valve is usually installed to give priority to the domestic hot water circuit (see Figure 8.10). That means it can only feed either the hot water or the central heating system at any one time. For this reason it should not be used where there is likely to be a high hot water demand during the winter months. The designed heating room temperature could drop below comfort level when the demand for hot water is high.

Fully pumped system using three-port mid-position valve (Y plan)

This type of system is common in new domestic properties and is designed to provide separate time and temperature control of the heating and domestic hot water circuits. To fully meet the requirements of the Building Regulations, time control must be managed via a programmer and TRVs, and an automatic bypass valve must be fitted (where required). The mid-position valve allows hot water and heating circuits to operate together.

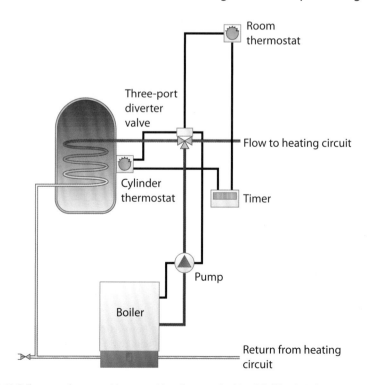

Figure 8.10: Fully pumped system with two-position diverter valve (simplified for clarity)

Did you know?

Although common in new properties, the Y plan system is not suitable for floor areas greater than 150 m^2. For larger areas, the S plan is recommended.

Be careful when working on existing systems: the three-port diverter valve and mid-position valve system look similar. However, the three-port valve and timing device are completely different for each system. The mid-position valve has more wires than the diverter valve.

Working practice

Ray qualified as a plumber last year. He now has his own apprentice and needs to explain to her the installation requirements and system operation for a fully pumped system using a two-position diverter valve.

- List all the points Ray needs to cover.
- What other information will Ray need to tell his apprentice?
- What work is the apprentice likely to do?

Activity 8.2

Think about how your own central heating and hot water system works and try to work out which plan it uses. Even if you have a combination boiler rather than a traditional boiler and cylinder system, it will still fit one of the plans.

Fully pumped system using 2 × 2-port valves (S plan)

This type of system is common in new domestic properties, particularly larger ones, and is recommended for floor areas above 150 m^2. The main reason for this is the limited capacity of a three-port valve installation to satisfy the heat demands of a larger system. The use of the 2 × 2-port valves also gives greater flexibility in system design, with additional valves being added to the system to zone separate parts of the building – for example, the upstairs from the downstairs – so that each area is controlled individually. The system provides separate temperature controls for heating and hot water circuits (see Figure 8.11). The features are similar to those of the other systems described earlier.

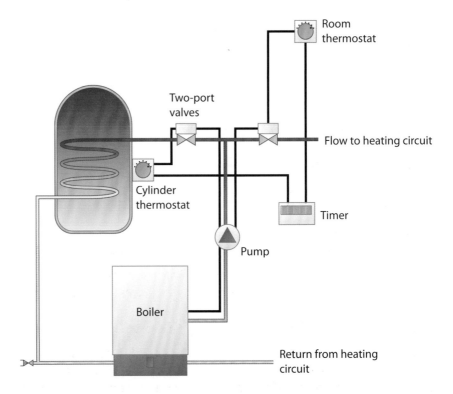

Figure 8.11: Fully pumped system with 2 × 2-port valves (simplified for clarity)

Progress check 8.01

1   In open-vented systems, what is the minimum diameter of the open vent pipe? What is the purpose of the open vent?

2   What is the purpose of an air separator in a central heating system?

3   What type of system would use an S plan?

4   A customer's heating system sometimes cools down when there is a demand for hot water to be heated. What type of motorised valve are you likely to find in this system? What 'plan' is this system likely to follow?

5   On some older systems, the circulating pump was fitted in the return pipe back to the boiler. What problem can this cause, particularly as the system ages?

KNOW THE DIFFERENT MATERIALS USED TO INSTALL DOMESTIC CENTRAL HEATING PIPEWORK

Modern plumbers have to work with various materials, so it is vital to the working life and efficiency of systems that they know and understand the application of these materials.

Identify the principle materials used in domestic central heating applications

The main materials you will come across and work with are copper, mild steel and various types of plastic. You can find out more about these and other materials used in domestic central heating applications in Chapter 4: *Common plumbing processes*.

Describe the use of plastic barrier tube for installing central heating circuits

When installing modern heating systems with plastic pipework, it is important to use barrier tube. This prevents air from permeating (passing through) the wall of the plastic pipe and into the heating system. Air in the system will quickly cause severe corrosion problems. Figure 8.12 shows the construction of plastic barrier pipe used in central heating systems.

The advantages of plastic pipe over more traditional materials such as copper include:

- it does not promote **electrolysis**
- it has a very smooth bore to help the flow of water through the pipe.

However, it should not be used on the surface (e.g. on skirting boards): because of its high **coefficient of expansion** it will sag and look unsightly.

Describe the advantages of insulating pipework

Pipe insulation works by trapping air in its enclosed cell-like structure. Trapped, still air is a poor conductor of heat, which makes it ideal for use in insulation. Even cold water contains some heat energy, and it is this energy that pipe insulation keeps in.

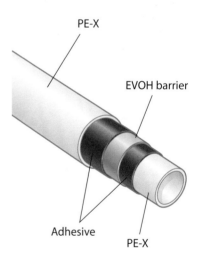

PE-X

EVOH barrier

Adhesive

PE-X

Figure 8.12: Barrier pipe construction

Key terms

Electrolysis (electrolytic action) – describes a flow of electrically charged ions from an anode to a cathode through a medium called electrolyte, usually water. (See Chapter 3, pages 99–100.)

Coefficient of expansion – the amount a material expands when heated. (See Chapter 3, pages 95–96.)

Figure 8.13: Polyethylene pipe insulation

Progress check 8.02

1 How does pipework insulation work?

2 State the different materials that pipework insulation can be made from.

3 Why must you use barrier pipe when installing plastic push-fit heating systems?

4 What would you calculate using Delta T?

5 When you are installing a radiator, what is the minimum distance you should allow between the bottom of the radiator and the floor? Why should you allow this distance?

Insulating pipework has several advantages:

- Water in the pipework will stay warmer for longer. This means that less water will be wasted where there are dead legs in hot water systems. It is also good for the environment, because the energy used to heat the water is being used more efficiently.
- Pipes are less likely to freeze and burst during cold weather, reducing the risk of water damage to the customer's property.
- It will help to control condensation problems on pipework. This in turn will help to reduce corrosion.
- It will help to control the noise level of water flowing through the pipework.
- It will help to protect the customer from burns if they touch any hot water or heating pipework.

State types of pipework insulation

Polyethylene pipe

Polyethylene insulation (see Figure 8.13) is often referred to as closed cell or foam pipe insulation and has several advantages:

- It is cost-effective as it is cheaper than nitrile rubber and foil-backed pipe insulation.
- It has a very low density with strong thermal, physical and chemical-resistant properties.
- It is effective over a wide range of temperatures and will prevent heat gain as well as heat loss.
- It can be installed in many different situations as well as domestic ones.
- It can be used with trace heating for areas that could be subject to very low temperatures.

Its disadvantages are that it is less flexible and more rigid than nitrile rubber and does not have strong fire resistance.

Foil-backed lagging

Foil-backed pipe fibreglass insulation comes in pre-formed section with a self-adhesive strip down one side for fast and efficient installation. A roll of foil tape is also used to help make joints at corners and where straight sections butt up against each other. It is designed for insulating pipework and equipment operating within a temperature range of about −50°C to 120°C. This makes it ideal for warm, hot and heating services and for cold, chilled and refrigeration services.

Nitrile rubber

Nitrile rubber pipe insulation has a fine, closed cell structure and works within a temperature range of around −45°C to 105°C. It is available in a full range of wall thicknesses and bore sizes and has superior thermal conductivity, making it one of the best pipe insulations in its class. Nitrile rubber is particularly suitable for:

- frost protection
- domestic heating
- ground source heat pump applications
- chilled pipework applications.

State bespoke tools used for the installation of domestic central heating systems

The specialist tools and equipment you will need for installing central heating systems are listed below and covered in more detail in Table 4.01 in Chapter 4:

- ½" (13 mm) hexagon radiator valve spanner (see page 124)
- radiator spanner (see page 124)
- adjustable water pump pliers (see page 128)
- radiator vent key (bleed key/air release valve) (see page 124)
- disposable pipe freezing kit (see page 134)
- reusable pipe freezing kit (see page 133).

In addition, you may need a ratcheted radiator valve spanner. This tool fits most radiator valves, nipples, lugs and cistern connectors, and the integral ratchet means that you do not need to remove the tool to turn it. You should always keep the ratchet lightly oiled and be careful not to let the hexagon key slip when you are using the spanner.

UNDERSTAND HEAT EMITTERS AND THEIR COMPONENTS

Heat emitters are the devices used within a heating system to heat up the space they are placed in. There are various types, and they range in style, material, cost and installation.

Identify the European Standard for the manufacture of radiators

BS EN 442

Radiators sold in Europe formerly had to meet standards set by each country's certification body, such as the BSI or the DIN in Germany. However, in 1996 BS EN 442: *Specification for radiators and convectors* replaced BS 3528 in the UK. Since then technology has moved on considerably in both manufacturing and testing. As from 1 July 1997 all radiators manufactured in Europe need to conform to the European standard BS EN 442.

Heat output performance is obviously the main requirement for radiators but the new standard covers several other important aspects not previously covered by BS3528 (and covers some more accurately). These are:

- minimum material thickness for all wet surfaces
- a detailed table of manufacturing tolerances
- periodic burst pressure tests to supplement the 100 per cent leak pressure tests
- pre-treatment and paint quality requirements for corrosion protection and resistance to impact damage
- requirements for product marking, labelling and catalogue data.

It is in the area of heat output performance, however, that the new standard has greatest immediate impact.

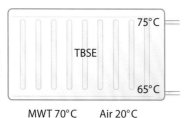

Figure 8.14: Radiator showing top bottom same end (how Delta T is calculated)

Example

For a water flow temperature of 75°C and a return temperature of 65°C, the mean water temperature (MWT) is 70°C. If we then subtract the room temperature of 20°C, we are left with a temperature difference (Delta T) of 50°C.

Remember

When choosing radiators, you should take note of manufacturers' fixing positions. It is often said that a radiator should be positioned beneath a window to reduce draughts. Curtains should finish 10 cm above the radiator.

Define Delta T

Delta T (▲T) is the difference between the mean (average) water temperature in a radiator and the ambient (surrounding) air temperature – so it is the change in temperature:

▲T= final temperature – initial temperature.

Under BS EN 442, radiators must be tested with a flow water temperature of 75°C and a return temperature of 65°C, in a test room with a consistent air temperature of 20°C. In addition, the flow and return connections should be connected at the same end, normally referred to as 'top bottom same end' (TBSE) (see Figure 8.14).

Identify different heat emitters used in domestic systems and explain their working principles

Heat emitters can be convectors, radiators or underfloor heaters, with variations within each type. This section looks at the most common types of heat emitter for domestic systems and how they work. Chapter 3 covers the scientific principles of heat emission, so refer to pages 96 and 114–116 as necessary.

Cast iron column radiators

Sometimes called hospital radiators, these are mostly found on older installations in buildings such as schools or village halls. However, some are now being installed in domestic properties as 'designer' decor (see Figure 8.15).

Panel radiators

Despite the name 'radiator', about 85 per cent of heat from radiators is given off by convection. The heat output of a standard panel radiator can be further improved by welding 'fins' (heat exchangers) onto the back. These increase the radiator's surface area as they become part of its heated surface. The design of the fins will also help convection currents to flow.

Types of panel radiator

Radiator design has developed dramatically as manufacturers aim to provide efficient radiators in a variety of styles. Figure 8.17 shows the most common types of steel panel radiator.

Figure 8.15: Cast iron column radiator

Figure 8.16: A typical panel radiator

- Manufacturers provide at least four height options, from 300 mm to 700 mm.
- Width measurements are from 400 mm, with increments of 100–200 mm, to a maximum of around 3 m.
- The recommended height from the floor to the base of the radiator is 150 mm (depending on the height of the skirting board). This allows adequate clearance for heat circulation and valve installation.
- Outputs will vary depending on design. You must ensure that the radiator's output will be sufficient to heat the room.

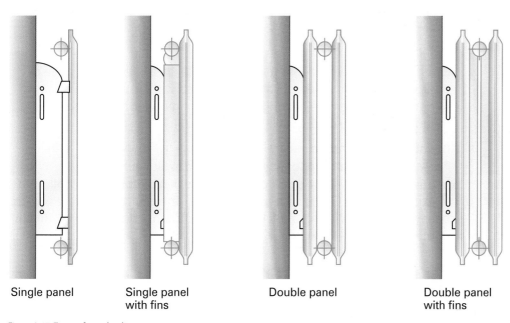

Single panel | Single panel with fins | Double panel | Double panel with fins

Figure 8.17: Types of panel radiator

Seamed-top panel radiator

This is currently the most commonly fitted radiator in domestic installations (see Figure 8.18). Top grilles are also available for this radiator.

Compact radiators

These have all the benefits of steel panel radiators, with the addition of 'factory-fitted' top grilles and side panels, making them more attractive to the consumer (see Figure 8.19).

Figure 8.18: Seamed-top panel radiator

Figure 8.19: Compact radiator

Figure 8.20: Rolled-top radiator

Rolled-top radiator

As the popularity of compacts has increased, the market for rolled-top radiators has declined. Some of the production seams from their manufacture can be seen following installation, making them less attractive to the customer (see Figure 8.20).

Combined radiator and towel rail

This design allows towels to be warmed without affecting the convection current from the radiator (see Figure 8.21). These are generally only installed in bathrooms.

Tubular towel rail

Often referred to as 'designer towel rails', these are available in a range of designs and colours. They can also be supplied with an electrical element option for use when the heating system is not required. They tend to be mounted vertically on the wall (see Figure 8.22).

Low surface temperature radiators (LSTs)

These were originally designed to conform to health authority requirements, where the surface temperature of radiators was not allowed to exceed 43°C when the system was running at maximum. LSTs are now becoming popular in children's nurseries, bedrooms and playrooms, and in domestic properties for occupants with disabilities.

Figure 8.21: Combined radiator and towel rail

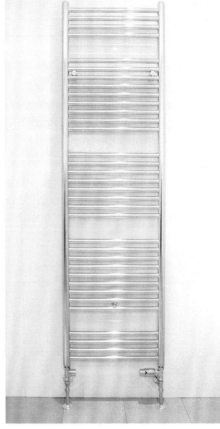

Figure 8.22: Tubular towel rail

Skirting convector heaters

These work on the principle of natural convection; the fins provide a large surface area for heat output. The panel is heated by conduction from the heating pipe (see Figure 8.23). Cool air then passes through it and is heated. This heated air rises and passes from the panel via louvres at the top. Skirting heating is no longer widely used in domestic properties because of the restrictions placed on output from the heater. However, you may still come across it in some older properties.

Figure 8.23: Skirting heater

Fan-assisted convector heaters

These work by forcing cooler air through the heating fins using an electric fan (see Figure 8.24). Therefore, they require connection to the central heating control system. This type includes include kick space heaters, which can be tucked away where space is limited (see Figure 8.25).

Activity 8.3

Heat emitters can be made of other materials, as well as steel. Using the Internet, text books and trade magazines, list as many of these materials as you can.

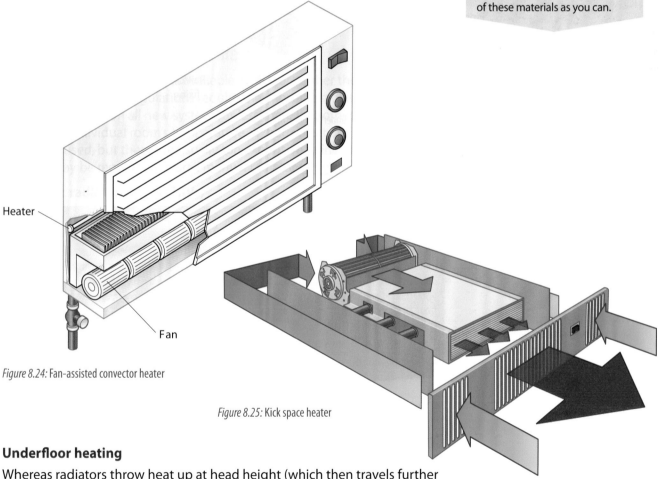

Figure 8.24: Fan-assisted convector heater

Figure 8.25: Kick space heater

Underfloor heating

Whereas radiators throw heat up at head height (which then travels further up to the ceiling only to travel downwards and return as a cold draught), underfloor heating provides a pleasant warmth for feet, body and head. This is because it produces only a very gentle air circulation.

> **Remember**
>
> When hanging radiators you should allow for the position of furniture. It has become the norm for radiators to be positioned under windows as not much furniture goes there.

Explain how to hang a radiator

In the days when windows were much less thermally efficient than today, radiators were positioned under them. Cold air from draughty single-glazed windows would be warmed as it passed over the radiator. These days, however, better glazing means that radiators can be sited wherever the customer wants.

The procedure for hanging a radiator is outlined below. Refer to pages 44–45 of Chapter 1 for advice on manual handling.

Hanging a radiator

① Check the job specification to make sure the radiator is the correct one for the room.

② Carefully unpack the radiator and inspect it for damage, making sure you have the vent plug, end plug and wall brackets.

③ Mark the centre position of the brackets. Use pencil so the marks can be removed easily afterwards.

④ Find the centre of the wall where the radiator is to be hung. Then lean the radiator against the wall and transfer the positions of the radiator brackets to the wall.

⑤ Lightly plumb down from these marks, to indicate where the brackets will be.

⑥ Measure the distance between the bottom of the bracket and the bottom of the radiator, then add this distance to the proposed height from the floor to the bottom of the radiator. (A minimum distance of 150 mm between floor and radiator is recommended, to allow for good air circulation.)

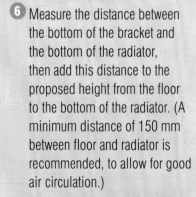

7 Mark where the bottom of the first bracket will be. Place the bracket against the wall on this mark and mark the screw positions. Then assemble the drill, ensuring all electrical safety checks are made, and secure the first bracket to the wall. Level across to position, mark and secure the second bracket.

8 Assemble the radiator with the valves required, using PTFE thread tape. It is standard to fit the flow and return valves at the bottom of the radiator, at opposite ends to each other. Fit the vent plug and end plug, then hang the radiator on its brackets and check for level.

Safe working

Radiators are heavy so you must use the correct manual handling techniques when lifting and handling them.

Did you know?

Some radiator manufacturers supply small plastic inserts for their brackets, to stop the radiator rattling. If these have been supplied, fit them before hanging the radiator.

Describe how to bleed a radiator

To bleed (vent) a radiator, you will need a vent key and a piece of cloth or other absorbent material to catch any water that is displaced from the valve.

- Before you begin, make sure the circulating pump is switched off. If the system is already full of water and you are only releasing a little trapped air, then more air may be drawn in if you try to bleed the radiator with the pump running. This will make the problem worse.
- If you have a system or combination boiler, check that there is enough pressure within the system to force out the air. This check can be made by reading the pressure gauge on the boiler: it should be at 1–1.5 bar. The gauge will drop as the air is released, and you will need to top up the system in order to get the gauge back to 1–1.5 bar. Use a filling loop to do this.
- If you have a more traditional open vented system, with the smaller F&E cistern in the loft alongside the main, larger, cistern, there will be enough natural head of pressure to force the air out. In addition, the system water will be topped up automatically via the ball valve as the air is released from the system.
- Hold a piece of cloth immediately under the vent pin and, with your other hand, use the vent key to turn the vent pin. You must never remove the pin completely as it is very small and very difficult to fit back in once the air has been removed and water is coming out.
- When all the air in the radiator has been vented, water will appear at the vent pin. Re-tighten the pin and test the radiator by turning on the heating system and placing your hand on top of the radiator. If the radiator warms up to the very top, you know that all the air has been removed.

Safe working

Do not attempt to bleed radiators when they are hot. If you do this, you may release scalding water, which could burn you.

Figure 8.32: Radiator plug and manual air vent

Key term

Centrifugal force – a force moving away from the centre of a body with circular motion.

Safe working

Ensure the discharge pipe from the pressure relief valve terminates where it will not harm anyone or damage the structure of the building.

Figure 8.33: Pump stripped down to show impeller

Figure 8.34: An A-rated domestic circulating pump

Figure 8.35: Ball-type pump valve

Figure 8.36: Gate-type pump valve

UNDERSTAND MECHANICAL CENTRAL HEATING CONTROLS

Vital to the efficiency of any modern central heating installation are the controls used in the system. This set of controls enables the user to operate the system to suit them and their lifestyle and to save on energy usage to reduce carbon emissions. This section looks at how these controls work together to minimise energy wastage and ensure the customer has plenty of heated water at the right temperature.

Describe the function of a domestic circulator pump

Pumps are fairly simple. They consist of an electric motor, which drives a circular fluted wheel called an impeller (Figure 8.33). This 'accelerates' the flow of water by **centrifugal force**. When installed, domestic circulating pumps are fitted with isolation valves to permit service and maintenance. Figures 8.34–8.36 show other types of domestic circulator pump.

It is the pump's job to circulate water around the central heating (and possibly the hot water) system, ensuring that the water is delivered at the desired quantities throughout the system components. Most pumps have three settings, and manufacturers provide performance data for each, which shows flow rate in litres per second, pressure in kPa and head in metres (m).

The flow rate should not exceed 1 litre per second for small-bore systems and 1.5 litres per second for microbore systems; anything higher can create noise in the system. Most pumps deliver 5 m or 6 m head. This is usually enough to overcome the flow resistance of the whole heating circuit, in particular the 'index' radiator circuit, which is the radiator circuit that offers the greatest frictional resistance to the flow of water.

It is good practice to position the pump so that it gives a positive pressure within the circuit. This ensures that air is not drawn into the system through microscopic leaks.

Describe the effects of the circulator pump in relation to feed and vent

The pressures developed in a central heating system vary according to the pump's position.

- The pressure on the outlet side of the pump is positive, as the water is being accelerated and expelled at a higher pressure from the pump.
- On the inlet side of the pump, the pressure is negative and the water is being drawn into the pump ready to be accelerated out.

Common sense tells us that there must be a point within the system when positive pressure created by the pump changes to negative pressure. This is known as the 'neutral point' and differs according to the type of system that is installed. In open-vented systems, it is where the cold feed enters the system. The only part of the system then under negative pressure is between the cold feed and the inlet of the pump.

The pressure developed by the static head from the feed and expansion (F&E) cistern will never vary (hence the term neutral point) provided the rules for installation of the F&E and the connection of the cold feed to the system are observed. In sealed systems, the neutral point will be where the expansion vessel connects to the pipework. The pump position is even more critical in a fully pumped system because of its position in relation to the cold feed and vent pipe (see Figure 8.06 on page 316).

State the differences between motorised valves

All motorised valves must provide the correct flow direction and priority to match the system. Diverter valves are used in modern combination boilers. When a hot tap is opened, the diverter valve operates to ensure that maximum heat input is given to heating the hot water system instantly and not to the central heating.

The type of valve used will depend on the system design, but the following are available:

- three-port diverter valve
- three-port mid-position valve
- two-port motorised zone valve.

Three-port diverter valve

This controls the flow of water in fully pumped central heating and hot water systems on a selective priority basis, normally for the domestic hot water (see Figure 8.37).

Three-port mid-position valve

This looks very similar to the diverter valve. It is used in fully pumped hot water and central heating systems in conjunction with a room and cylinder thermostat. It provides full temperature control of both the hot water and heating circuits, which can operate independently of each other or both at the same time.

Two-port motorised zone valve

A single-zone valve is used in gravity domestic hot water and a separate one is used to control pumped central heating systems to enable separate temperature control of both the heating and hot water circuits. Motorised valves are also used in fully pumped systems to provide separate control of heating and hot water circuits. They can be used to zone different parts of a building. Figure 8.38 shows a two-port motorised zone valve.

> **Key term**
>
> **Synchron motor** – synchronous motor. An AC motor where the shaft rotation is synchronised with the frequency of the supply current.

Figure 8.37: Three-port diverter valve

Figure 8.38: Two-port motorised zone valve

Describe the process to exchange a synchron motor

① Ensure safe isolation of the motorised valve.

② Slacken the screw which holds the cover in place and remove the cover.

③ Cut the wires which lead to the motor.

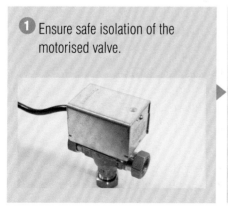

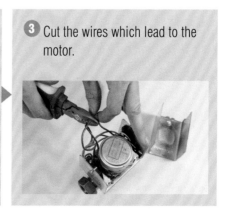

4 Undo and remove the small screw securing the motor to the valve. Take care not to drop and lose the screw.

5 Remove and discard the old motor and replace it with the new one.

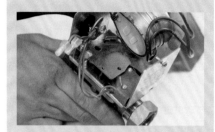

6 Use new connectors to reattach the wires, taking care not to leave any of the conductor exposed. Refit the cover and tighten the screw securing the cover in place.

Knowledge check

1 What is the outside design temperature when calculating radiator sizes?

 a +1°C
 b 0°C
 c −1°C
 d −2°C

2 If a central heating system forms a continuous loop around the dwelling to supply each radiator, what type of system is it?

 a full-gravity
 b two-pipe
 c one-pipe
 d semi-gravity

3 What is the name of the electronic device used to control the room temperature inside a dwelling?

 a cylinder thermostat
 b thermostatic radiator valve
 c boiler thermostat
 d room thermostat

4 If the floor area of a dwelling exceeds 150m², what type of system must be installed?

 a two-port zone valve
 b three-port diverter valve
 c three-port mid-position valve
 d semi-gravity with two-port valve

5 By how much will the water in a central heating system expand when it is up to operating temperature?

 a 1 per cent
 b 2 per cent
 c 3 per cent
 d 4 per cent

6 What is the minimum diameter of the cold feed from the F&E cistern?

 a 8 mm
 b 10 mm
 c 15 mm
 d 22 mm

7 What is the minimum height the F&E cistern should be installed above the primary flow to the cylinder in a fully pumped system?

 a 500 mm
 b 1 m
 c 1.5 m
 d 1.75 m

8 With regard to an F&E cistern, how far should the open vent extend above the water level in the cistern?

 a 450 mm
 b 475 mm
 c 500 mm
 d 550 mm

9 The pump position in relation to the open vent and cold feed is important in an open vented fully pumped central heating system. What does it prevent?

 a air from being drawn in and water from being pumped over
 b water from being drawn in and air from being expelled
 c cool water from circulating around the system
 d the boiler from cycling on and off

10 What type of boiler recycles the flue gases before expelling them to atmosphere?

 a regular boiler
 b condensing boiler
 c system boiler
 d combination boiler

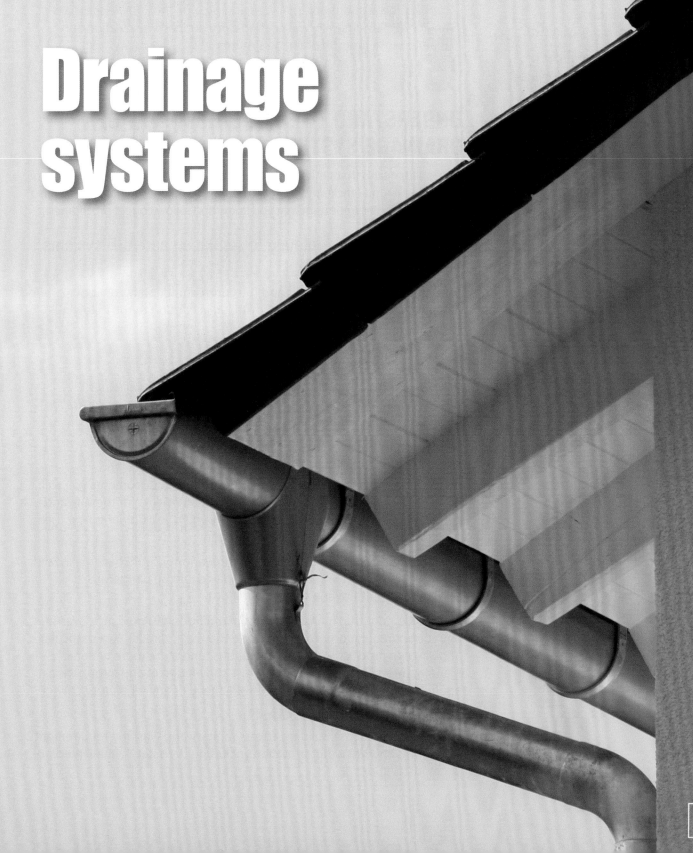

Chapter 9

Drainage systems

This chapter will cover the following learning outcomes:

- **Understand the requirements of drainage systems**
- **Know the types of traps and associated requirements**
- **Know the procedures for soundness testing and commissioning above ground systems**
- **Know the requirements of rainwater systems and associated guttering**

Introduction

This chapter will look at the requirements for installing above ground drainage (AGD) systems, including the relevant legislation, equipment and procedures. It will also introduce you to common sanitary installations and practices and provide the knowledge, skills and understanding you will need to carry out basic installation and maintenance tasks on drainage systems.

UNDERSTAND THE REQUIREMENTS OF DRAINAGE SYSTEMS

It is vital for health that sanitary above ground drainage systems remove human waste cleanly and efficiently. Your own health and safety as a plumber are also paramount, so it is important that you know and follow correct procedures.

Identify documents relating to sanitation and AGD systems and components

To ensure good workmanship and the correct functioning of above ground drainage systems, they should be installed according to the following:

- manufacturer's instructions
- Building Regulations Parts G and H
- BS 6465: *Sanitary installations*
- BS EN 12056:2000: *Gravity drainage systems inside buildings*
- BS 8000: *Workmanship on building sites Part 13: Code of practice for above ground drainage.*

Manufacturer's instructions

When it comes to installation, servicing and maintenance, the manufacturer's instructions are the most important documents you will have available to you. They set out all the requirements for the safe and correct use and installation of their equipment. If you do not follow their guidance, a number of things could happen:

- The product or appliance guarantee/warranty could become invalid.
- The installation or maintenance repair could be dangerous.
- You may contravene some regulations.

You should note that manufacturer's instructions and guidance override any current regulations. This is because they refer directly to a particular appliance or product, whereas regulations tend to be more general. Manufacturers also have the very latest knowledge of and information about their products, which can be well ahead of any regulatory change.

Building Regulations Parts G and H

Part G: Hygiene of the Building covers sanitary and washing requirements, bathrooms and hot water provision, including the safety of unvented hot water installations.

- Part G1 covers sanitary conveniences and washing facilities.
- Part G2 covers bathrooms.
- Part G3 covers hot water storage.

Soil and vent pipe systems can be installed inside or outside but the design and installation must comply with Building Regulations Part H1. Generally speaking, Part H1 states that the foul water system must:

- convey the flow of water to a foul water outfall (a foul or combined sewer, a cesspool, septic tank or settlement tank)
- minimise the risk of blockage or leakage
- prevent foul air from the drainage system from entering a building under working conditions
- be ventilated
- be accessible for cleaning blockages.

BS 6465: Sanitary installations

BS 6465 is in four parts and concerns the design of sanitary facilities and the scale of the sanitary provision that is required. It covers new buildings and buildings undergoing major refurbishment, and also covers the provision of portable toilets.

- **BS 6465-1:** Code of practice for the design of sanitary facilities and scales of provision of sanitary and associated appliances
- **BS 6465-2:** Code of practice for space requirements for sanitary installations
- **BS 6465-3:** Code of practice for the selection, installation and maintenance of sanitary and associated appliances
- **BS 6465-4:** Code of practice for the provision of public toilets

BS EN 12056: Gravity drainage systems inside buildings

This gives guidance on the minimum standards of work and materials to be used. Its requirements cover:

- wastewater drainage systems that operate under gravity
- drainage systems in houses, commercial, institutional and industrial buildings
- correct installation and calculation of drainage systems in the UK.

Identify different types of above ground drainage system

Modern pipework in domestic dwellings are one-pipe systems of one of four types:

- primary ventilated **stack**
- ventilated branch
- secondary ventilated
- stub stack
- grey water recovery.

Primary ventilated (single-stack) system

This is the most common system and is often specified for domestic dwellings because, unlike other one-pipe systems, it does not need a separate ventilating pipe. This means it costs less in terms of materials and installation time, which is good for both business and the environment. Figure 9.01 on page 340 shows a primary ventilated stack with appliances.

Safe working

Remember to wear the correct PPE when handling used sanitary pipework (see page 345).

Key term

Stack – the vertical discharge pipe that carries waste to the main sewer below ground.

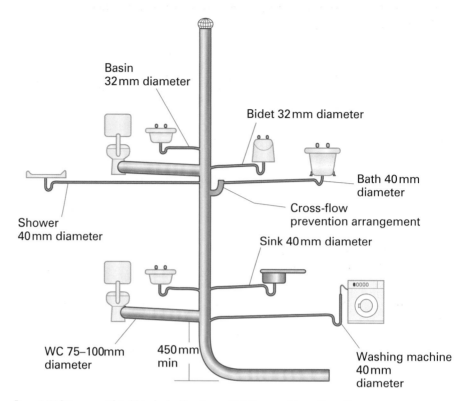

Basin
32 mm diameter

Bidet 32 mm diameter

Bath 40 mm
diameter

Cross-flow
prevention arrangement

Shower
40 mm diameter

Sink 40 mm diameter

WC 75–100 mm
diameter

450 mm
min

Washing machine
40 mm
diameter

Figure 9.01: Primary ventilated (single stack) system, with full range of domestic appliances

Building Regulations Part H1 sets out a number of rules about the design of this system. The rules governing pipe diameter and the minimum depth from the lowest connection above the invert of a drain (450 mm) are shown in Figure 9.01. The invert level is the lowest part of a drainage pipe on the *inside* bore of the pipe, not the outside. It refers to the height above/below a benchmark of the lowest part of the pipe channel at a given point on the drainage system (see Figure 9.02).

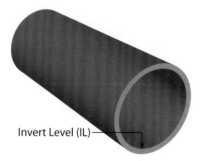

Invert Level (IL)

Figure 9.02: Invert level of a drainage pipe

In addition, there are limits to the maximum lengths of the **branch connections** and their gradients. These are shown in Table 9.01.

	Pipe size (mm)	Maximum length (m)	Minimum slope (mm fall per metre run)
Basin	32	1.7	18–20
Bath	40	3.0	18–90
Shower	40	3.0	18–90
WC	100	6.0	18

Table 9.01: Limits to length and slope of pipes (pipework gradients)

The appliances in a primary ventilated stack system must be grouped closely together. There is some flexibility, however. For example, if you install a shower with a 50 mm waste fitting, it can be located up to 4 m away from the stack, as opposed to 3 m if using 40 mm pipe. The branch pipes should always be at least the same diameter as the trap. The maximum pipework lengths and gradients may be exceeded but only when additional ventilation is provided within the system by a device such as an anti-siphon trap or a self-sealing valve.

Ventilated branch

In this discharge system, the location of a branch pipe in a stack should not cause **cross-flow** into another branch pipe. Cross-flow can be prevented by working to the details shown in Figure 9.03. This will prevent trap seal loss, which is covered in more detail on pages 354–356. It is, however, permissible to have connections from two WCs in opposing positions.

Key terms

Branch connection – where a discharge pipe is connected to the main discharge stack.

Cross-flow – occurs when two branches are located opposite each other.

Chapter 9

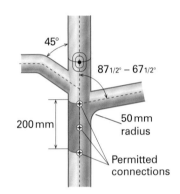

Figure 9.03: Cross-flow prevention

Ventilated discharge branch system

Secondary ventilated stack system

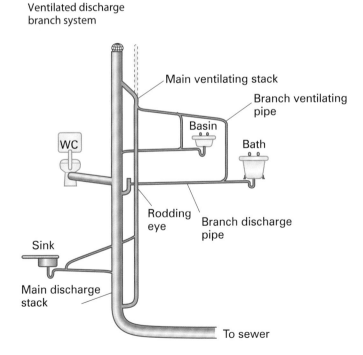

Figure 9.04: Ventilated discharge branch and secondary stack system

Chapter 9

On some installations you may find it easier to run the kitchen sink waste pipe into a gully rather than a pipe to the stack. This is allowed as long as the pipe end finishes between the grating or sealing plate and the top of the water seal. Where these conditions cannot be met, separate ventilation will need to be installed. This can be done in two ways:

- by ventilating each appliance into a second stack – the ventilated discharge branch system
- by directly ventilating the waste stack – secondary ventilated stack system.

The trap sizes and seals required are described in the section on traps on pages 349–354 and in Table 9.03 (page 353).

Branch ventilating pipes

A way of avoiding excessively long pipe lengths or steep gradients is to ventilate the pipework system using separate ventilating pipework (see Figure 9.05). This does not happen on many new installations, as anti-siphon traps or self-sealing valves (see page 351) are now commonly used as a modern alternative, reducing material costs and installation times.

There are some factors to consider.

- The branch vent pipe must not be connected to the discharge stack below the spill-over level of the highest fitting served.
- The minimum size of vent pipe to a single appliance should be 25 mm.
- If it is longer than a 15 m run or serves more than one appliance, it must be 32 mm minimum.
- The main venting stack should be at least 75 mm. This also applies to the 'dry part' of the primary vented stack system.

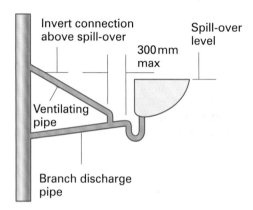

Figure 9.05: Branch ventilating pipes

Stub stacks

Stub stacks reduce the amount of ventilating pipework in a ventilated discharge stack. They also avoid the need for weatherings on internal stacks. Requirements for stub stack systems are covered on pages 353–354.

Grey water recycling

As mentioned in Chapter 5, grey water is waste water generated by activities such as bathing, laundry and dishwashing. Up to half of the water used in a house can end up as grey. For more on grey water recycling, refer back to Chapter 5, page 198.

General discharge stack requirements

- All stacks should have access for cleaning and clearing blockages.
- Rodding (unblocking) points and access fittings should be placed to give access to any length of pipe that cannot be reached from any other part of the system.
- All system pipework should be easy to access for repair purposes.

Explain the installation considerations for primary ventilated stack systems

Volume of waste

Primary ventilated stack systems must cope with a high volume of waste at peak demand times without excessive pressure fluctuations within the discharge system. In some circumstances, it is worth considering oversizing the discharge pipework to prevent these fluctuations.

Branch connections

The location and number of branch connections are also important. This has been covered in detail on pages 340–341.

Self-cleansing gradients

In addition, the gradient of all pipework must be sufficient so that the pipes are 'self-cleansing'. A self-cleansing gradient does not flow too quickly, which would allow solid debris to be left behind in the pipe. If this was allowed to happen, eventually the pipe would become blocked. The aim of a self-cleansing gradient is to allow the waste water to flow away and carry all the debris with it to the main drainage system. Minimum pipework gradients are listed in Table 9.01 on page 341.

Inspection eyes

Pipework must include adequate and easy access for clearing obstructions inside the pipe. Inspection eyes are pipe fittings equipped with a plug that can be removed to allow examination or cleaning of the pipe run. These points should allow rodding of pipes that cannot be accessed by removing the appliance trap. Inspection eyes should be fitted to every change in direction in the pipe. Figure 9.06 shows various inspection eyes.

> **Remember**
>
> Remember to consider access. Check whether an access pipe fitting will be needed.

Figure 9.06: Typical inspection eyes: access pipe, plug and junction

Describe the requirements of a stub stack system

The stub stack is commonly used to connect the ground floor appliances in a property containing more than one bathroom, where the high point in the drainage run includes a ventilating pipe to fully ventilate the drainage system (see Figure 9.07).

- The highest waste connection allowed is 2 m above the invert of the drain and 1.3 m to the base of the WC.
- The length of the branch drain from the stack is 6 m for a single appliance connection and 12 m if there is more than one.

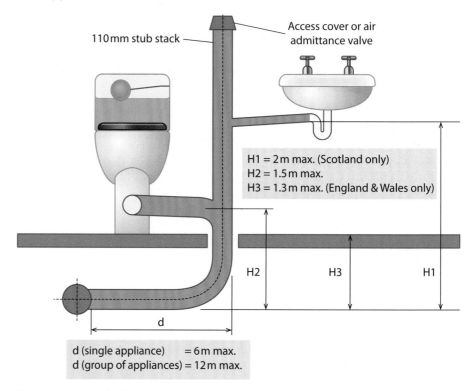

Figure 9.07: Waste heights for a stub stack system

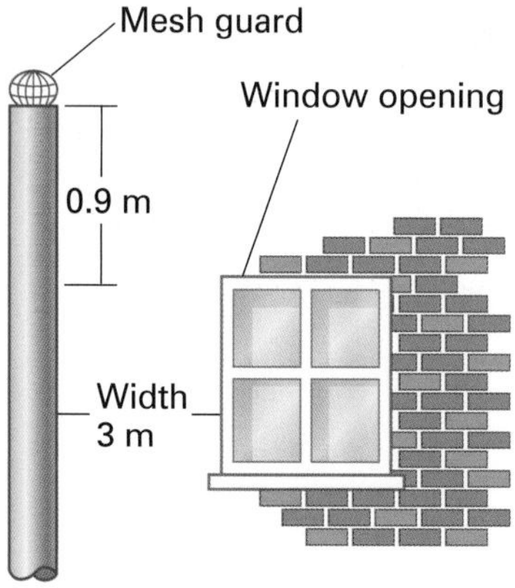

These dimensions can be exceeded when an air admittance valve is used to allow air to enter the system. The valve, which operates on negative pressure (the open valve lets air in) and positive pressure (the closed valve contains the smells), can be located in the roof space or the pipe-boxing arrangement to the WC, subject to certain requirements (see Figure 9.08).

Figure 9.08: An external stack must be terminated as shown

Identify terminals associated with stacks and stub stacks

Air admittance valves

Discharge systems can be terminated inside buildings when they are fitted with an air admittance valve (AAV), as shown in Figure 9.09. This valve avoids the need for the ventilating part of an internal discharge system to penetrate through the roof.

- Only AAVs that carry the British Board of Agreement certificate should be used.
- Before you install an AAV on to the ventilation pipe, you must ensure that the drainage system is independently ventilated, or that there is a

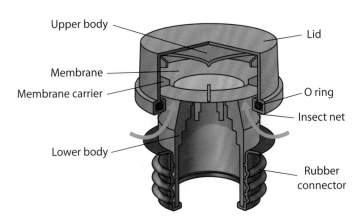

Figure 9.09: Construction and operation of an air admittance valve

second open soil vent. Failure to do so will cause a dangerous build-up of sewer gas within the pipework.

- They can be used on buildings with up to five storeys or on a ground floor for a stub stack system.
- The AAV comes within a polystyrene case; the top half of this case should be kept and fitted to the valve when the discharge system is complete. This will prevent the valve from freezing closed in cold weather conditions.
- AAVs are not meant to be fitted externally as this exposes them to weather conditions for which they are unsuited.
- They are not required if the ventilation part of the discharge system is terminated correctly.

External terminals

A terminal mesh guard (see Figure 9.10) should be fitted to prevent birds nesting on the stack.

A vent cowl (see Figure 9.11) could be fitted where the stack is sited in exposed, windy conditions.

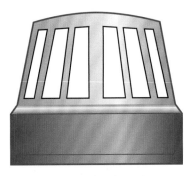

Figure 9.10: A terminal mesh guard

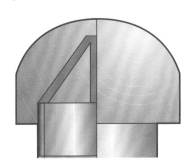

Figure 9.11: A vent cowl

Describe the differences between permanently and temporarily decommissioning above ground drainage systems

From time to time, it will be necessary to either permanently or temporarily decommission a discharge system. Decommissioning above ground drainage systems will normally involve stripping out old appliances and pipework to replace them with new.

- A system is **permanently decommissioned** if it is never to be used again – once decommissioned it can never be put back into use.
- **Temporary decommissioning** may occur when remedial or refurbishment work is being carried out. Existing appliances need to be removed and new ones fitted when the rest of the work is completed; or the customer might require the old appliances to be refitted.

When removing appliances, take care not to damage them. Removing a cast iron bath, particularly if it has to be carried downstairs, requires careful handling. Some plumbers break the bath into four pieces for easier removal. A club hammer is the best tool for this, but correct PPE should be worn: full face protection, ear protection and gloves. The bath is tipped upside down to

 Safe working

Personal protective equipment (PPE) is very important. Always wear a hard hat for these jobs and other PPE as required with particular tools and equipment.

Use rubber gloves when handling old sanitary pipework and appliances. Contact with their contents could lead to health problems.

Safe working

Old pipework systems could be made of lead, so you must take the usual precautions when handling this material.

Many soil and vent pipes used to be made from asbestos cement, so always follow the handling rules for this material. (See Chapter 1, page 8.)

reduce the risk of being hit by shards of enamel, which will shatter off. Once the appliance has been removed, it should be stripped of any scrap metal to be taken for recycling.

On externally mounted pipework you will have to chop out the mortar between the pipe and the masonry. Take care when doing this as it will mean less making good after you have installed the new pipework.

Taking down cast iron soil and vent pipes can be dangerous due to their weight, so they need careful handling. It is best to try to take down short sections by partially cutting them with an angle grinder and then tapping the pipework with a hammer, which will cause it to shear. Tie a rope to each section in turn so that they can be lowered to the floor. Make sure no one is in the work area. Fixing lugs can be broken from the joint, and the nails prised out using a wrecking bar.

Once the stack is removed, make sure the joint to the drain is covered or capped. This will prevent:

- anyone tripping on it
- debris entering the drain
- obnoxious smells escaping from the drain.

A range of blanking plugs for decommissioning work is invaluable.

Working practice

Colin is asked to replace and upgrade an old two-pipe system of above ground drainage to a detached property. The system is a mixture of cast iron and asbestos cement. The whole system is showing signs of leakage and general deterioration.

Bearing in mind the requirements of Building Regulations Part H1 and BS EN 12056 Part 2, Colin decides on the primary ventilated (single-stack) system.

Discuss the following points:

- What steps must Colin make to safely remove the redundant pipework? Include any legislation that he must abide by and any PPE required.
- What criteria must Colin work to? Make a comprehensive list covering all the points required under the legislation.
- Colin has noticed a problem: the bedroom next to the bathroom has a washbasin installed 2.5 m away from the stack. What could Colin do to overcome this problem?
- Does Colin need to notify anyone of the changeover from the old system and, if so, who?
- How can Colin determine which type of below ground drainage system he will be connecting the new system to?

Describe below ground drainage systems

Plumbers very rarely work on below ground drainage systems. This job tends to be done by the builder on smaller projects or ground workers on large contracts. However, plumbers do need to know the basics of these systems and how to connect the soil and vent pipe to the drain.

Essentially, you must ensure that **foul water/drainage** cannot enter the **surface water** system. This is because, with certain types of system, surface water is discharged cheaply via separate pipework to streams, etc. Foul water drainage should never be discharged to such pipework as it could be a potential health hazard. The simple rule here is: always know what you are discharging into – then you can't go wrong.

Key terms

Foul water/drainage – anything discharged from a sanitary appliance such as a WC, bath, basin or sink.

Surface water – water collected via the rainwater system.

There are three main types of below ground drainage system: combined, separate and partially separate, as outlined in Table 9.02.

System	Description	Advantages	Disadvantages
Combined svp = soil vent pipe rwp = rainwater pipe fwg = foulwater gully *[diagram: fwg, svp, rwp, fwg, rwp, rwp, foul and surface water sewer]*	Foul water from sanitary appliances and rainwater go into one sewer.	• Cheap and easy to install • Gets a good flush out during heavy rain	• More costly to treat the water at the sewage works • Inadequately sized drains could overflow during heavy rain • Air gullies must be trapped
Separate *[diagram: Inspection chambers, fwg, svp, rwp, rwp, rwp, fwg, foul water sewer, surface water sewer]*	Foul water runs into one sewer and rainwater runs into a separate sewer for surface water.	• No need for water treatment of surface water • No need for trapped gullies on surface water drains	• Danger of cross-contamination – foul to surface water
Partially separate *[diagram: fwg, svp, rwp, rwp, rwp, fwg, soakaway or ditch, foul water sewer, surface water sewer]*	This system still uses two pipes, but some of the surface water is discharged into a watercourse, soakaway or drainage ditch.	• Greater flexibility with system design	• Danger of cross-connection

Table 9.02: Summary of different below ground drainage systems

Working practice

Gary is a Level 3 plumber. His boss has sent him to deal with a complaint from the tenant of a house. Her landlord had employed a plumber six months previously to install a new bathroom because of the state of the old one. The tenant has complained time after time to the landlord that things aren't right with the new bathroom. She thinks there must be a blockage as there are often smells, and waste water appears in the bath when the washbasin is emptied.

Gary has a quick look round and soon identifies the problem. It is nothing to do with a blockage.

Discuss the following points:

- What is the probable cause of the problem if it is not a blockage?
- What term is used to identify this type of trap seal loss?
- Should Gary carry on and repair the problem?
- What are the consequences if he does continue and repair the problem?
- Who would be responsible for paying for the work to be carried out?

Identify health hazards when working with drainage systems

Weil's disease

This is transmitted through infected animal urine – mainly from rodents, cattle and pigs – normally via infected water. It can enter the human body in a variety of ways: through the nose, mouth, eyes and cuts and grazes. Mild cases of Weil's disease (leptospirosis) affect millions of people every year. Of these, only a very few suffer severe cases, but Weil's disease is responsible for killing two or three people every year. In severe cases, failure of major organs (the liver and kidneys) leads to jaundice and dialysis, and recovery time is many months. In milder cases, a person develops flu-like symptoms such as headache, chills and muscle pain.

Hepatitis

Hepatitis is a virus that can be contracted from infected people or human faeces. The three most common types are A, B and C. (You may also encounter hepatitis D or E, but these strains are much rarer.) Hepatitis A is generally the most common and is transmitted via poor sanitation and sewerage. To contract the virus you must put something in your mouth that has been in contact with human faeces from someone who has the virus. This disease could easily be contracted by a plumber who does not maintain high standards of hygiene and personal protection when working on sanitation systems. Hepatitis C is the most common type in the UK, and is transmitted through bodily fluids (blood, saliva, etc.) and intravenous drug use.

Dermatitis

Dermatitis is inflammation of the skin and can be very uncomfortable and unsightly. It can be caused by contact with a substance in the environment or workplace – this is known as contact dermatitis. Detergents and solvents can strip the skin of its natural oils, and repeated exposure will leave the skin very dry, sore and cracking unless precautions are taken. This type of dermatitis is common to people who work in the wet trades and it particularly affects their hands.

Case study

Two plumbers are called out to a public toilet that won't flush as it is jammed. The toilet has a high-level cistern and the chain cannot be pulled to flush it. One plumber uses a stepladder to safely work on the cistern. After removing the lid he reaches in to see if he can unjam the siphon. After feeling around he finds an object that has been drawn into the siphon from a previous flush.

Both the plumbers are shocked to find it is a syringe that had been hidden in the cistern, probably by a drug user. The syringe could easily contain many diseases such as HIV (which can lead to AIDS) and hepatitis, another serious illness. If the needle had pricked their fingers, this could have led to very serious consequences.

Safe working

Be careful when dealing with blocked WCs and pipework as these can be collection points for potentially dangerous items such as hypodermic syringes left by drug users or incorrectly disposed of feminine hygiene products. Don't just go exploring with your hands; make a visual inspection first.

Progress check 9.01

1   In a primary ventilated stack system, what is the minimum depth allowed between the lowest connection and the invert of the drain?

2   How can the maximum recommended pipework lengths and gradients be exceeded in a primary ventilated stack system?

3   What is the important requirement if a kitchen sink waste pipe is terminated into a gully rather than the stack pipe?

4   Subject to certain requirements, where can air admittance valves be located internally?

5   What is a combined system of below ground drainage?

6   What is a separate system of below ground drainage?

7   What is a partially separate system of below ground drainage?

KNOW THE TYPES OF TRAPS AND ASSOCIATED REQUIREMENTS

The purpose of a trap is to retain a 'plug' of water to prevent foul air from the sanitation and drainage pipework entering a room. They are used on above ground discharge system pipework and appliances. This section looks at a range of traps for sinks and other appliances.

Identify different traps used for sanitary appliances

Traps are usually made of plastic (polypropylene to BS EN 274) – as shown in Figure 9.12. They are also available in brass for use on copper pipework, where a more robust installation is required, and can be chromium-plated to provide a pleasing appearance.

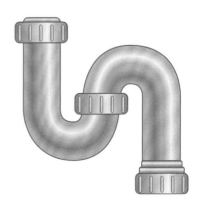

Figure 9.12: A selection of plastic traps

P and S traps

These get their name from their shape, as you can see in Figure 9.13 (although here they are turned to show how they fit into the appliance). They are available in a tubular design or with a joint connection, which allows more options when installing pipework and fittings.

Figure 9.13: P and S traps

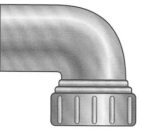

Figure 9.14: Swivel elbow

P traps are often used where the waste pipe is installed directly through a wall from the appliance and into a drain or directly into a stack. P traps and bottle traps (see below) can be converted to S traps using swivel elbows like the one shown in Figure 9.14.

S traps are used where the pipe has to go vertically from the trap through a floor or into another horizontal waste pipe from another appliance (with other constraints being used to avoid induced siphonage, of course). S traps can be a real problem, as the fall of pipe from the basin is generally too steep, which can lead to trap seal loss and obnoxious smells entering the building.

Tubular P or S traps are the best type for kitchen sinks, as these have a better self-cleaning flow through them. Food waste from a kitchen sink tends to collect at the bottom of bottle traps and will in time block the trap.

Bottle traps

Bottle traps are often used because of their neat appearance and because they are easier to install in tight situations (Figure 9.15). Figure 9.16 shows what the trap looks like inside and how the depth of seal is measured. They should be avoided on sinks and, to a lesser extent, on baths and shower trays, as they can be prone to blockage from the accumulation of food or soap deposits.

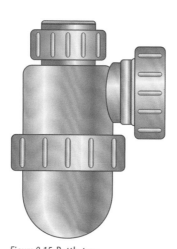

Figure 9.15: Bottle trap

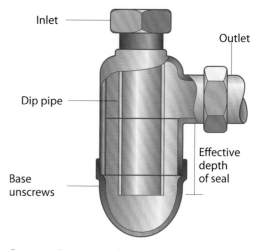

Inlet

Outlet

Dip pipe

Effective depth of seal

Base unscrews

Figure 9.16: Cross-section of bottle trap

Sink traps

Sink traps can be combined for use with more than one appliance, as shown in Figures 9.17 and 9.18.

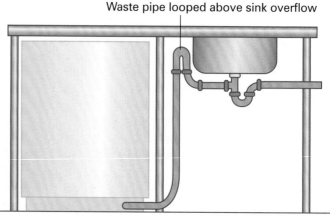

Waste pipe looped above sink overflow

Figure 9.17: Combined sink trap fittings

Figure 9.18: Combined sink/washing machine trap

Self-sealing valve

This is also known as a waterless trap and can be used as an alternative to a traditional trap. It works on the simple principle of using an internal membrane as a seal (Figure 9.19). The membrane allows water to flow through it when the water is released, then closes to prevent foul air from entering the building (see Figure 9.20). The valve can be used on systems meeting BS EN 12056 Part 2. It is ideal for fitting behind pedestals and under baths and showers, and is supplied with a range of adaptors so that it can be used in various situations. The valve has the potential to revolutionise the installation of above ground systems, the requirements for which are covered on pages 339–342 and 353–354.

Figure 9.19: Self-sealing valve

Chapter 9

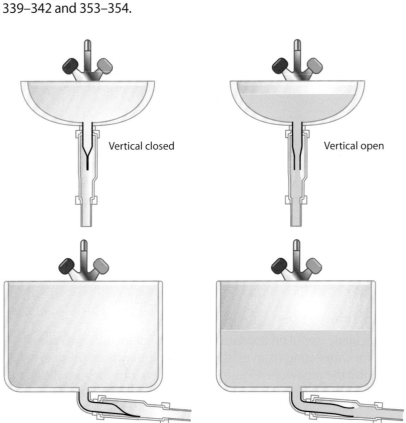

Vertical closed

Vertical open

Horizontal closed

Horizontal open

Figure 9.20: How a self-sealing valve (waterless trap) works

Running traps

Running traps are used in public toilets, one running trap serving a range of untrapped washbasins (see Figure 9.21). In domestic installations, a running trap could be used where a P or S trap arrangement is not possible. Running traps are sometimes used with washing machine or dishwasher waste outlets, although specialist traps are available for these appliances.

Anti-siphon and resealing traps

A well-designed and installed above ground discharge system should prevent loss of trap seal. Some reasons for loss of trap seals are covered on pages 354–356, but anti-siphon traps are designed to prevent seal loss due to the effects of siphonage. These types of traps may be specified or fitted in situations where normal installation requirements cannot be met.

Figure 9.21: Running trap

Did you know?

The Hepworth valve provides a very similar function to the anti-siphon trap, although it is less complicated.

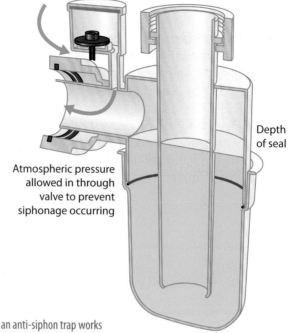

Figure 9.22: How an anti-siphon trap works

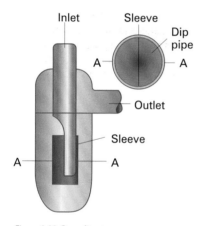

Figure 9.23: Resealing trap

The anti-siphon trap uses an anti-vacuum valve located on the top of the trap. If the pressure drops inside the pipework, the valve is activated, allowing air to enter the system to equalise the pressure (see Figure 9.22).

The resealing trap has a bypass within the body of the bottle. The dip pipe allows air to enter the trap via the bypass arrangement. As the seal is lost due to siphonage, air is allowed into the trap, breaking the siphonic effect (see Figure 9.23).

Tubular swivel traps

These are particularly useful on appliance replacement jobs, as they allow more options for connecting to an existing waste pipe without using extra fittings or altering the pipework (Figure 9.24). On new jobs they are often used on sinks with multiple bowls, again because of their multi-position options.

Low-level bath traps

These are designed to fit into tight spaces under baths and shower trays (Figure 9.25). This type of trap is available with a 50 mm seal depth, making it suitable for connection to a soil and vent stack. Note that the 38 mm seal depth version is only suitable for waste connection into a ground floor gulley.

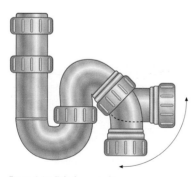

Figure 9.24: Tubular swivel trap

Remember

Whatever type of trap is fitted, it is important that you can access it for cleaning. Some traps have cleaning eyes; others can be split at their swivel joints to enable a section of the trap to be removed.

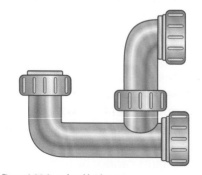
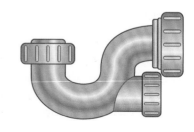

Figure 9.25: Low-level bath traps

352

Straight-through traps

These are used as an alternative to an S trap where space is limited. They are also easier to hide behind pedestal basins. The main problem with this design is the two tight bends, which slow down the flow of water (see Figure 9.26).

Describe the purpose of trap seals

Traps are used on above ground discharge system pipework and appliances. As mentioned on page 349, traps retain a 'plug' of water to prevent foul air from the sanitation and drainage pipework entering a room. The depth of this plug of water will depend on where it is to be used.

Part H of the Building Regulations details the minimum trap diameter and connecting pipe diameter for a range of sanitary appliances. It also includes the requirements for minimum depth of seal. You must ensure that the minimum trap seal depths are maintained, as shown in Table 9.03.

Figure 9.26: Straight-through trap

> **Remember**
>
> If you don't follow Building Regulations requirements on trap sizes and seal depths, noxious smells and unsanitary conditions may be created in the building.

Appliance	Diameter of trap (mm)	Depth of seal (mm of water or equivalent)
Washbasin[1] Bidet	32	75
Bath[2] Shower[2]	40	50
Food waste disposal unit Urinal bowl Sink Washing machine[3] Dishwasher[3]	40	75
WC pan – outlet <80 mm WC pan – outlet >80 mm	75 100	50 50

[1] The depth of seal may be reduced to 50 mm only with flush grated wastes without plugs on spray tap basins.
[2] Where these appliances discharge directly to a gully, the depth of seal may be reduced to no less than 38 mm.
[3] Traps used on appliances with flat bottom (trailing waste discharge) and discharging to a gully with a grating may have a reduced water seal of no less than 38 mm.

Table 9.03: Minimum trap sizes and seal depths

> **Activity 9.1**
>
> Why is it necessary to install a trap? List the different types of trap available and state where you would use each one.

Describe design considerations

When designing an above ground drainage system, you must think carefully about the pipe size, gradient (slope) and bends. You should then choose the pipework to ensure the maximum efficiency of the system without setting up conditions that may result in trap seal loss. Table 9.04 on page 354 shows pipe design considerations for different appliances.

	Pipe size (mm)	Maximum length (m)	Minimum slope (mm fall per metre run)
Basin	32	1.7	18–20
Bath/sink	40	3.0	18–90
Shower	40	3.0	18–90
WC	100	6.0	18

Table 9.04: Pipe design considerations

Failure to meet the requirements will result in poor performance and noise. Figure 9.27 sets out the requirements for ensuring correct functioning of the system.

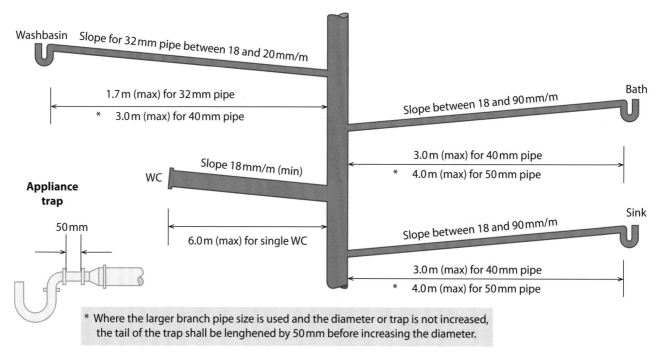

Unvented branch connections to stacks

Washbasin Slope for 32 mm pipe between 18 and 20 mm/m

1.7 m (max) for 32 mm pipe
*   3.0 m (max) for 40 mm pipe

Bath

Slope between 18 and 90 mm/m

3.0 m (max) for 40 mm pipe
*   4.0 m (max) for 50 mm pipe

Slope 18 mm/m (min)

WC

Appliance trap

50 mm

6.0 m (max) for single WC

Sink

Slope between 18 and 90 mm/m

3.0 m (max) for 40 mm pipe
*   4.0 m (max) for 50 mm pipe

* Where the larger branch pipe size is used and the diameter or trap is not increased, the tail of the trap shall be lenghened by 50 mm before increasing the diameter.

Figure 9.27: Requirements for good drainage system design

Describe the reasons for trap seal loss

Trap seals can be lost as a result of:

- poor practice – caused by self-siphonage, induced siphonage or compression
- not following the regulations – details of which have been covered above
- natural causes – such as **capillary action** or evaporation.

Key term

Capillary action (capillarity) – the process by which a liquid is drawn up through a small gap between the surfaces of two materials.

Self-siphonage

This is most common in washbasins, as their shape allows water to escape quickly. As the water discharges, a plug of water is formed, creating a partial vacuum (negative pressure) in the pipe between the water plug and the basin (see Figure 9.28). This is enough to siphon the water out of the trap. Ensuring that the waste pipe is within the lengths allowed and to the correct fall, or that it is ventilated, should prevent self-siphonage. Resealing traps would also avoid this problem.

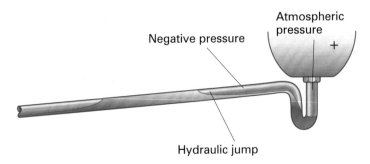

Figure 9.28: Self-siphonage

Induced siphonage

This is caused by a discharge of water from an appliance that is connected to the same waste pipe as other appliances. As the water plug flows past the second appliance connection, negative pressure is created between the pipe and the appliance that siphons the water out of the trap (see Figure 9.29). This arrangement is not acceptable on a primary ventilated stack. Fitting a branch ventilating pipe between the two traps would solve the problem – as would fitting a resealing trap.

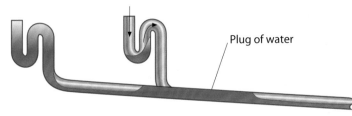

Figure 9.29: Induced siphonage

Compression

As water is discharged from an appliance into the main stack (usually a WC at first floor level), it compresses at the base of the stack, causing back pressure (see Figure 9.30). This can be enough to force the water out of the trap, thus losing the seal. Regulations advise the use of large radius bends and a minimum 450 mm length between the invert of the drain and the lowest branch pipe in order to prevent this.

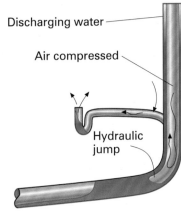

Figure 9.30: Compression

- Ensure that the ladder is secure before attempting to climb and work from it.
- Keep your working periods up the ladder to a minimum.
- Take regular rest breaks so that fatigue does not set in.

Access

You should consider access to the rainwater system carefully, as well as making provision to avoid damaging the customer's property.

Report any existing damage to the customer straight away before work commences.

- If a ladder is to be used to access the gutter system, you should pay particular care and attention to the location of the gutter system.
- Any garden areas and lawns should be protected and a ladder should not be allowed to sink into soft earth, as it could make the ladder topple over.
- Use a ladder stand-off. It will help prevent the ladder from slipping and will protect the customer's property.
- The area surrounding the ladder or scaffold should be sealed off from other people.

Personal protective equipment (PPE)

When installing or maintaining gutter systems, you should always wear standard PPE including:

- steel-toed (capped) boots or shoes
- strong trousers with knee pads
- well-fitting upper body clothes
- suitable gloves
- a hard hat – for you and anyone working beneath you.

Safe working

If you are working on gutter systems, nobody should be working or crossing directly under you.

Describe the installation method for rainwater systems

Installing guttering

The first job is to set out the gutter brackets. Gutters can be laid level or to a slight fall. The fall should be 1 in 600. For a 6 m length of gutter, this would be 10 mm. This is hardly noticeable once the gutter is fixed in position.

Once you have worked out the fall over the total length of the installation, you should fix the first bracket at the highest level on the run and then fix the last bracket to give the amount of fall required.

Then fix a string line between the two brackets. This can be used to position the rest of the brackets: offer each bracket up to touch the string line and ensure an adequate fall is allowed. Brackets should usually be spaced at 1 m centres, unless the manufacturer states otherwise.

The gutter can then be positioned in the brackets. On the first length, depending on the roof shape, you should fit the stop end (or corner bracket) and union bracket. You will then deal with the next joint in position.

Fixing a gutter

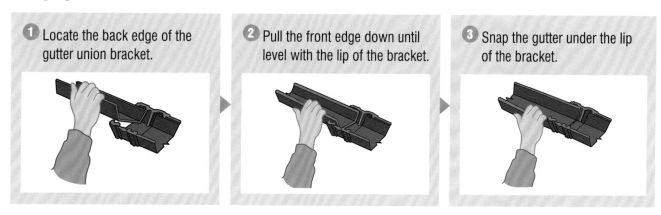

1 Locate the back edge of the gutter union bracket.

2 Pull the front edge down until level with the lip of the bracket.

3 Snap the gutter under the lip of the bracket.

Expansion and contraction

Make sure you allow for thermal movement. On a plastic gutter, a raised fixing mark is usually found on the inside joint. If expansion marks are not on the gutter joints, you should allow 3 mm for every 1 m run. Expansion takes place because of temperature differences and cannot be avoided, so you must allow for it in an installation. If allowance is not made, the guttering could expand and force joints to fracture.

Installing downpipes

Once the gutter system is complete, you can work on the downpipe. Mark out the brackets using a plumb line dropped from a masonry nail driven into the nearest mortar joint below the running outlet. Centre the brackets with the plumb line and mark the wall through the fixing holes.

You will usually have to install an offset (sometimes known as a swan neck) to clear the width between the building wall and the fascia. The underside of the protruding roof at this point is known as the soffit. The easiest way to do this is to install the top section of downpipe with the first offset on, placing the second offset against the outlet of the running outlet and taking the measurements in position (see Figure 9.44). Finish off the fall pipe at the base with a rainwater shoe or connection to the drain.

> **Did you know?**
>
> All screws used for fixing rainwater systems should be alloy or stainless steel so that they do not rust.

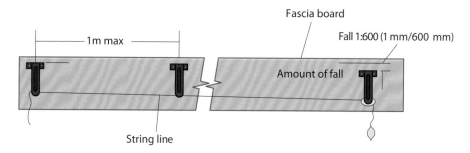

Figure 9.44: Positioning the brackets

Labels: Fascia board; Fall 1:600 (1 mm/600 mm); 1 m max; Amount of fall; String line

> **Remember**
>
> Make sure the roof felt is tucked into the gutter to prevent wind-driven rain entering the property. This also allows any rainwater that leaks through broken or damaged roof tiles to run into the gutter. The old sand-covered bitumen felt will degrade and break down where it is exposed to the elements. In this case, it may need replacing first before you can install the gutter system.

Clip spacings should be every 1.5 m and you should make allowance for thermal movement: leave 6–8 mm from the end of the pipe to the inside shoulder of the fitting. Place clips where the pipes are jointed.

You should cut cast iron downpipes and gutters using an angle grinder. Do not attempt to use an angle grinder unless you have been trained to do so, and ensure that you wear the required PPE for this operation at all times.

Knowledge check

1 What is the recommended fall (gradient) for gutter systems?

 a 1:600
 b 600:1
 c 1:500
 d 500:1

2 What potentially dangerous material might you come across while maintaining rainwater systems?

 a plastic
 b copper
 c cast iron
 d asbestos

3 What gutter profile is this?

 a ogee
 b square
 c half-round
 d deep flow

4 What gutter fitting is shown?

 a gutter union
 b external corner
 c running outlet
 d external stop end

5 What is the maximum recommended fascia board clipping distance for a gutter?

 a 900 mm
 b 950 mm
 c 1000 mm
 d 1050 mm

6 What is the purpose of the 'install gutter to here' mark on the inside of gutter fittings?

 a to save on materials
 b to allow room for expansion of the gutter
 c to make the gutter easier to fit
 d to make the gutter look better when it is complete

7 What fitting is used at the bottom of a downpipe when the rainwater discharges over a gully?

 a angle
 b pipe end
 c open end
 d shoe

8 What tool is used to ensure all the gutter fascia brackets are installed at the same level?

 a plumb line
 b straight line
 c string line
 d level line

9 Where would an offset be used on a rainwater system?

 a at the end of a downpipe above the floor gully
 b around the roof in the guttering
 c a special fitting to go from cast iron gutter to plastic gutter
 d a series of bends from the gutter outlet to the top of the downpipe

10 What type of bracket is this?

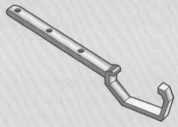

 a fascia bracket
 b downpipe bracket
 c metal rafter bracket
 d special union bracket between gutter profiles

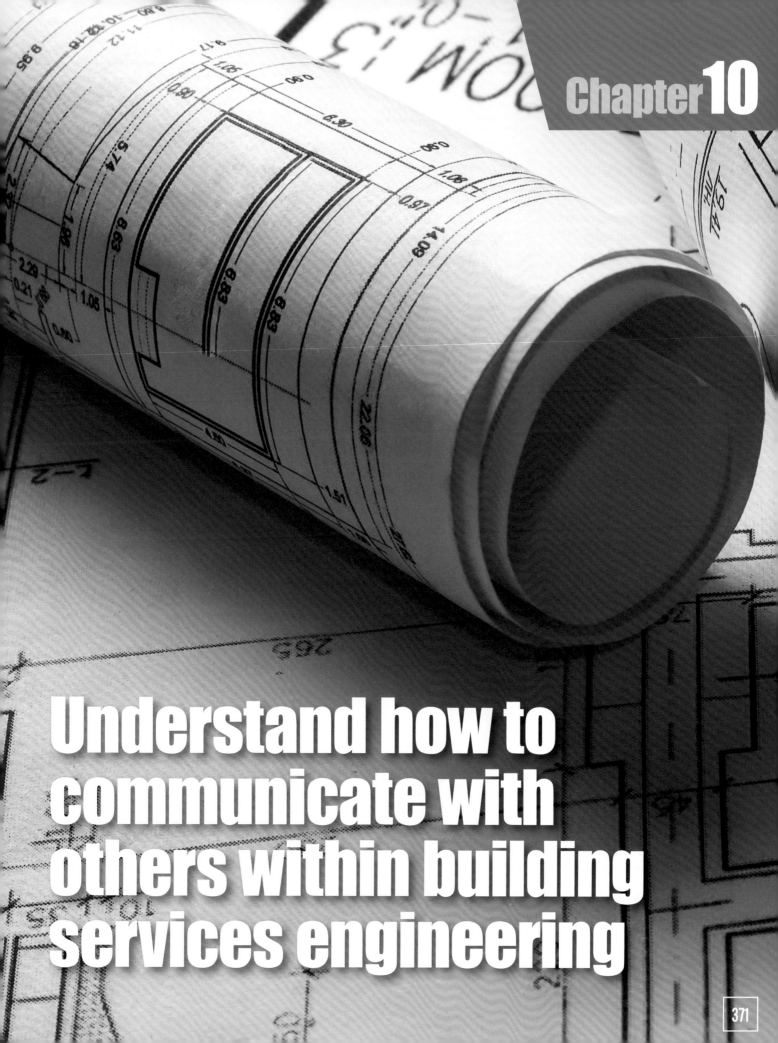

Understand how to communicate with others within building services engineering

This chapter will cover the following learning outcomes:

- **The construction team members and their role within the building services industry**

- **Applying information sources in the building services industry**

- **Communicating with others in the building services industry.**

Introduction

For a building services engineer to work in a consistent, efficient way with others on site, it is essential to be able to communicate well. This may mean understanding exactly who is on site but also what specific role each person has. It is easy to make assumptions about a person's role but this could be misleading and end up in an expensive misunderstanding, or even danger to those on site.

In this chapter you will look at the various methods of communication that an engineer in the building services industry can expect to use.

THE CONSTRUCTION TEAM MEMBERS AND THEIR ROLE WITHIN THE BUILDING SERVICES INDUSTRY

You will come into contact with many people when you are working on site; it is very important to understand exactly what they do and what they do not do.

Key roles of the site management team

A construction project is a very complex and involved process that requires managing all the way through from initial planning to final handover to the customer. The site management team has overall responsibility on site for the people working on the project. This is because, on every construction site, there is a team of people with different roles and responsibilities. Each part of this site management team manages an aspect of the installation and may have other teams below them with further managers looking after specific tasks.

Figure 10.01: A construction team

However, the client has overall responsibility as they hire the architect and sign the contract with the main **contractor**.

It is the responsibility of the main contractor to employ sub-contractors and specialist companies to build what the client wants.

Architect

Buildings need to be designed so they can be used for specific functions. The architect's responsibility is to carry out the design based on a client's request but also make sure that it is within the rules and regulations set out for that type of building. An architect may also call upon the services of more specialist design engineers or consultant engineers if the project is very large or complex. He or she will be employed directly by the client and may also oversee the whole construction project on their behalf.

Key term

Contractor – an individual or a company that has a specific job function within a project.

Chapter
10

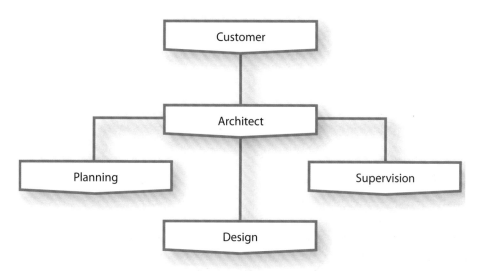

Figure 10.02: An architect may need to work with several teams

Figure 10.03: An architect must liaise with many people

Project manager or clerk of works

These two roles can be quite similar but generally a project manager will have a specific project management qualification such as PRINCE 2 or APM (Association of Project Managers). These formal qualifications may be required to progress to senior roles such as senior project manager, project director or to work on specific contracts within large corporations. Some government contracts will insist on the project manager holding this level of formal qualification. A highly trained project manager will be able to run most types of project without being a specialist in the subject they are delivering – solid project management principles are all they require.

A clerk of works will have day-to-day responsibility for quality assurance. This may involve monitoring the quality of workmanship and materials. The architect may also give them responsibility to issue instructions for changes. For very large construction projects, there may be a clerk of works/project manager for different aspects such as electrical work or air conditioning/ventilation. There may also be a need for an overall project director who manages a team of project engineers/managers.

Structural engineer

Not only must a new building look good but it must be designed so that it is safe and able to withstand the weather and continual use that it is meant for. It is the responsibility of the structural engineer to make sure this is the case. Structural engineers:

- provide technical advice
- select the most appropriate materials for the job
- inspect the property to check the condition
- inspect foundations
- inspect damage.

Structural engineers are highly trained, the highest level being a chartered structural engineer.

Surveyor

A surveyor will work closely with an architect and structural engineer to assess the condition of a building. The surveyor will have knowledge of, and be able to advise on, the construction and materials used in a building. He or she will also be aware of the up-to-date regulations and commercial aspects of the building such as market value for insurance or rebuild requirements.

Building services engineer

Imagine a fully functioning building. All the services that exist within that building have been designed, installed and maintained by a building services engineer. The range of services is shown in Figure 10.4.

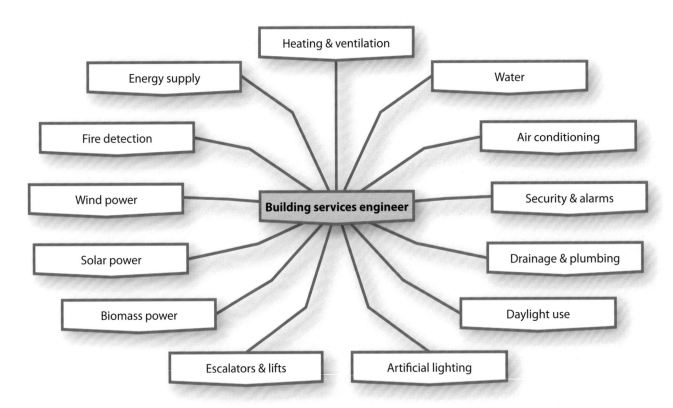

Figure 10.04: A building services engineer is responsible for many services in a building

More information can be found on the website of the Chartered Institution of Building Services Engineers: www.cibse.org.

Quantity surveyor

The costs of a new installation can vary but it is the job of the quantity surveyor to monitor, control and take action if costs change. The quantity surveyor will be involved at the beginning of a project by creating an initial bill of quantities from the design plans. This will then be used to set the budget for construction and will be monitored during the project roll-out.

Buyer

Professional buyers are often part of a large contract process. The buyer's responsibility is to source the most cost-effective products for the job. Increasingly this may also involve other issues to meet customer demands; for example a customer who is very environmentally aware might demand only ecologically sustainable products or locally sourced products to reduce the carbon footprint of the building. Customers might also specify that only carbon neutral products can be used.

Estimator

Before a contract can be agreed between a customer and the contractor doing the work, initial costs need to be decided. The estimator will be responsible for producing the costs that will go into the customer tender response.

Case study

During the 2012 Olympics, 10 million litres of water were needed to fill the pool at the Aquatics centre.

Use the Internet to find at least five other facts about the plumbing or heating of the London 2012 Olympic stadium.

Figure 10.05: Ten million litres of water were needed to fill the pool at the London 2012 Aquatic centre

Contracts manager

Large contracts require a contracts manager. A contracts manager will be involved at the beginning of a contract when it is being set up and at the end when it is handed over as completed to the customer. The contracts manager is also responsible for the day-to-day management of the contract engineers. This can involve managing the ongoing contract engineer costs of the project, making sure they are closely monitored against any budgets.

Construction manager

Sometimes also referred to as a construction project manager (CPM), the construction manager will have overall responsibility for planning, coordinating and controlling a major building project. This responsibility includes the start of the build project and only ends once the building has been handed over, completed to the client. This role is different from that of a project manager (see page 373), who only has responsibility for the construction project.

Principal contractor

On construction sites, work that takes over 30 days or 500 man hours is considered 'notifiable work' under the Construction Design Management (CDM) Regulations 2007. With construction work of this size, a principal contractor must be involved and as soon as they are appointed, details need to go to the Health and Safety Executive (HSE).

The principal contractor has legal duties for certain aspects of project management, contractor engagement and workforce engagement. An example of this says: 'Contractors must provide the workforce with the required training and information to carry out work safely. The principal contractor must check that this happens and every worker is provided with an induction on health and safety and training.'

Progress check 10.01

1   On a large construction project, who is the customer's main point of contact?
2   What are the main responsibilities of the contracts manager?
3   What does a quantity surveyor do?

Key roles of individuals who report to the site management team

Sub-contractor

For specialist work a contractor may need another company to come in and complete specific tasks for them. These sub-contractors may be specified by the client due to their personal preferences or they may have won a tender process issued by the main contractor. Clients often have specific specialist companies in mind due to their reputation, specific skills or if they have worked successfully with them before. Plumbing and heating companies will generally be sub-contracted.

Site supervisor

On all major building projects there will be a site supervisor to oversee all day-to-day activities. Their role will be to carry out regular inspections of security and the environment, taking action where required. The site supervisor may have several other roles that are unique to each site, but main responsibilities may include:

- ensuring the site is run in compliance with all company and statutory health and safety regulations
- induction of all temporary workforce and contractors on site including visitors
- running team meetings and liaising with different groups and management on site.

Trade supervisor

Within the specific trade areas a day-to-day supervisor is required to oversee that particular area. For plumbing and heating, the trade supervisor will be a fully qualified Level 3 plumbing and heating engineer. The trade supervisor will be aware of what stock and materials are on site and available as the job progresses. An amount of liaison will form part of their role to make sure any special instructions from the contracts engineer or architect are implemented once the relevant paperwork (such as variation orders) has been processed.

Trades

Groundworker – at the beginning of a construction contract, particularly on a new building, there is a requirement to prepare the ground. Groundworkers are responsible for preparing the site foundations and ground before other trades come in to complete their tasks. This could involve digging foundations, trenches, footings or simply clearing an area in preparation for construction.

Figure 10.06: Groundworkers must prepare the ground at the beginning of a construction contract

Figure 10.07: A joiner at work

Bricklayer – following on from groundwork, the bricklayer is responsible for building the structure before other trades start their work.

Joiner – on most construction sites, wooden structures need to be built, including door frames, doors, stud work and window frames. This is the responsibility of the joiner, a specialist carpenter.

Plasterer – ideally, when the first fix pipework for all plumbing and heating systems has been installed and tested, a plasterer will cover any of the pipework that will be concealed. The plasterer may need to put up plasterboard first and then it is their responsibility to finish off the wall to a good level so that decorators can sand, paint or put up wall covering. Careful liaison between trades is required at this stage to make sure the pipework connection points are clearly marked, or they may disappear forever!

Tiler – tiles are another type of finish for a wall. It is the responsibility of the tiler to fit the tiles near the end of the construction stage. This stage will often coincide with the decoration stage when all the walls have been finished. Liaison between plumbers and tilers is also important. For example, the tiler may need to work around sanitary appliances such as baths or shower enclosures.

Electrician – installation electricians test and install wiring systems (power, lighting, security, fire) and equipment. Depending on the job, they may work on building sites or in domestic properties.

The electrician's role is very similar to that of the plumber, but the electrician connects electrical cabling to electrical components, rather than pipework to plumbing appliances. There are many rules and regulations which all electricians must abide by.

Heating and ventilation (H&V) fitter – heating and ventilation fitters are specialists that install systems to control the **ambient temperature** in a building. As technology changes and becomes more available, more buildings are moving towards climate control. The developing technology has seen trades getting closer and closer together, with many of the most sought-after engineers in building services having multiple qualifications and skills.

Gas fitter – responsible for installing, testing and repairing equipment connected to a gas supply. The gas may be natural gas or liquefied petroleum gas (LPG). Gas engineers must be specifically qualified for this work and their competence must be checked regularly under the Gas Safety (Installation and Use) Regulations 1994.

A gas fitter must, by law, be a Gas Safe registered installer (operative). Many plumbers and heating engineers will also qualify for Gas Safe registration, either during or after their NVQ Level 3 qualification.

Decorator – responsible for painting, filling in cracks in plaster, sanding window frames, doors and any wood finishes. The decorator will mainly be involved in the last stage of an installation and will generally have to work around all the other trades, negotiating when he or she can come in and finish off an area.

> **Key term**
>
> **Ambient temperature** – the temperature of the air where equipment or cable is installed.

Chapter 10

Figure 10.08: A decorator will be involved in the later stages of an installation

Figure 10.09: Visitors to a construction site could be doing various tasks

Key roles of site visitors

Visitors to site are commonplace and may come from different local authority or government departments, interested in various aspects of the construction work. Even though these visitors may be wearing a suit and seem important, they should always be challenged and asked to go through the same site induction program and signing-in process as other contractors or visiting trades. All visitors should expect this as the site will become unsafe if there are two sets of rules. If an incident occurs on site you will need to know who is there and who can be accounted for by the emergency services.

Building control inspector – building control inspectors provide advice on request to help building projects comply with the building regulations. Inspectors normally work directly for the local authority. An inspector will come to the site at the beginning of the build project to make sure all the regulations have been considered and planned for. Inspectors can visit a building site as many times as required to satisfy any questions they may have but will generally come again at the completion stage to sign off the installation.

Water inspector – very occasionally you might see a water inspector from the Drinking Water Inspectorate (DWI). These inspectors are in place to maintain the quality of 'wholesome' water or drinking water provided by the water companies. They have the power to go to water companies and test at source but they also have the power to go on site and measure at the tap. They will be testing for microorganisms, chemicals and metals, as well as what the water looks and tastes like.

HSE inspector – health and safety law is enforced by inspectors from the Health and Safety Executive (HSE) or inspectors from the local authority. Inspectors have the right to go into any workplace without notice to look at the management of health and safety and compliance with the law. Inspectors will come to a site if an incident has been reported or they can just simply turn up unannounced. Inspectors have the right to close a site down for very serious health and safety breaches or issue a notice for improvement.

More about the process of HSE inspections and the consequences is covered in Chapter 1: *Health and safety in building services engineering*.

Case study

During the Olympics over 40,000 people worked on site, with a peak of 13,000 at any one particular moment. The Olympic Delivery Authority set about with the aim of being the safest construction project ever. At one point, the project went over 3 million hours without a single reportable injury.

Electrical services inspector – approved companies registered with, for example, the **NICEIC**, **NAPIT** or **ECA** will have visitors from time to time to inspect the level of work. Domestic installers who self-certify their own work under self-certification schemes will require periodic checks. These inspectors are there to check on the quality and adherence to the wiring regulations. Electrical services inspectors are qualified in the electrical trade with the experience to make judgements on installation standards and workmanship.

For a company to comply with a scheme they must be inspected on a regular basis depending on the terms of the scheme. Local authority inspectors may also be involved and come to a site at various stages of a build to make sure the regulations are followed, or for final sign-off just before completion and handover.

Key terms

NICEIC – the National Inspection Council for Electrical Installation Contracting.

NAPIT – the National Association of Professional Inspectors and Testers.

ECA – the Electrical Contractors' Association.

Activity 10.1

1 You are a self-employed plumbing and heating engineer and you have been injured in an accident at work. The injury has meant you are unable to work for eight days. Go to the HSE website and find out what you must do about the accident and injury.

2 You are the manager of a construction company and you are going to build five houses on the open space next to a college. Go to the HSE website and download and print a 'Notification of construction project' form. Complete it for the construction of the five new properties.

Progress check 10.02

1 What are the main responsibilities of the site management team?

2 When can a building control inspector visit a site?

3 What are the responsibilities of a building control inspector?

4 What powers do HSE have when visiting a construction site?

APPLYING INFORMATION SOURCES IN THE BUILDING SERVICES INDUSTRY

Types of statutory legislation and guidance information that apply to the industry

Legislation means literally 'to make law'. Laws exist to protect the rights of individuals and businesses. Each country has different legislation and processes but the UK is heavily influenced by European law. An area that is particularly affected is employment law. In the UK the laws that govern what you must and must not do are influenced by three main sets of legislation. These are:

- common or contract law
- UK legislation
- European legislation and judgements from the European Court of Justice.

Some of the relevant legislation in this country that sets out the minimum rights and responsibilities of an employer and employee are given below.

The Race Relations Act 1976 and Amendment Act 2000

When originally passed, the Race Relations Act 1976 made it unlawful to discriminate on racial grounds in relation to employment, training and education, the provision of goods, facilities and services, and certain other specified activities. The 1976 Act applied to race discrimination by public authorities in these areas, but not all functions of public authorities were covered.

The 1976 Act also made employers vicariously (explicitly) liable for acts of race discrimination committed by their employees in the course of their employment, subject to a defence that the employer took all reasonable steps to prevent the employee discriminating.

The Commission for Racial Equality (CRE) proposed that the Act should be extended to all public services and that vicarious liability should be extended to the police. The main purposes of the 2000 Act were to:

- extend further the 1976 Act in relation to public authorities, thus outlawing race discrimination in functions not previously covered
- place a duty on specified public authorities to work towards the elimination of unlawful discrimination and promote equality of opportunity and good relations between persons of different racial groups
- make Chief Officers of police vicariously liable for acts of race discrimination by police officers
- amend the exemption under the 1976 Act for acts done for the purposes of safeguarding national security.

The Sex Discrimination Act

The Sex Discrimination Act 1975 makes discrimination unlawful on the grounds of sex and marital status and, to a certain degree, gender reassignment. The Act originated out of the Equal Treatment Directive, which made provisions for equality between men and women in terms of access to employment, vocational training, promotion and other terms and conditions of work.

The Equal Opportunities Commission (EOC) has since published a Code of Practice. While this is not a legally binding document, it does give guidance on best practice in the promotion of equality of opportunity in employment, and failure to follow it may be taken into account by the courts.

The Sex Discrimination Act was amended to ensure compliance with the Equal Treatment Directive, with all changes being effective from April 2008. The definition of harassment is extended so that if, for example, a male supervisor makes disparaging comments about women, it is no longer a defence to show that he makes similar comments about men. In addition if someone witnesses sexual harassment of a colleague, they can bring a claim of harassment themselves if they felt it made their work environment intimidating.

Employer liability has also been extended to make organisations liable if they haven't taken reasonable steps to prevent harassment by a third party such as a visitor or customer.

Employment Relations Act 1999 & 2004

The 1999 Act is based on the measures proposed in the White Paper: Fairness at Work (1998), which was part of the Government's programme to replace the notion of conflict between employers and employees with the promotion of partnership.

As such it comprises changes to the law on trade union membership, to prevent discrimination by omission and the blacklisting of people on grounds of trade union membership or activities; new rights and changes in family-related employment rights, aimed at making it easier for workers to balance the demands of work and the family and a new right for workers to be accompanied in certain disciplinary and grievance hearings.

The Employment Relations Act 2004 is mainly concerned with collective labour law and trade union rights. It implements the findings of the review of the Employment Relations Act 1999, announced by the Secretary of State in July 2002, with measures to tackle the intimidation of workers during recognition and de-recognition ballots and provisions to increase the protections against the dismissal of employees taking official, lawfully organised industrial action.

The Human Rights Act 1998

The Human Rights Act 1998 covers many different types of discrimination – including some not covered by other discrimination laws. However, it can be used only when one of the other 'articles' (the specific principles) of the Act applies, such as the right to 'respect for private and family life'.

Rights under this Act can only be used against a public authority (such as the police or a local council) and not a private company. However, court decisions on discrimination will generally have to take into account what the Human Rights Act says.

The main articles within this Act are: right to life, prohibition of torture, prohibition of slavery and forced labour, right to liberty and security, right to a fair trial, no punishment without law, right to respect for private and family life, freedom of thought, conscience and religion, freedom of expression, freedom of assembly and association, right to marry, prohibition of discrimination, restrictions on political activity of aliens, prohibition of abuse of rights, limitation on use of restrictions on rights.

The Employment Rights Act 1996, Employment Acts 2002 & 2008

Subject to certain qualifications, employees have a number of statutory minimum rights (such as the right to a minimum wage). The main vehicle for employment legislation is the Employment Rights Act 1996 – Chapter 18. If you did not agree certain matters at the time of commencing employment, your legal rights will apply automatically. The Employment Rights Act 1996 deals with many matters such as:

- right to statement of employment
- right to pay statement
- minimum pay
- minimum holidays
- maximum working hours
- right to maternity/ paternity leave.

The Employment Act 2002 amended the 1996 Act to make provision for statutory rights to paternity and adoption leave and pay. The Employment Act 2008 makes provision for the resolutions of employment dispute including compensation for financial loss, enforcement of minimum wage and of offences under the Employment Agencies Act 1973 and the right of Trade Unions to expel members due to membership of political parties.

The Race Relations Act 1976 (Amendment) Regulations 2003

The Race Relations (Amendment) Regulations 2003 modify the Race Relations Act 1976.

- Indirect discrimination on grounds of race, ethnic origin or national origin is extended to cover informal as well as formal practices.

- The concept of a 'Genuine Occupational Requirement' is introduced for situations where having a particular ethnic or national origin is a genuine requirement for the employment in question.

- The definition of discriminatory practices is extended to cover those who put particular groups at a disadvantage, rather than only those where there is proof that a disadvantage has been experienced.

- The Act is extended to give protection even after a relationship (such as employment in an organisation, or tenancy under a landlord) has finished.

- The burden of proof is shifted, meaning an alleged discriminator (such as an employer or landlord) has to prove that he or she did not commit unlawful discrimination once an initial case is made.

Case study

Avneet works in a small plumbing and heating company that specialises in work for the healthcare sector. Over the last year, whenever Avneet has gone to the depot to collect supplies she has suffered verbal racial abuse from her manager. Because this is a small company and she loves her work so much, she says and does nothing. After a year the comments become more frequent and Avneet eventually puts in a grievance against her manager. The company upholds the grievance and admits that racial abuse has occurred. The manager is reprimanded but no further action is taken by the company. Avneet later finds herself on a poor performance target list and is put forward for redundancy even though her work has always been highlighted as 'good practice' in her yearly reviews.

1 What laws may have been broken here?

2 What course of action could have been taken?

Racial and Religious Hatred Act 2006

The Racial and Religious Hatred Act 2006 makes inciting hatred against a person on the grounds of their religion an offence in England and Wales. The House of Lords passed amendments to the Bill that effectively limit the legislation to 'a person who uses threatening words or behaviour, or displays any written material which is threatening... if they intend thereby to stir up religious hatred'. This removes the abusive and insulting concept, and requires the intention – rather than just the possibility – of stirring up religious hatred.

Employment Equality (Religion or Belief) Regulations 2003

These regulations make it unlawful to discriminate against, harass or victimise workers because of religion or religious or similar philosophical belief. They are applicable to vocational training and all aspects of employment, recruitment and training.

Equality Act 2006

This amends the Sex Discrimination Act and places a statutory general duty on employers when carrying out their functions to have due regard to the need to eliminate unlawful discrimination and harassment, and also to promote equality of opportunity between men and women.

Equality Act 2010

From 1 October 2010, the Equality Act replaced most of the Disability Discrimination Act (DDA). However, the Disability Equality Duty in the DDA continues to apply. The Equality Act 2010 aims to protect disabled people and prevent disability discrimination. It provides legal rights for disabled people in the areas of:

- employment
- education
- access to goods, services and facilities including larger private clubs and land-based transport services
- buying and renting land or property
- functions of public bodies, such as the issuing of licences.

The Equality Act also provides rights for people not to be directly discriminated against or harassed because they have an association with a disabled person. This can apply to a carer or parent of a disabled person. Also people must not be directly discriminated against or harassed because they are wrongly perceived to be disabled.

Protection from Harassment Act (PHA) 1997

Harassment is defined as any form of unwanted and unwelcome behaviour (ranging from mildly unpleasant remarks to physical violence) that causes alarm or distress by a course of conduct on more than one occasion (note that it doesn't need to be the same course of conduct).

The PHA is the main criminal legislation dealing with harassment, including stalking, racial or religious motivation and certain types of antisocial behaviour such as playing loud music. Significantly, the PHA gives emphasis to the target's perception of the harassment rather than the perpetrator's alleged intent.

Employment Equality (Age) Regulations 2006

The Employment Equality (Age) Regulations 2006 is a piece of legislation that prohibits employers from unreasonably discriminating against employees on grounds of age.

Data Protection Act 1998

Information about people can also be subject to abuse, so the Information Commissioner enforces and oversees the Data Protection Act 1998 and the Freedom of Information Act 2000. The Commissioner is a UK independent supervisory authority reporting directly to the UK Parliament. It has an international role as well as a national one.

The principles put in place by the Data Protection Act 1998 aim to ensure that information is handled properly. Data must be:

- fairly and lawfully processed
- processed for limited purposes
- accurate, adequate, relevant and not excessive
- not kept for longer than is necessary
- processed in line with your rights
- secure and not transferred to other countries without adequate protection.

'Data controllers' have to keep to these principles by law.

Regulations

Regulations are the practical part of a law that tell the participants exactly how they are meant to act. The regulations give control or rights and also allocate responsibilities. A regulation can put legal restrictions on a company, contractor or individual that will be monitored by a government body. Regulations can also be self-regulated by industry such as a trade organisation.

Examples of regulations that exist to control and allocate responsibilities are the Electricity at Work Regulations (EAWR). The 33 regulations are specifically designed to reduce the risk of death or injury from electricity in the workplace. The EAWR covers voltages over 1000 V a.c. and 1500 V d.c.

BS 7671: Requirements for Electrical Installations (The IET Wiring Regulations)

The 'regs', as they are known by technicians in the electrical industry, are not actually regulations and they differ slightly in their approach when compared to the **Electricity at Work Regulations**. Whereas EAWR cover all electrical work, BS 7671 only covers specific areas of electrical installation and maintenance. BS 7671 is periodically reviewed and new editions are brought out every few years depending on the scale of the changes and amendments required. The early editions over 100 years ago consisted of only a few pages with details of what could and could not be done. The latest 17th edition is very different and relies heavily on the technician to have the skill, knowledge and understanding of the document. A fully qualified electrician will be able to read and interpret the wiring requirements. BS 7671 has been developed this way to allow flexibility with all the developing technologies and products available but still stay within the guidelines dictated by EAWR.

British Standards

Technical committees and specialist boards set up by the British Standards Institute (BSI) approve British Standards – of which there are currently 27,000. Having a British Standard means that certain specifications are met while encouraging manufacturers to standardise the way they produce products. The Kitemark can be used to indicate something is certified by the BSI but it is mainly used for safety and quality. A competent electrician can also apply for a Kitemark from BSI.

Harmonisation with Europe has meant that some BS numbers have been superseded by the EN number (**BS EN**), showing that a British Standard has been replaced by the European Standard.

Examples of the standards related to electrical installation work can be found in Appendix 1 of BS 7671, Requirements for Electrical Installations. Some of the relevant standards are shown in Table 5.3.

Codes of practice

The main purpose of a code of practice is to provide extra help and guidance at work when applying current legislation. Codes of practice are specifically aimed at areas that require more help and a little more detail.

Activity 10.2

Look up the Kitemark website and write down the main advantages a customer has when employing a Kitemark electrician.

Figure 10.10: The British Standards Kitemark

Key term

BS EN – British Standards European Norm.

The scope, purpose and requirements of the work to be undertaken are usually described using drawings. You need to ensure that the installation meets industry standards and the requirements as set out in the contract specification.

Materials used in plumbing installations should be to the relevant EN or BS number. British Standards also make recommendations on design and installation practice. In addition to British Standards, the following legislation places statutory responsibilities on plumbers:

- Water Supply (Water Fittings) Regulations 1999
- Electricity Supply Regulations 1998 and Electricity at Work Regulations 1989
- Building Regulations 2000
- Health and Safety at Work etc. Act (HASAWA) 1974.

The approved codes of practice are government-approved advice that is sanctioned by both the HSE and the Secretary of State. Although failure to comply does not mean you are officially breaking the law, if it came to a criminal court case and you were found not to be following them, you would have to prove you had an equally good alternative. In some cases statutory regulations may be accompanied by codes of practice approved under Section 16 of the HASAWA 1974. These codes do have a legal status and this is defined fully in Section 17 of HASAWA.

Although codes of practice do not have to be British Standards, most of them are. These codes of practice provide guidelines about topics such as:

- dangerous substances and explosive atmospheres
- legionella
- asbestos
- gas safety
- hazardous substances
- workplaces
- management and health and safety
- pipelines.

Case study

On 11 December 2005 a number of large explosions occurred at Buncefield oil storage depot in Hemel Hempstead. One of the explosions registered 2.4 on the Richter scale and to this day counts as Britain's most costly disaster. Following the prosecution of five companies, the main reasons for the disaster were shared in a report by the HSE so that other companies could learn from the experience. Poor maintenance and management were among the main factors that led to the explosion and large fire.

Figure 10.11: Explosions at the Buncefield oil storage depot

Guidance from manufacturers

Manufacturers' documentation and guidance notes are a very valuable source of information to a building services engineer. This information must always be fully read and understood before installation of specific equipment. Manufacturers' guidance will be supplied with all the equipment a building services engineer has to work with. A manufacturer has a duty of care to provide all the installation fitting details that might be required. This may include any specific details on the positioning of equipment, load and weight restrictions, fittings, heating or cooling requirements or simply what the product contains and is made of. Manufacturers' guidance is essential reading for the engineer and cannot be skipped over, even though it might take several hours to read and digest fully. Reading all the manufacturers' information may save hours, days and a considerable amount of money so you should get used to taking the time to do this.

Two key sets of information are generally produced by manufacturers in the form of technical information or data sheets and operational/functional instructions. The data sheet will contain as much information as the manufacturer believes you will need. Operational/functional instructions are used in the installation process but will also be useful for the customer after the installation has been completed. This type of information will form part of the handover pack to the client.

With technology developments and the boundaries becoming blurred between the different specialisms and trades, installation work has become very complex. Often, further advice will be required and engineers will no longer be able to rely on basic information and guidance. Engineers should expect to have regular contact with manufacturers. There will be cases when the installation specification cannot be completely matched and advice about tolerances on the installation method will be required. Any extra information you receive as the installing engineer is potentially valuable to the customer and must be kept and presented to the client as part of the handover pack. Remember you may not be the next engineer that works on site. This extra information will be very important if your company is to look good and maintain a good relationship with the client.

Working practice

As part of a large installation, a building services heating engineer has been contracted to run in the heating main pipework for a telecommunications room in a major department store as part of a refurbishment. The engineer has been given basic instructions, a set of drawings and some manufacturers' equipment guides. He has fitted this type of equipment before and decides to save time and start straight away with the installation work in the next available project management slot while the store is closed. The installation work is completed and the job is handed over to the test and inspection team. It is found that the equipment specification has changed from the previous installation and there is now a requirement for the heating mains to be connected to the boiler house separately from the existing pipework. The shop has to be closed so the pipework can be refitted correctly. The project go live date has to be postponed a week and the customer is not happy as customer service is affected and the store director receives complaints. The building services company has to provide compensation under the terms of their contract.

1   What makes a contract binding?

2   What else can happen if a contract is broken?

3   Who was at fault here: the customer, the building services company or the heating engineer?

The purpose of information used in the workplace

The amount of documentation that a building services engineer can expect to see on site is vast. A lot of companies use standard format documents and forms such as risk assessments but many companies create their own to meet their specific needs or ways of doing things. For large projects documentation must be controlled so that all the engineers and trades involved on site have the same common information. For this reason a project manager or project director will issue **controlled documentation**.

Any changes that are required will then have to go through the project manager in a process called 'change control' (note that project changes also go through a change control process). As an engineer you will receive your instructions via documentation generally. You will also have to produce or complete documentation throughout the job as it is very rare that you will be working in complete isolation. You will always have to pass information on to either the customer or to other groups of engineers/site operatives so they can continue and finish their stage of the installation or repair. For some types of information it is acceptable to pass the detail on verbally – for other information that needs to be tracked, a written document is the preferred way of communication. Wherever possible it is a good idea to write information down – this is considered best practice.

Job specifications

When an installation is planned, certain details are required so that any engineer or operative knows exactly what it is they are expected to do. A job specification will give that detail. Job specifications will also be used in conjunction with many other types of documents.

A specification can be in many forms based on the purpose. You might have a general specification that details the materials, construction, weight, colour and power requirements of a particular product. A performance specification will be slightly different and may include detail on how well a product performs in different conditions.

Plans/drawings

As an engineer in the building services industry you will come into contact with many different types of drawings, plans and diagrams. Each one will give slightly different information to build up the whole picture of what needs to be achieved. Some of this information will only be required during the installation and some will be required on an ongoing basis for maintenance and installation growth. The need for engineers to be familiar with all types of plans and drawings is key to a successful installation or finding and repairing faults.

Assembly drawings

Assembly drawings are very common and are used for complex machines to show how something is put together. A control panel or a motor might have an assembly drawing. An assembly drawing is also sometimes known or referred to as an exploded diagram and is very good at showing all the components and the order in which they are put together. As you can imagine, if you had to take a combination boiler to pieces to replace a component, knowing the order and how the boiler is put together and

Key term

Controlled document – a changing document that is reviewed and then reissued by the project manager so that all the people involved in a project have exactly the same information and are using the most current version.

Chapter
10

taken apart will save a considerable amount of time and effort. An assembly drawing is also a very handy way to identify why you have one bolt left after you have put something together!

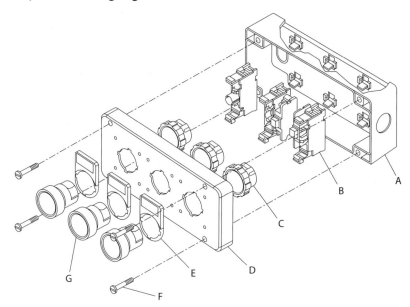

Figure 10.12: An assembly drawing for an electrical control box

Wiring diagrams

Although wiring diagrams do not specifically use circuit symbols, they do show the actual layout of a circuit with the physical connections. For this reason they are a very common diagram that you as a building services engineer will use.

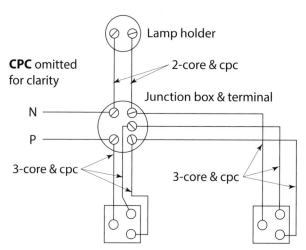

Figure 10.13: A wiring diagram

Layout drawings

A layout drawing shows exactly how everything in the installation is laid out. This is often used in combination with other forms of diagram to give more information that might help clarify the installation. A layout drawing will be based on the site drawings provided by the architect. A common practice is to use the standard BS 1192 graphical symbols for construction drawings. However, if these diagrams are produced by a computer-aided design software package there can be some variation in the symbols used. Layout drawings are drawn to scale so that measurements can be taken off for installation positions of specific components and equipment. The main symbols you will use are shown in Table 4.14 in Chapter 4 (page 177).

As fitted drawings

As an installation progresses, changes will occur due to new information or customer contract variations. An example may include moving a radiator to another wall if the side on which a door opens is changed by the customer to enable easy access for new furniture – the drawing will be updated or notes added. At the end of the job all these changes are put together on the drawing and they become the 'As fitted drawings'. These are the most up-to-date records of the installation and will be used for ongoing work such as maintenance or upgrades.

Work programmes

An important part of installation work is understanding what is required in stages. A work programme will give you all the work required in a construction project and show the interrelationships between the different construction activities. A works programme is also referred to as a programme of works. This information can feed into a project plan or be part of the main project planning process.

Job sheets

Job sheets can also be known as job cards or works instructions. A typical job card is shown in Figure 10.14.

A job card gives details about a specific task. The job card will be given out to an engineer to perform a task, typically giving the time allocated and most of the information required for successful completion of the job. Once the job is finished, the job card is completed and handed back to the relevant supervisor. Job cards are often automated computer applications held on laptops or handheld devices. A job is defined, issued to an engineer, completed and then handed back to the job controller. This type of process allows larger tasks to be broken down, allocated to a range of differently skilled operatives and then carefully managed.

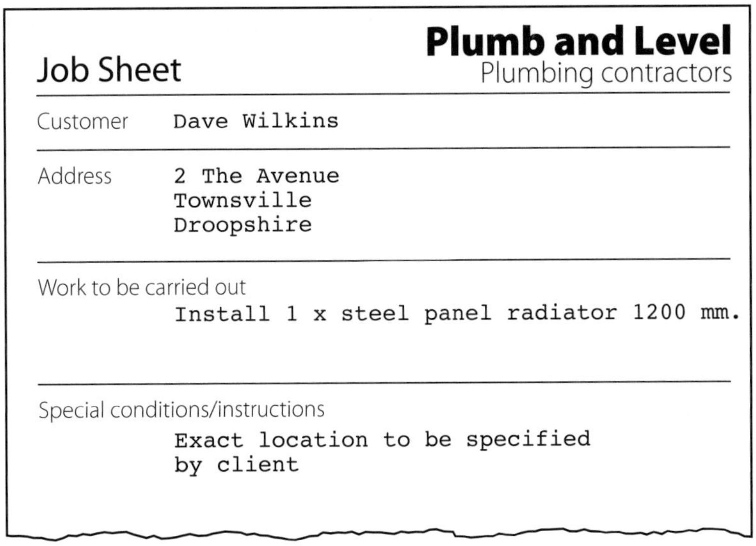

Job Sheet	**Plumb and Level**
	Plumbing contractors
Customer	Dave Wilkins
Address	2 The Avenue
	Townsville
	Droopshire
Work to be carried out	
	Install 1 x steel panel radiator 1200 mm.
Special conditions/instructions	
	Exact location to be specified by client

Figure 10.14: A typical job sheet

Delivery notes

Communicating with suppliers and wholesalers effectively makes the difference between success and failure. Arranging for stock and materials to turn up on site at a specific time slot can be fraught with issues. On a large construction site, deliveries are constantly coming in and if the received goods are not carefully checked, a particular part of the project might not go ahead on time and contract penalties will be issued to

Delivery note		**S. BENDS** *Plumbing supplies*
Order No.		Date

Delivery address	Invoice address
2 The Avenue	Plumb and Level
Townsville	Plumbing Contractors
Droopshire	

Description	Quantity	Catalogue No.
Steel panel radiator	1	
1200 mm		
Comments		
Date and time of receiving goods		
Name of recipient		Signed

Figure 10.15: A purchase order for plumbing equipment

Architect's Instruction

Issued by: Ivor Kingston Associates   Job Reference: IK/AL/001

Address: Kingston Road

Employer: Mr A Waterside

Address: 60 Well Lane   **Variation Order No:** 001

Contractor: A.Leak Plumbing and Heating   **Issue date**: 11th September 2012

Address: 58 Well Lane

Works: New dwelling   Sheet 1 of 1

Situated at: 60 Well Lane

Under the terms of the above mentioned contract, I/we issue the following instructions

	Office use: Approximate costs	
	£ omit	£ add
1. Re-route cold water supply pipework services	0.00	200.00
2. Re-route hot water distribution pipework services	0.00	200.00
1) Approximate totals	0.00	400.00
2) Signed: Ivor Kingston		

Figure 10.16: A typical variation order

sub-contractors. A method of checking the delivered stock is the delivery note. This needs to be checked off against the order list or purchase order by the authorised electrician or supervisor. Any variations between the delivery note and the purchase order need to be confirmed with the delivery person and then the wholesaler, so that any missing materials can be arranged, reordered or reworked.

Variation orders

The customer is always right, but on the odd occasion when things do not completely go to plan, extra work may need to be ordered. This extra chargeable work is over and above the original contract agreement and must be paid for by the customer. A variation order is a good way to make sure there are no misunderstandings or conflicts when the final bills are due to be paid as all extra work is accounted for. Variation orders are controlled by a senior member of the site team that may include the lead project manager or architect. Any work over and above an agreed contract must be requested through a controlled process and a variation order issued for the work.

Working practice

A team of plumbing and heating engineers start work on a school dormitory refurbishment at the beginning of the summer holidays. The programme is fairly tight with only a six-week time slot to remove all the old pipework from the original antiquated plumbing and heating systems. Within the six weeks, the team also has to install new radiators and pipework and new bathrooms and WCs on each floor. All pipework is to be run at high level in the corridors and will later be concealed by a new false ceiling. All pipework must be tested and approved before commissioning and handing over to the other trades. The other trades include:

- carpenters, who are putting in new door frames and skirting
- plasterers, who are giving the whole building a skim
- tilers, who are retiling the four bathrooms
- electrical engineers who are rewiring and fitting new socket outlets, light switches and bulb holders
- ceiling contractors for the new ceilings throughout the three-storey building
- carpet fitters, who are levelling all floors before fitting special lino and carpet.

The customer site manager is in constant contact with all the trade supervisors and trades. When the site manager inspects one of the student dormitory rooms she decides there are not enough radiators for the number of beds in the room and for the two large radiators in each dormitory to be replaced with smaller ones, while still retaining the calculated heat input requirement for each room. The site manager discusses this with the supervisor and promises to sort it out with the project manager. This never happens but the work is still done by the plumbing and heating engineers. When the time comes for the central heating stage to be handed over to the ceiling specialist, the contract is two weeks later than agreed. The ceiling contractors had allowed two weeks to complete the task before they move onto a contract somewhere else. A new contractor is required at short notice. The project is late and more expense has been incurred. Because no variation order was issued and no contractual evidence was available, the plumbing and heating sub-contractor had to pay the penalty as they had breached their contract.

1 From the trades listed, in what order should the work be carried out?

2 What programme should have been produced originally and then amended?

3 Who should have overall responsibility to make sure the project is delivered on time and within budget?

Time sheets

Time Sheet				**Plumb and Level**		
				Plumbing contractors		
Employee			Project/site			
Date	**Job No.**	**Start time**	**Finish time**	**Total time**	**Travel time**	**Expenses**
Mon						
Tue						
Wed						
Thu						
Fri						
Sat						
Sun						
Totals						

Employee's signature _____ Supervisor's signature _____

Date _____

Figure 10.17: A typical time sheet

As a member of staff you will often be required to complete a time sheet. Larger companies run electronic schemes but the principles are the same. As an engineer you will often be working away from your official company office. You will need to keep your employer informed of your work on site. A time sheet helps your employer keep track of your work on site, including hours, extra expenses incurred, any travel time and the actual job address you are working at. Time sheets are regularly completed and handed in to the employer so that contracts can be tracked and project costs amended to reflect what is actually happening on site. The time sheet is the main method by which an employee gets paid. Different types of installation work might also mean overtime work – work outside of the normally contracted hours. Extra work may mean special payments are made and again the time sheet is the method by which these payments are made. Some companies will not actually pay an employee until the time sheet has been completed, so treat them with respect and always complete them when told to!

Policy documentation

Policy documentation can come from many different sources. A government policy comes from a law being passed and then instructions follow detailing how something must be done or not done. Alternatively, a company will have their own policy documentation to guide employees.

Most successful companies are unique because they have a unique proposition that makes them marketable. This leads to many companies having their own unique policies. These policies define how the company operates in different areas of their business or how they want their staff to work. A policy statement may state how a safe working environment is achieved or there could be a company policy stating that only certain ecologically friendly products can be used.

As a contractor or sub-contractor, your manager, supervisors or company directors will be aware of specific customer policies and will manage or negotiate how you work with these.

The purpose of information given to customers

Quotations

For a contract to be won, a quotation must have been given to a customer. This can be via a tender process or simply a request from a customer for a quote for a specific piece of work. Large contracts often go through a formal tender process that involves a customer writing out a requirement formally and then asking a list of companies to bid for the business. The response to tender will not only give the costs in the form of a quotation but highlight any extra reasons why that particular company should win the bid. The quotation contained within the tender response must be as close to the actual requirement as possible as this will generally form the basis of the resulting contract.

The tender process can be a very formal process with set response timescales and quote formats. Some companies keep the process very simple and have a set list of preferred suppliers that get invited to tender. Once the tender process has been completed, and the customer is happy with the quote, there could be a further process to refine the detail down to a more focused quote.

An example of a simple quotation would be an electrical company asking their wholesaler for a best price quotation on a list of required stock. This quotation may arrive in the form of a letter or an email directly from the wholesaler to the company director.

Estimates

A less formal process of getting costs for a job or stock involves asking for an estimate. A potential customer could ask you for an estimate. Although this is only an estimate, it is intended to give the customer a general idea of costs so they can set expectations for others in their company or possibly set a budget. Estimates do need to be as accurate as possible but they can only be as good as the information they are based on. A medium-sized plumbing and heating company would have a person or people with the necessary skills to go to a customer's site and estimate the costs of a job. The estimate might be the basis of the contract or used within a tender response. The customer might also be happy with the estimate and sign a contract for the work without further negotiation.

Invoices/statements

An invoice is used to help finalise payment. It is sent to the individual or company that is paying for the work to be completed.

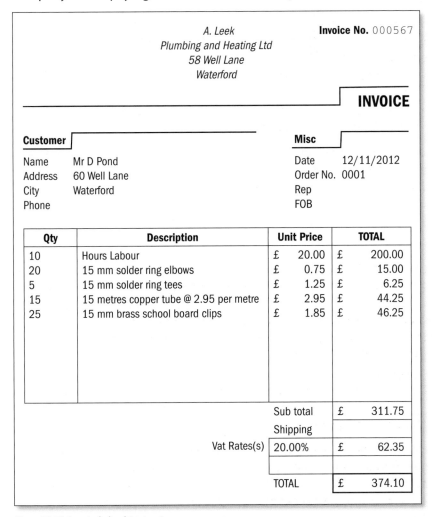

A. Leek
Plumbing and Heating Ltd
58 Well Lane
Waterford

Invoice No. 000567

INVOICE

Customer

Name	Mr D Pond
Address	60 Well Lane
City	Waterford
Phone	

Misc

Date	12/11/2012
Order No.	0001
Rep	
FOB	

Qty	Description	Unit Price	TOTAL
10	Hours Labour	£ 20.00	£ 200.00
20	15 mm solder ring elbows	£ 0.75	£ 15.00
5	15 mm solder ring tees	£ 1.25	£ 6.25
15	15 metres copper tube @ 2.95 per metre	£ 2.95	£ 44.25
25	15 mm brass school board clips	£ 1.85	£ 46.25

Sub total	£	311.75
Shipping		
Vat Rates(s) 20.00%	£	62.35
TOTAL	£	374.10

Figure 10.18: A typical plumbing invoice

> **Key term**
>
> *Contract* – a legally binding agreement between two or more parties. For a contract to exist there has to be an offer, acceptance and consideration by both parties. Contracts do not need to be written.

A customer might require a fully itemised statement for their accounts department. This will show all of the costs of the project and be dated. Invoices are generally sent to a company for payment and the period for payment is agreed. Many companies have a stated payment period after being invoiced that is controlled by the company accountant. Failure to meet the payment date after invoicing can result in legal action or a review of the company credit rating.

Statutory cancellation rights

When you buy goods or services from a supplier or trader, you are entering into a contract with them. Contract law in the UK states that under this **contract** you now have a set of implied rights known as statutory rights. The word 'statutory' means laws are involved and these include the Sale and Supply of Goods to Consumers Regulations 2002 and the Unfair Contract Term Act 1977. These laws state that the consumer has the right to goods that are deemed to be of satisfactory quality, fit for the purpose they are intended for, free from faults, safe and which have a satisfactory finish and appearance.

Contracts can be cancelled at any point up until the offer has been accepted but the contractor must notify the customer of their intention to withdraw the offer. Some contracts are often time limited. This means if the offer is not accepted within a certain time, the contract is not valid. Other reasons for a contract not becoming legally binding are rejection of the contract by the customer and death of a contractor before the offer is accepted. However, it must be proved that the customer was informed of the contractor's death for the contract to be made invalid.

When a contract has actually been made, it is difficult to get out of this legally binding deal unless a breach of contract can be proved. In certain circumstances, the customer is given the right to cancel a contract over a specific period of time. This time is called the 'cooling off period' but it depends on what was purchased and how. In the UK, buying goods or services online, by phone or by mail is subject to the Distance Selling Regulations, which give the consumer a seven-day cooling off period in which they can cancel without financial penalty. There are also certain circumstances that mean the cooling off period can be extended by a further three months.

Handover information

With every completed installation, a certain amount of information will need to be given to the customer so they are able to operate any new equipment or carry out basic maintenance. The handover pack should be presented to the customer and each aspect run through individually with them until they understand all the various aspects of the installation. Handover information will typically include:

- as fitted diagrams
- drawings
- operation manuals
- flow diagrams/system block diagrams
- test certificates
- maintenance contracts
- technical data sheets
- any health and safety specific information.

This list is not exhaustive and will be different for each installation and customer.

> **Activity 10.3**
>
> You have just installed a complete plumbing and central heating installation on a three-bedroom, detached house, using a gas-fired boiler connected to an unvented hot water storage cylinder. Write a list of what you believe would be included in a handover pack.

The importance of company policies and procedures that affect working relationships

Company working policies and procedures

As mentioned before, all companies are unique and some companies will have written down versions of their company policies. These may cover all aspects of the company operation. If your company has a set of policies, it is your responsibility to make yourself aware of the detail. A company may have induction days that include detailed training on how the company policies are put into operation.

Typical company policies may include:

- behaviour expectations on and off site
- timekeeping
- acceptable dress code
- contract of employment.

Some policies are legal responsibilities such as health and safety and equal opportunities but others may be dictated by the nature of the business, for example an environmental energy provision company which will only use 100 per cent recyclable products.

If you have signed a contract with a company then you need to understand the company policies and work within them.

Working within your limits

If you are trained to be able to complete a specific job and you have all the relevant equipment (including safety kit) then it is not unreasonable for you to complete the job in hand. If, however, you do not have the training or the correct equipment then you are working outside your limitations. If you are asked to complete work that you know you are not qualified to do, you should not continue with the job and clarify your position with your supervisor. If something goes wrong, ultimately it will be you who pays the price – this could be a very costly experience for the sake of not clarifying a point!

Working practice

An apprentice is put in a team of plumbers who are contracted to repair leaking guttering on a three-storey factory unit. The only way to work on the gutters is to use a cherry picker-type lift from the back of a rented truck. The apprentice is told that he must complete all the repairs before 4 p.m. that day while the rest of the team prepare stock for another project the following day.

The apprentice and one other plumber get to work but the cherry picker gets stuck and the apprentice is trapped 40 ft in the air – the rest of the team who are fully trained cannot be found. As the day comes to an end the apprentice decides to climb down but he falls, breaking his ankle. The HSE is called and the company receives a prohibition notice, a large fine and is closely monitored for the next two years for repeat offences and breaches of the Health and Safety at Work Act.

1 What training courses are available for working at height?

2 What does the HSE recommend about working at height?

3 Regarding the injury to the apprentice, who was at fault in this situation?

Supervisor and management responsibilities

Specific roles are allocated to specific members of staff within a company. Generally they are in that role because they are the most qualified and suitable for the tasks that come with their responsibilities. A supervisor in a team of plumbing and heating engineers will have the relevant technical qualifications and also the experience to make decisions about day-to-day activities. A manager is appointed following either specific management training or years of experience in the business. Some managers are there because it is their business but, either way, clear responsibilities for each role within a company exist, based on the company job description. A job description will define the role and responsibilities and the staff member will have been interviewed with those in mind.

A manager will understand the nature of the company's business. He or she will also be in a position to make decisions that affect the company, company contracts and its employees. The supervisor will take their instructions from the manager. These instructions will then be passed down one stage further to the employee completing the work. The supervisor will then monitor the work and the manager will monitor the supervisor's performance and the contract. If any amendments to a contract are required as it runs, the employee may need to pass information up the chain via the supervisor to get something changed. This chain of command should always be followed unless there is an immediate health and safety risk.

Progress check 10.03

1 Describe how a contract is made.

2 What might be included in a customer handover pack?

3 Give three examples of when a variation order could be used.

COMMUNICATING WITH OTHERS IN THE BUILDING SERVICES INDUSTRY

Methods of communication available

Communications skills are a highly prized skill in industry. A person's ability to communicate effectively can mean the difference between success and failure on a job. Some people are natural communicators and others take a long time, if ever, to pick up the skills.

Oral communication

You are sure to know someone who communicates well verbally – so what is their secret? It is the ability to know their audience, and to understand quickly the best words, phrases and type of language to use. Some people prefer very short instructions whereas others require a full discussion to go into finer detail. But remember not to get too complicated or you will lose your message.

Engineers can be the worst offenders at using shortened terms, phrases and acronyms for tools and commonly used products. This is a steep learning curve for all new staff so make sure you ask for an explanation when a term comes up you are not familiar with. It could be very costly if you do not understand.

Activity 10.4

Some experienced tradesmen have their own terms for tools or materials that have been passed down to them. Some tradesmen use their own terms that have come about due to not listening, mishearing or misinterpreting. This is usually as a result of poor verbal communication and you may hear some bizarre things!

Use the experience of the people around you to discuss acronyms and terms that you have been too afraid to ask about, or the names given to plumbing products, and write a list with definitions.

Written communication

Written communication is good for putting down information that needs to be referred to at a later date. This can be in the form of a letter, report, text message, fax or an email. Because of the advances in readily available technology, increasing numbers of tradesmen have advanced phones or portable data devices. These devices can process the same information as personal computers. Some manufacturers are also developing software for phones, especially for the building services industry, which makes information a lot more accessible. The different forms of written communication were covered on pages 391–398.

Communicating effectively with others

Building services and construction have a very multicultural environment and you can expect to work with a diverse mix of people. There is a wide range of barriers that need to be addressed when working and communicating with different people. Common sense must prevail to overcome any awkwardness and reservations.

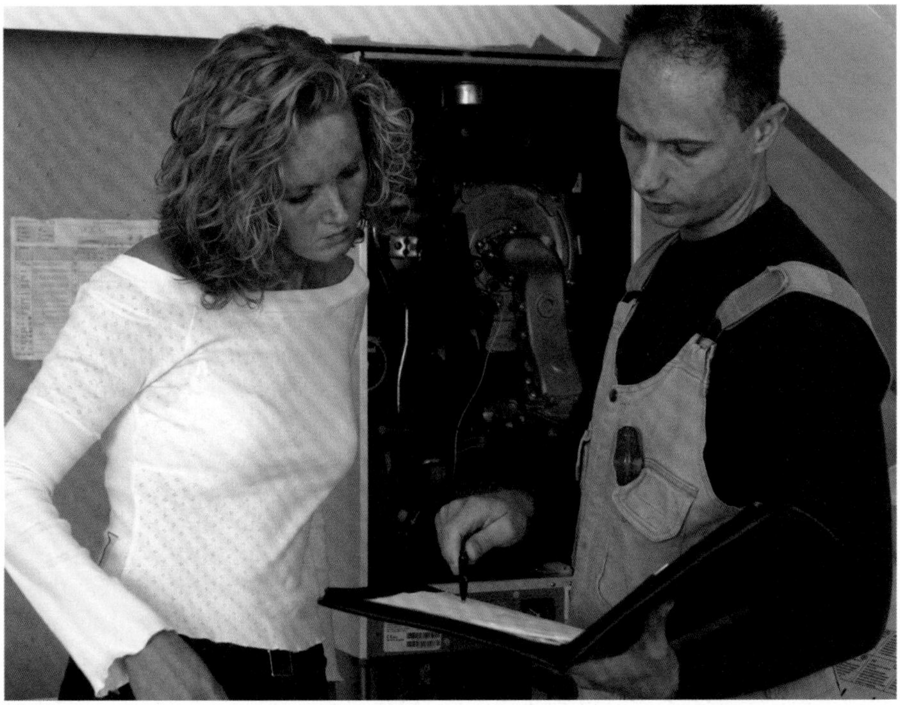

Activity 10.5

Write a brief instruction sheet for a new recruit on how to change a hacksaw blade in a large frame hacksaw.

Figure 10.19: A diverse mix of people work in building services and construction

Language differences

If you travel abroad for work, you may find yourself in an isolated position. You cannot understand your colleague and they cannot understand you. Signs, symbols and pointing often work – don't be tempted just to speak louder! Speak clearly, concisely and slowly as a lot of European languages have the same common roots – some words will be understood both ways. If you find yourself working with a team, and English is not the first language, take the opportunity to pick up some new words and develop your language skills. If accent or dialect is the barrier, ask your colleague to slow down a little and repeat the word.

Physical disabilities

Communicate with people with physical disabilities in the same way as you communicate with anyone else. You will be led by the person with the disability but listen to the feedback you are given. There are, however, some common-sense points to think about.

- If you offer someone assistance, wait until your help is accepted before you carry out the action.
- If someone has a visual impairment, make sure you identify yourself and others and use names if talking in a group.
- Call people with disabilities by their first names only if you are on first name terms with the rest of the group.
- If someone has a speech impediment, wait for them to finish speaking before you start – don't interrupt or speak over them.
- Speak directly to someone rather than going through a work colleague or interpreter.
- If someone has a physical disability, still offer your hand to shake.

Above all, relax and do not be embarrassed if you accidentally say, 'Have you seen this?' or 'Have you heard about this?'

Learning difficulties

Communicating with someone with learning difficulties may have its own unique challenges which depend on the complexities of the individual's needs. For example, communicating with a person who has profound and multiple learning disabilities (PMLD) may mean that using speech, symbols or signs will not be adequate. Specialist help and training may be required in this instance. The main considerations are for you to think about your tone, voice and body language and again take guidance from the individual you are trying to communicate with. Communication is a basic human right and allows an individual to interact with other people, show their feelings and make decisions that affect their lives. A considerable amount of guidance can also be found from bodies such as Mencap or disability-specific organisations.

Dealing with conflict in the workplace

Customer conflict

Conflict in the workplace is not good, although sometimes it can clear the air between people. Occasionally it cannot be avoided as there will always be people who fundamentally disagree with the way something is being done or they simply do not get on. However, if the conflict is between the customer and an operative, this can cause complications that are bad for the contract and business.

Contractual disputes are the biggest source of conflict between customers and contractors. If a contract has not been set out in enough detail, and the detail has not been discussed and agreed formally in writing, there may be conflict. Issues like this are always best settled by a meeting with all the parties involved and a chair person who is independent of both groups if possible.

If the conflict cannot be solved by simple 'round the table' discussion, a mediator might be required. The Ministry of Justice is responsible for developing Alternative Dispute Resolution (ADR) to avoid costly use of the court system. Mediation is where both parties come together and discuss their needs and concerns in the presence of an independent person (the mediator) to reach an agreement. This type of approach is good for small claims, civil disputes and contractual disputes. However, if the conflict is contractual and cannot be agreed in this way, lawyers can be involved.

The best way to avoid conflict with a customer is to keep them fully informed at all times, stay professional and bring any issues to their attention as quickly as possible. If there is a problem, go to the customer and tell them about it, but also offer solutions as the conflict could simply arise due to frustration. If you offer a solution at the same time you may be able to avoid conflict.

Conflict with co-workers

Conflict between co-workers can be due to individuals not getting on, personal competitiveness, jealousy, lack of information, lack of training or a whole range of other possibilities. Conflict between workers on site will not reflect well on the company, as the last thing a customer wants to see on site is arguments. This would be bad for their business and it will obviously affect productivity and potentially the quality of the work being done.

Conflict with supervisors and operatives

At some point in your working life, you may encounter a manager who you do not get on with. This can be for a number of reasons but this situation can become very difficult and lead to a lot of unhappiness in the workplace. Your manager or supervisor may be asking you to do something you are not trained to do. You may not want to admit that you do not know how to do it as you want to protect your job. Your manager is frustrated because every time he asks you to do something, you take too much time and don't seem willing to ask questions. This can lead to bad feeling on both sides.

A situation like this can arise when people are not honest with each other and do not take the time to discuss matters. If this occurs, you need to call a meeting with your supervisor and discuss the points. You could set up some ground rules that both parties sign up to, focusing on better ways of working together. Some examples are described in Table 10.1.

Activity 10.6

Conflict is a part of life that everyone has to deal with at some time but it is your ability to manage conflict without escalation that can set you apart from others. Discuss with another learner an example of conflict that you have managed successfully.

Employee commitment	Manager/supervisor commitment
I will ask questions when I need clarification.	I will not get angry when I am asked questions.
I will ask for feedback after work.	I will make time at the end of the day to review work and give constructive comments.
I will attend all the training offered to me.	I will look at training requirements of individuals and offer advice and courses.
I will attend monthly reviews.	I will arrange monthly reviews.

Table 10.1: Ground rules for working together

If disputes cannot be resolved by talking together, a third party might be required. If a more senior manager or supervisor cannot resolve the issue then the use of an alternative dispute resolution service such as the Advisory, Conciliation and Arbitration Service (ACAS) might be required. As part of ACAS, a third party is invited into the meeting. The level of dispute resolution will be determined by the level of control the two parties are willing to give up. Obviously if the dispute cannot be resolved in this way, and the courts are involved, the final decision will be bound by the law and neither party will have a say.

Method	Descripton
Conciliation	This is where both parties retain all the power. There is a facilitated discussion where all points, needs and concerns are discussed and all reach an agreement that is 'honour bound'.
Mediation	Some of the powers are given up by both parties and this is agreed beforehand and signed up to. The rest of the process is the same as conciliation.
Arbitration	Both parties give up their power to the independent arbitrator beforehand and sign up to this. Although arbitration is voluntary, once signed up to both parties have to stick to the decision.

Table 10.2: Alternative methods of dispute resolution (ACAS)

A lot of arguments start from a simple misunderstanding between workers. To avoid this situation, simply ask your colleague for clarification, stay calm and, if necessary, check with another colleague or manager.

Effects of poor communication on an organisation

Effects of poor communication between operatives

Poor communication can cause a number of issues. In a team working to tight deadlines, good communication becomes even more essential. Sometimes there is only limited time to give and get instructions on site. If these instructions are passed down the line of management to a team, and then the operative carries out the activity, he or she has to be confident that it is correct. If you are not completely sure about an instruction that you have been given, do not start work – check again until you feel confident. If you don't, it could cost the company a great deal of money or damage the company's reputation.

Effects of poor communication between operatives and management

With large contracts, instructions can be passed through a management chain. These must be written down to ensure the message does not get changed or misinterpreted. Any official changes will be controlled by a project manager, principal contractor or architect. These instructions will be written and again, if there is any doubt, no one will mind if you stop to check.

Case study

John was given his first installation job as part of a very small team involved in the redevelopment of a warehouse that was being converted into offices. He was given the task of running pipework along the corridors at high level and into each of the 15 rooms ready for connection. The plumbing supervisor and another apprentice were in another part of the warehouse. They were tracing existing pipework and removing it as required. John had a meeting with the senior plumber at the beginning of the first day and was given verbal instructions – he did not write them down or ask any questions as he did not want to appear stupid. The instructions were to run a 22 mm copper tube into each room, terminating each pipe at low level with a machine made bend. John was not totally sure but thought he would figure it out – after all, he had seen this in college.

The senior plumbing and heating engineer got caught up in a difficult draining off problem, due to a faulty isolation valve, and didn't manage to speak to John at length until the end of the second day. By this time, John had run a single 15 mm tube to each room. This cost the company 600 m of 15 mm copper tube and two man days, plus a third day to salvage as much of the 15 mm tube as possible for reuse. John wasn't dismissed but it took a few months before he was allowed to take on this kind of responsibility again.

The case study above is real so how could this have been avoided?

1 Describe a set of actions that could have been taken to avoid this situation.

2 Write a set of 'open' questions that could have been asked to avoid this kind of poor communication between a manager and apprentice.

Consequences of poor communication with a customer

Poorly managed installations with no clear lines of communication are certain to end up in difficulties. A customer pays the contractor and has the final say in what is to be done on site. Regular structured conversations and communication are required between a company and customer to make sure the installation work stays on track, within budget and finishes on time. If the customer is not kept informed of changes, difficulties, setbacks, stock issues or workforce issues, the contract could finish late. If the customer is unable to start business on time because of this, it could lead to a legal case being brought against the company.

Formal project management is one very good method of controlling communication in a structured way. Meetings should be held on a regular basis and any changes to instructions should be written down in minutes and a change of control process put in place.

Assessment assignment

You are in the process of trying to win a tender for a large installation job in a private faith school. The potential customer has asked for more information before they are willing to award the contract to you. If your company wins, you will be the sub-contractor responsible for a complete refurbishment of the plumbing and heating system. Write a short report (no more than 500 words) explaining to the customer your company's health and safety policy and how you will carry out the work maintaining a safe environment. You must take into account any considerations and requirements for a faith school (you can choose the particular faith of the school and be prepared to debate your assumptions).

Knowledge check

This chapter is assessed via online multiple choice questions. Please attempt the following knowledge check questions. If you do not get all the questions correct first time, read the relevant parts of the chapter again and re-attempt the questions until you are confident of the answers.

1 What does APM stand for?

  a Associate Post Manager
  b Alternative Project Management
  c Association of Project Managers
  d Alternative Practical Measure

2 Under the CDM Regulations, on construction sites, 'notifiable work' has to be registered when what conditions are reached?

  a A project is 30 days or 500 man hours long
  b A project is 15 days and 50 man hours long
  c A project is 10 days and 10 man hours long
  d A project is 5 days and 5 man hours long

3 What does ECA stand for?

  a Electrical Contractors Academy
  b Edwards Contractors and Artisans
  c Electrical Contractors Association
  d Elevator Contractor Association

4 Which of the following is not one of the three main sets of industry statutory legislation and guidance?

  a common or contract law
  b UK legislation
  c European legislation and judgements from the European Court of Justice
  d The Human Rights Act

5 Part 5 of BS 7671 covers what specific subject area?

  a Definitions
  b Protection for safety
  c Scope
  d Selection and erection of equipment

6 What is the role of a clerk of works on a construction site?

  a to ensure operatives have lunch and tea breaks
  b to ensure quality assurance
  c to ensure work schedules are adhered to
  d to ensure materials arrive on site at the correct time

7 In which year was the Sex Discrimination Act amended to include the Equal Treatment Directive?

  a 1974
  b 1975
  c 2008
  d 2011

8 What diagram uses BS 1192 symbols?

  a Layout drawing
  b Schematic diagram
  c Circuit diagram
  d Wiring diagram

9 What is the best diagram to show you simply how a system works?

  a Block diagram
  b Schematic diagram
  c Wiring diagram
  d Assembly drawing

10 What does ACAS stand for?

  a Alternative Conciliation and Service
  b Advisory, Conciliation and Arbitration Service
  c Alternative Combat Advisory Service
  d Advisory Contact Arbitration Service

Glossary

Ambient temperature – the temperature of the air where equipment or cable is installed.

Ampere (amp/A) – the measure of electrical current; the flow of 1 coulomb in 1 second.

Anode – the negative electrode of an electrochemical current source while being discharged.

Back EMF (electromotive force) – a voltage produced within electromagnetic windings (such as those in motors and transformers). As well as producing the voltage required, the magnetic field in the iron core produces this second, opposing voltage. The current from back EMF flows against the supply current. This opposition is called reactance, which reduces the amount of power produced. This inefficiency causes a low power factor.

Backflow – water flowing in an unwanted/unintended direction, which could lead to foul water contaminating a drinking supply.

Bonding – connecting non-current carrying metalwork (metallic devices) together and into the earthing system to protect from electric shock.

Branch connection – where a discharge pipe is connected to the main discharge stack.

BS EN – British Standards European Norm.

BSPT – British Standard Pipe Thread.

Calorific value – the amount of heat energy a substance releases when burned.

Capillary action (capillarity) – the process by which a liquid is drawn up through a small gap between the surfaces of two materials.

Catalyst – a substance that increases the rate of a chemical reaction while itself remaining unchanged.

Cathode – the positive electrode of an electrochemical current source while being discharged.

Centrifugal force – a force moving away from the centre of a body with circular motion.

Chamfer – smooth a cut edge by slightly bevelling or angling it.

Circuit – a network of components (e.g. battery, wire, switch) joined in a loop to allow the flow of electricity.

Circuit breaker – an automatic switch that opens when its rated current is exceeded.

Coefficient of expansion – the amount a material expands when heated.

Coincidence of demand – when many consumers try to use the same services (e.g. water, gas or electricity) at the same time. For example, there may be surges in electricity demand during the commercial breaks in popular television shows, as large numbers of people try to make tea at the same time. This may cause problems to the providers of these services.

Combustible – able to burn or be burned (flammable).

Competent person – BS7671:2008 and the associated guidance notes and codes of practice define a competent person as someone with the technical knowledge or experience to carry out electrical work without risk of injury to themselves or others.

Compound – a mixture of two or more elements.

Conductor – an electrical conductor is a material that allows electricity to flow through it, such as metal. (Thermal conductors allow heat to pass through them.)

Conduit – a pipework system designed to carry wiring and protect it from mechanical damage.

Connectivity – a way of describing how good an electrical connection is. A loose connection will have poor or low connectivity and a secure connection will have good or high connectivity.

Contract – a legally binding agreement between two or more parties. For a contract to exist there has to be an offer, acceptance and consideration by both parties. Contracts do not need to be written.

Contractor – an individual or a company that has a specific job function within a project.

Controlled document – a changing document that is reviewed and then reissued by the project manager so that all the people involved in a project have exactly the same information and are using the most current version.

Co-products – by-products of some industrial processes which can be used as biomass fuel.

Corrosion – the gradual damage or destruction of a hard material through chemical action.

Cross-flow – occurs when two branches are located opposite each other.

Cross-sectional area – the area of a cable conductor's face. Look at a conductor end-on and that is the area to be measured.

Crutch head – the handle of a stop tap.

Dangerous occurrence – a near miss. This is when an accident occurs that could have caused serious injury but didn't, for example, if a scaffold collapses after working hours when the entire workforce is off site.

Dead head – a situation in which the pump is running but all heating controls are closed, preventing the flow of water around the heating circuit. This can damage the pump impeller and lead to pump failure.

Dead leg – a water pipe with no draw-off point. In the case of hot water supply, a pipe over the recommended length.

Degradation – breakdown of a material caused by exposure to ultraviolet (UV) light, usually in sunlight.

Dezincification – the breakdown of zinc in brass fittings by the electrolytic corrosion between zinc and copper, making the components porous. This occurs in some soil and water conditions, especially in soft water areas.

Differential valve – a valve that opens and closes automatically in response to the flow of water passing through the unit.

Dressing – plumbers often refer to 'dressing the suite'. It means installing the taps, wastes and – in the case of baths – the cradle frame or feet. It also includes installing float valves, overflows or flushing valves and the handle assembly to the WC cistern.

Ductile – able to be drawn out into wire.

Earth/earthing – connecting exposed conductive parts of an installation to its main earthing terminal.

Earth continuity – all exposed metalwork in a building is bonded together and connected to the earthing block in the consumer unit.

ECA – the Electrical Contractors' Association.

Electrical conductivity – how well or poorly a material conducts electricity.

Electrolysis (electrolytic action) – describes a flow of electrically charged ions from an anode to a cathode through a medium called electrolyte, usually water.

Electromagnetism – the relationship between electric currents or magnetic fields.

Fabric – what the building is made of: brickwork, concrete, blockwork or timber.

Fascia/fascia board – the piece of wood or plastic fixed to the rafter; timber/plastic board (or other flat pieces of material) covering the ends of rafters.

Ferrous – metals that contain iron.

Ferrule – a metal fitment used to connect to a main to allow isolation.

Finite – available in limited amounts, and will therefore run out eventually.

Fit for purpose – using the right tool for the right job and the right environment for the work. For example, try assembling delicate electronic components in a dirty, tumbledown barn. The environment is simply not right for that job. It is not 'fit for purpose' and will lead to a faulty product.

Flux – a paste (or liquid) used in soldering techniques.

Foul water/drainage – anything discharged from a sanitary appliance such as a WC, bath, basin or sink.

Friction – force that opposes motion, as when two surfaces rub together.

Fuse – a conductor inside a cartridge or holder, designed to melt and break when its rated current is exceeded.

Galvanise – to coat iron or steel with a layer of zinc to prevent oxidisation (rusting). The term comes from the 18th-century Italian scientist Luigi Galvani.

Gooseneck – looking down into the trench, the pipework should be in the shape of a letter P.

Gravity circulation – cold water is heavier than hot water. This means that gravity exerts a stronger pull on cold water, drawing it down and allowing the hot water to rise through the system.

Hazard – a situation that poses a threat. For example, a drill with a damaged power lead is a hazard because if you attempt to use it, the drill could give you an electric shock.

Hazardous malfunction – this occurs when a tool or item of equipment goes wrong and does not injure anyone, but could have caused injury to the person using it or to people in the vicinity.

Head – the difference in height (vertical distance) between the top and bottom of a water pipe, or between a reservoir and the consumer outlet. This determines the water pressure at the consumer outlet. Water is stored at a higher level than the consumer outlet and is delivered by gravity in most areas, rather than being pumped.

Heat emitter – a device in a heating system that heats up the space it is placed in.

Immersion heater – an electric element fitted inside a hot water storage vessel. It can be controlled by a switch and thermostat.

Impervious – does not allow fluids to pass through.

Joule – International System of Units (SI) unit of energy.

Kinetic – energy produced by motion or movement, as in wind- or water-driven power stations.

Lockshield valve – used to balance the system when it is installed or maintained. It is also used as a service valve and is solely intended for use by the installing or maintenance plumber and not by the consumer.

Malleable – able to be hammered into shape.

Mechanical damage – physical damage to cables or electrical equipment. Sheaths, enclosures and containment, cable armouring and toughened casings are all called mechanical protection.

Monobloc – a mixer tap with a single mounting hole.

Motorised valve – a valve in a water pipe that can be turned on or off by an electric motor. It is one of the heating controls in a system and a valve can be two- or three-port.

Multipoint – an instantaneous water heater that can supply water to more than one outlet.

NAPIT – the National Association of Professional Inspectors and Testers.

National Grid – a network of nearly 8,000 km (5,000 miles) of overhead and underground power lines that link power stations together throughout the UK.

NICEIC – the National Inspection Council for Electrical Installation Contracting.

Ohm – the SI unit of resistance, denoted by the Greek letter omega (Ω).

Packing-gland nut – a nut used to compress packing to make valve spindles watertight.

Palatable – pleasant to taste.

PAT – stands for portable appliance testing and is a legal requirement under health and safety legislation.

Permeation – the penetration of a substance, such as a gas or liquid, through a solid material such as a pipe wall (for example, air entering the water supply through the external wall of a pipe).

Porous – allow fluids or air to pass through.

Prospective fault current (PFC) – a large fault current that will flow from the source of supply to the fault itself when an electrical fault occurs.

Proving unit – a low-voltage, inverted DC testing device.

Radial circuit – a single cable is taken from the electricity mains into the consumer's property and terminates at the consumer's meter.

Rainwater pipe – a pipe attached to the side of a building to collect and disperse rainwater into the drain. Note this is also called a downpipe or fall pipe in different parts of the UK.

Renewable energy source – an energy source that will not be used up, such as wind power.

Resin particles (zeolites) – microporous molecule minerals, whose structure resembles a cage made up of many cavities and channels that are interconnected to form a framework.

Ring circuit – ensures continuity of the electricity supply to all consumers, even at peak demand times. As one area requires more energy, another area will probably require less.

Service reservoir – where cleaned and treated water is stored ready for use by the consumer. The water is protected from contamination, including sunlight.

Short circuit – occurs when electricity flows along an unintended path.

Single point – an instantaneous water heater supplying one outlet.

Snaked – the pipe is laid in a zigzag formation, using more than would be needed to transfer water directly from point A to point B.

Sparge pipe – a horizontal pipe that connects to the flush pipe. It is mounted on the face of the urinal and has pre-drilled holes (urinal spreaders), which are used to wash the face of the slab and channel.

Stack – the vertical discharge pipe that carries waste to the main sewer below ground.

Statutory – statutory documents are those that have been debated and issued by the government. The word 'statutory' comes from the fact that they are on the Statute Book, which means that they are part of the law of the land.

Sterilisation – purification by boiling, or, within the industry, dosing the supply with chlorine or chlorine ammonia mix.

Structural engineer – a professional who analyses, and contributes to the design of, any structure that will support a load. For example, a structural engineer would be heavily involved in the design of a new bridge, as the main purpose of the construction is to take the weight of the traffic that passes over it.

Surface water – water collected via the rainwater system.

Swaging – where the end of a tube is flared out using a swaging tool or drift. It may be necessary to anneal the end of the tube to allow this procedure, particularly on larger diameter tubes.

Synchron motor – synchronous motor. An AC motor where the shaft rotation is synchronised with the frequency of the supply current.

Tendering – the tendering process is the stage of a project when companies are invited to submit a price and a proposal to the customer. The customer will choose the most suitable offer and appoint the company to carry out the work.

Thermal – a type of power station where water is heated so it turns to steam (heated gas turns water to steam in nuclear power stations), which then drives a turbine.

Thermal conductivity – how well or poorly a material conducts heat.

Thermostat – device which senses the temperature in the system and opens or closes the respective zone valve.

Venturi – a constricted piece of tube that has the effect of reducing fluid or liquid pressure when a fluid flows through it.

Vitreous china – a material produced from a solution called slip, or casting clay. Slip has the consistency of pouring cream and contains ball clay, china clay, sand, fixing agent and water. Vitreous products have a glasslike appearance.

Voltage – the pressure that pushes electricity around a circuit, known as electromotive force (emf). Voltage may be supplied by a battery or a mains supply.

Water governor – pressure-reducing valve.

Water hammer – loud noise in pipework caused by a valve being closed suddenly.

Water mains – the network of pipes that supply wholesome water to domestic and commercial properties.

Water table – the natural level of water in the ground.

Water undertaker – the legal term for a company that supplies domestic water.

Watt (W) – the SI unit that indicates the amount of power an electrical item consumes.

Whetting – sharpening using friction or grinding.

Wholesome water – good-quality water for human consumption.

Index